Monitoring and Modelling Dynamic Environments

(A Festschrift in Memory of
Professor John B. Thornes)

Monitoring and Modelling Dynamic Environments

(A Festschrift in Memory of Professor John B. Thornes)

EDITED BY

Alan P. Dykes
Mark Mulligan
John Wainwright

WILEY Blackwell

Library of Congress Cataloging-in-Publication Data applied for.

Hard Back: 9780470711217

A catalogue record for this book is available from the British Library.

Wiley also publishes its books in a variety of electronic formats. Some content that appears in print may not be available in electronic books.

Cover image: Badlands in the Neogene – Quaternary sediments of the Mula basin near the town of Campos del Rio, Murcia, Spain. Photograph courtesy Asunción Romero Díaz.

Set in 8.5/12pt Meridien by SPi Global, Pondicherry, India
Printed and bound in Singapore by Markono Print Media Pte Ltd

1   2015

This book is dedicated to the memory of Professor John B. Thornes – geographer, geomorphologist, leader, mentor, colleague and friend.

Photograph courtesy Rosemary Thornes.

Contents

List of contributors

Ashraf Afana
Department of Geography
Durham University
South Road, Durham, UK

Irasema Alcántara-Ayala
Instituto de Geografía
Universidad Nacional Autónoma
de México (UNAM)
Coyoacan, Mexico

Gabriel del Barrio
Estación Experimental de Zonas Áridas (EEZA)
CSIC, Cañada de San Urbano
Almeria, Spain

Louise J. Bracken
Department of Geography
Durham University
South Road, Durham, UK

Jane Brandt
Fondazione MEDES
Sicignano degli Alburni (SA), Italy

Denys Brunsden
Vine Cottage
Seatown, Dorset, UK

Tim Burt
Department of Geography
Durham University
South Road, Durham, UK

Artemi Cerdà
SEDER – Soil Erosion and Degradation
Research Group
Department of Geography
University of Valencia
Valencia, Spain

Thomas J.B. Dewez
Direction of Risks and Prevention – BRGM – French
Geological Survey
Orléans-la-Source, France

Alan P. Dykes
School of Civil Engineering and Construction
Kingston University
Penrhyn Road, Kingston upon Thames, UK

Carolyn F. Francis
Halcrow Group Ltd,
a CH2M HILL Company
Swindon, UK

Nichola Geeson
Fondazione MEDES
Sicignano degli Alburni (SA), Italy

Antonio Giménez-Morera
Departamento de Economía y Ciencias Sociales
Escuela politécnica superior de Alcoy
Universidad Politécnica de Valencia
Alicante, Spain

Ken Gregory
Department of Geography
University of Southampton
Southampton, UK

Janet Hooke
Department of Geography and Planning
School of Environmental Sciences
University of Liverpool
Liverpool, UK

Antonio Jordan
MED_Soil Research Group
Department of Crystallography, Mineralogy and
Agricultural Chemistry
University of Seville
Seville, Spain

Saskia Keesstra
Soil Physics and Land Management Group
Wageningen University
Wageningen, the Netherlands

Mike Kirkby
School of Geography
University of Leeds
Woodhouse Lane, Leeds, UK

Andrew T. Lowe
Halcrow Group Ltd, a CH2M HILL Company
a CH2M HILL Company
Warrington, UK

Jenny Mant
River Restoration Centre
Cranfield University Campus
Bedford, UK

Matthew Marsik
Land Use and Environmental
Change Institute
University of Florida
Gainesville, FL, USA

Jorge Mataix-Solera
Environmental Soil Science Group
Department of Agrochemistry
and Environment
Miguel Hernández University
Alicante, Spain

Mark Mulligan
Department of Geography
King's College London
London, UK

Agata Novara
Dipartimento dei Sistemi Agro-ambientali
University of Palermo
Palermo, Italy

Paulo Pereira
Department of Environmental Policy
Mykolas Romeris University
Vilnius, Lithuania

Marvin Quesada
Departamento de Ciencias Sociales
Universidad de Costa Rica Sede Occidente
San Ramón, Costa Rica

Asunción Romero Díaz
Department of Geography
University of Murcia
Murcia, Spain

José Damián Ruíz Sinoga
Department of Geography
University of Málaga
Málaga, Spain

Leszek Starkel
Department of Geomorphology and Hydrology
Institute of Geography,
Polish Academy of Sciences
Krakow, Poland

Iain S. Stewart
Centre for Research in Earth Sciences, School
of Geography, Earth, & Environmental Sciences
University of Plymouth
Plymouth, UK

John Wainwright
Department of Geography
Durham University
South Road, Durham, UK

Glenn Watts
Evidence Directorate, Environment Agency
Bristol, UK
and
Department of Geography
King's College London
London, UK

Peter Waylen
Department of Geography
University of Florida
Gainesville, FL, USA

About the editors

Alan P. Dykes is currently a senior lecturer in civil engineering at Kingston University, London. He obtained a geography degree from the University of Bristol in 1990, and he went on to undertake research for his PhD under John Thornes' supervision, transferring to King's College London when John moved to King's. His PhD project, *Hydrological controls on shallow mass movements and characteristic slope forms in the tropical rainforest of Temburong District, Brunei*, was part of the 'Brunei Rainforest Project' organised by the Royal Geographical Society (RGS) with the Universiti Brunei Darussalam. Throughout several years at Huddersfield University and more recently Kingston University, Alan has used his wide-ranging expertise in fieldwork, geotechnical laboratory testing and computer modelling to focus his research on landslides/slope instability and hydrological systems in landscape contexts as diverse as Ecuador, Indonesia, Iran, Ireland, Italy, Malaysia, Malta and Mexico and is now a leading authority on peatland instability and failure.

Mark Mulligan is currently a reader in geography at King's College London. He obtained a geography degree from the University of Bristol in 1991, which included fieldwork with the 'Brunei Rainforest Project', then moved to King's College London to undertake a PhD on '*Modelling hydrology and vegetation change in a degraded semi-arid environment*', supervised by John Thornes. Mark remained at King's College London as a permanent member of academic staff following his PhD, apart from a year on research secondment at the Istituto di Botanica, Università di Napoli, Italy, and in 2004 was awarded the *Gill Memorial Award* of the Royal Geographical Society – Institute of British Geographers for 'innovative monitoring and modelling' of environmental systems. Mark works on a variety of topics in the areas of environmental spatial policy support, ecosystem service modelling and understanding environmental change, at scales from local to global and with a particular emphasis on tropical forests in Latin America and semi-arid drylands in the Mediterranean.

John Wainwright is a professor in physical geography at Durham University. He completed a PhD on '*Erosion of semi-arid archaeological sites: A study in natural formation processes*' at the University of Bristol, supervised by John Thornes, in 1991. After brief spells at Keele and Southampton universities, John W. took up a permanent position at King's College London, becoming a professor within 10 years and in 1999 being awarded the *Gordon Warwick Award* of the British Geomorphological Research Group for his innovative work in the understanding of past and present semi-arid environments. John subsequently moved to Sheffield University, including 3 years as a visiting professor at Université Louis Pasteur, Strasbourg, then to Durham University in 2011. He has undertaken extensive field research, especially in drylands in the Mediterranean, the US Southwest and sub-Saharan Africa, and links the data to theory using computer modelling combined with results from laboratory experiments on slope and channel processes.

Like John Thornes, Alan Dykes and John Wainwright are originally from the county of Yorkshire in the north of England. Mark is from a little further south.

Acknowledgements

The preparation of this book has been supported by many members of the global geomorphological community. We are grateful to a large group of people who were kind enough to give their time to review the chapters submitted for inclusion in this book (in alphabetical order): Concepcion Alados, Irantzu Lexartza Artza, John Boardman, Julia Brown, Nick Chappell, Stephen Darby, Caitlin Douglas, Ian Douglas, Bob Evans, Fai Fung, Jeff Herrick, Martin Hurst, Sandra Junier, Neil McIntyre, Helen Moggridge, Bhopal Pandeya, Dorthe Pflanz, Maria Concepcion Ramos, David Smith, Elina Tarnavsky, Arnout van Soesbergen, Stephen Wise, Adrian Wood and Leonardo Zurita. We thank the publishers for their patience and assistance, and we (APD in particular) are grateful to the chapter authors for bearing with us through difficult periods when the 'day job' prevented any editorial progress for months at a time. Several figures were redrawn for this book by Chris Orton.

Specific copyright acknowledgements: t.b.c.

CHAPTER 1

Introduction – Understanding and managing landscape change through multiple lenses: The case for integrative research in an era of global change

Alan P. Dykes[1], Mark Mulligan[2] and John Wainwright[3]

[1] School of Civil Engineering and Construction, Kingston University, Kingston upon Thames, UK
[2] Department of Geography, King's College London, London, UK
[3] Department of Geography, Durham University, Durham, UK

The twenty-first century is when everything changes. And we must be ready.

(BBC 2006)

The world is changing. It has always been changing, as internal geophysical processes drive the tectonic systems that slowly rearrange the distributions of land masses and oceans and external influences such as orbital eccentricities alter the Earth's climate. Over hundreds of millions of years, the physical and ecological environments at any given location on land have been shaped and reshaped by natural then, from the late Pleistocene onwards, increasingly by human processes – so much so that some argue for a new geological epoch (the Anthropocene) in recognition of this (see Steffen et al. 2011; Brown et al. 2013). Our recognition of different kinds of impacts and implications of landscape changes has developed in parallel with the increasing rates of industrialisation of the last 250 years (Hooke 2000; Wilkinson 2005; Montgomery 2007). Humans are highly inventive in providing technological 'solutions' to feeding, watering and providing energy for ever-increasing populations, but this inventiveness does not always produce resilient or sustainable solutions. Indeed, it is often only when we are highly dependent upon these technologies that we begin to understand their negative effects on the environment that also sustains us. This is a reflection of both scientific advancement throughout the period and the increasing impacts on landscapes of the expanding industrial activities, including the industrialisation of agriculture, water and energy provision. However, although we know some of the impacts of these technologies on natural landscapes and the 'ecosystem services' that they provide, many aspects of the natural functioning of these landscapes remain uncertain.

The continually increasing population of the Earth has inevitably led to the expansion of anthropic modification of landscapes into increasingly marginal (in terms of primary productivity) or unstable (in terms of ecology and/or geomorphology) lands. The associated risks have driven much of the agricultural and geomorphological research that has provided the

Monitoring and Modelling Dynamic Environments, First Edition. Edited by Alan P. Dykes, Mark Mulligan and John Wainwright.
© 2015 John Wiley & Sons, Ltd. Published 2015 by John Wiley & Sons, Ltd.

basis for the management of landscapes under-going such changes. However, towards the late 20th century, there developed a greater aware-ness of the potential for long-term catastrophic losses of productive agricultural lands as a consequence of inadequate or inappropriate management (e.g. Montgomery 2007), usually stemming from inadequate relevant scientific knowledge – or the lack of communication, or application, of that knowledge in policy formu-lation. The early years of the 21st century have also focused attention on the potentially increasing risks of natural hazards arising from regional manifestations of anthropic, global cli-mate change. Extreme climate events lead to floods and landslides and wind storms, all of which cause losses of life, property and liveli-hood, but they may also lead to more chronic adverse impacts on societies through losses of land productivity – and these effects may be exacerbated by inappropriate land management strategies and agricultural practices.

John B. Thornes recognised the importance of good basic and applied geomorphological knowledge for the sound management of land-scape change relatively early. Many of his ideas were developed from his early research experi-ences in central Spain in the mid-1960s, although his work on the particular problems of semi-arid environments did not begin until 1972 (e.g. Thornes 1975). Fundamental to his work was the development of modelling approaches for understanding geomorphological processes and systems in parallel with the implementation of intensive field data collection methodologies to support the parameterisation of the models. Perhaps crucially, he identified the need to measure not only 'descriptive' parameters such as morphology or particle size – typical of early quantitative geomorphology – but also land-scape properties and processes that would allow quantification of the key geomorphological issues of spatial *and temporal* scale and variability. As such, he was a key player in the adoption of both monitoring and modelling approaches in geomorphological research and a leader in the explicit integration of these approaches and in the application of the findings to management policy and practice. For John, geomorphology included climate, hydrology and vegetation and their interactions with geomorphological pro-cesses and thus forms and their dynamics. As well as the management of modern processes, this interaction between process and form also informed John's understanding of past environ-ments, particularly in his work on palaeohy-drology and geoarchaeology.

Monitoring, modelling and management

Monitoring in environmental science is the process of keeping track, over time, of how one or more material properties or system states behave, based on repeated observations and measurements. It underpins the empirical understanding of environmental processes by observation of their consequences. For example, investigation of micro-scale controls on the pro-cess of soil matrix throughflow could require monitoring of water temperature (material property), soil water content and instantaneous flow velocity (static and dynamic system states, respectively). In geomorphology, many material properties are effectively constant at the time-scale of a research project, but for other varia-bles, the monitoring data required and hence the method by which they are obtained depend on the spatial and temporal scales of the prob-lem being investigated. Technological advances over the last 30–40 years have increasingly expanded the range of what can be measured (i.e. parameter/variable type and scale) and at what frequency, although in some cases there remain constraints. For example, satellites cannot yet provide both very high temporal frequency and very high spatial resolution data – although most applications of this approach such as monitoring land-use change do not require high frequencies of repeated observations, and rapid developments in UAV techniques are addressing some of these limitations at local scales. Most field monitoring takes place at points, and the cost of equipment and labour to maintain field monitoring 'stations' means that there are often relatively few of these, so that significant interpolation is required for landscape-scale analysis. However, it must be remembered that remotely sensed observations are proxies of the properties or system states of interest and must be grounded in field measurement.

Modelling is the process of representing or displaying something that cannot otherwise be experienced, through abstraction of reality and representation in the form of a conceptual, mathematical or physical model. John Thornes was one of the pioneers of mathematical models in geomorphology, which are increasingly used as representations of geomorphic systems. They are usually used as a way of evaluating conceptual models of components of a system such as a specific process or a topographic subsystem representing a set of processes; furthermore, today's office PCs can typically carry out hundreds of simulations using highly sophisticated models and produce the results before the morning coffee break. However, as with any investigation, the type and formulation of the model necessarily depends on the purpose of the research and, critically, on the quality and quantity of data that may be available to set up the model and evaluate its outputs (Mulligan and Wainwright 2004). John always understood that the geomorphological processes of greatest interest to him existed within a complex context. Over his career, through his own research and that which he supervised, collaborated with or helped to fund, he tried to ensure that as much of the relevant contextual complexity as was useful, was included. In his modelling of erosion, he started by defining and representing the process itself (Embleton and Thornes 1979). Subsequently, he introduced interactions with terrain, climate and geology (Thornes and Alcántara-Ayala 1998), then vegetation (Thornes 1990) and finally animals (Thornes 2007). All of

this was done with a clear focus on the socio-economic context and the policy outcomes (Brandt and Thornes 1996; Geeson et al. 2003) and an understanding of the role of time (Thornes and Brunsden 1977) and of history (Wainwright and Thornes 2003).

The degradation of agricultural lands in the Mediterranean region has been investigated at all spatial scales, from small experimental plots on hillslopes to remote sensing of the entire region. One of the major challenges has been to integrate not only the monitoring data with the modelling of erosion processes and changing state of the landscape system but also the findings from different scales of investigation into regional-scale, operational management tools. John Thornes was quick to embrace the possibilities presented by GIS technology (particularly digital elevation models) and satellite remote sensing and the integration of such techniques with smaller scale field studies and process or system models. Indeed, combining field and remote sensing-based monitoring with modelling was one of John's key foci during the latter part of his career, particularly in the European Mediterranean. He then led the development of the conceptual framework that ultimately led to the GIS-based and web-based, management-focused tools that came later. These will undoubtedly continue to be refined, not least to accommodate the threats to sustainability deriving from new technologies and/or revealed by ongoing monitoring programmes.

Management is used here in its broadest sense to refer to all levels of application of research findings by different agents, ranging from individual farmers and local communities through to regional managers and national/international policymakers such as the US federal government or the European parliament. We also use it to include all types of management, from details of how an individual field with particular soil types and rainfall patterns should be ploughed and then planted up to multi-decadal plans for new water resource infrastructure. The outputs of several of the desertification research projects initiated by John Thornes and outlined in this book provided outputs that are designed to be applied at all scales of application including EU-scale funding decisions. Application of research so as to have tangible effects on individual farmers, governmental policies and/or any other level of governance or commercial activity would constitute what is now generically referred to as 'impact' in the UK research context. 'Impact' is now a critical element of any research funding proposal from public funds (e.g. the UK Research Councils) and research quality assessments that determine how much money is given to each individual university to support research. It is arguably the case that measurable or evidenced research 'impact' is more likely to arise from more integrated and synthetic research projects that incorporate and integrate a greater number and type of research 'elements', that is, methodologies/techniques, disciplines, commercial or other organisational partners (providing end-user perspectives).

The strength of geomorphology as an academic discipline lies in its fundamental importance to so many aspects of society, thus being well placed to succeed in the increasingly selective and output-driven research funding environment in the United Kingdom and many other countries. Multinational engineering companies have built up substantial and interdisciplinary geomorphology teams in recent years, and insurance companies as well as national governments continually seek improved evidence of probabilities of damaging natural hazards occurring requiring hydrological, ecological, geomorphological and socio-economic research. The latter example provides a practical application of geomorphological research to the issue of tectonic hazards, such as interpretations of new evidence for frequency, nature, scale and thus possible impacts of earthquakes, volcanic eruptions and tsunami in particular locations. It is perhaps unfortunate that insurers probably tend to use the higher end of any range of probabilities of occurrence (i.e. more likely), whereas policymakers may be keen to emphasise the lower end in order to defer having to commit to public spending on mitigation measures! However, as more different types of evidence are obtained and integrated, often from different disciplines and using different approaches and techniques, so the uncertainties are reduced, weights of evidence increased and overall credibility in any headline message is increased among the potential end users. John Thornes showed how geomorphology could integrate and structure information from other disciplinary areas and recognised the need for understanding complex systems in a non-reductionist way (e.g. Thornes 1985; Wainwright 2009). His collaboration with Antonio Gilman demonstrated the value of geomorphology to archaeology (Gilman and Thornes 1985), and there are many other examples of interdisciplinary research and commercial applications in ecology (Nortcliff and Thornes 1988), hydrology (Francis and Thornes 1990; Dykes and Thornes 2000), climate (Thornes and Alcántara-Ayala 1998) and society (Trimble and Thornes 1990; Thornes 2005).

Aims, purpose and structure of this book

The aim of this book is to demonstrate the lasting significance of the integrated monitoring and modelling philosophy of geomorphological research pioneered by John Thornes, by reporting recent and ongoing research and applied work that utilise this approach. This collection of work includes a wide range of types of research and applied studies undertaken using this integrated approach and, as such, serves to demonstrate the value and effectiveness of adopting such an investigative framework. The authors of all of these works are academics and practitioners who were either taught by John Thornes as postgraduate (and in one case undergraduate) students or who worked with him on one or more research projects, and their present colleagues and research partners. John's inspirational teaching and leadership is reflected in the enthusiasm with which the contributors agreed to write for this book, which should also serve as a lasting tribute to his academic career as a leading geomorphologist. The book is structured in two parts. The first is made up of studies inspired and informed by the philosophy outlined earlier, while the second reflects more directly on John's own contributions to the discipline.

Part 1 of the book is made up of chapters that vary in emphasis from data collection (monitoring) at one extreme to the application of models to policy development at the other. The first six chapters cover this range and, as such, broadly parallel the changing focus of John Thornes' research in Spain and, later, most of southern Europe. Romero Díaz and Ruíz Sinoga (Chapter 2) examine the measurement of soil erosion in Spain, starting with methods initiated by John Thornes. Through their examination of scale effects and other factors relating to the measurement and monitoring techniques used, they highlight the potential difficulties of obtaining data that can be used reliably to quantify soil loss and, thus, set up and indeed validate models of soil erosion for varying soil, slope and vegetation conditions in particular. Cerdà et al. (Chapter 3) report findings from a later stage of experimental fieldwork that examined the influence of shrubs on the hydrological and erosional conditions of Spanish hillslopes, demonstrating an effect identified and argued by John Thornes long before relevant monitoring was undertaken to provide data that could enable the effect to be modelled. Hooke and Mant (Chapter 4) further highlight the difficulties of parameterising models in their presentation of combined flood and vegetation change data for ephemeral channels in Spain. In this study, the integration of two themes – channel morphology and vegetation dynamics – appears to have greatly increased the apparent level of complexity inherent in the connectivity between these landscape components.

More emphasis on the value of modelling is provided by Wainwright (Chapter 5) who discusses conceptual frameworks for attempting to analyse the nature of the complexities of past land use and associated geomorphological change in the Mediterranean region. The key element here is the development of a non-linear modelling framework that integrates not only vegetation and erosion processes but also the evidence of patterns of human settlement, effects of agricultural practices and, often, later abandonment. In Chapter 6, Brandt and Geeson explain how the scientific findings of the desertification research relating to contemporary environmental and socio-economic pressures, that is, incorporating anthropogenic factors integrated with knowledge of the range of geomorphological and ecological processes, were made available to end users. This chapter demonstrates the importance of engaging with all types of end users from practitioners to managers, as well as other scientists, while undertaking the research and designing the outcomes of the research to be as user-friendly as possible across the board. Mulligan (Chapter 7) takes many of these ideas a further step forward by presenting the historical timeline of a web-based modelling framework that simulates the effects of particular policy decisions relating to management of sensitive and highly variable landscapes in the Mediterranean and beyond. With provenance in the early monitoring and modelling of Mediterranean desertification under John's supervision, this framework is now a sophisticated spatial policy support system in wide use at various levels of decision-making around the world.

Subsequent chapters present hydrological, geomorphological and environmental management studies undertaken in a variety of global contexts, reflecting the very wide range of John's interests and activities beyond his core focus on the Mediterranean region. His particular concern for water resources and water management in semi-arid environments and elsewhere, as well as natural hydrological processes and systems, provide a common theme for Chapters 8–10. In Chapter 8, Francis and Lowe examine the studies behind the preparation of a strategic integrated regional plan for sustainable development in the Rift Valley Lakes Basin of Ethiopia and in particular the assessment of environmental impacts of any natural resource exploitation. There is an emphasis on water resources throughout a programme of work that necessarily and explicitly followed an 'integrated philosophy'. Watts, on the other hand (Chapter 9), focuses on the modelling of water resource planning for the United Kingdom, highlighting both the modelling and management issues driving the approaches used and the application of the research findings. Marsik et al. (Chapter 10), concerned with changing river flows potentially reducing water supplies in Costa Rica, present research findings that attempt to explain the changing discharges in terms of catchment-wide land-use and/or climate change. They highlight the importance of combining multiple approaches in order to adequately investigate the problem.

The next three chapters focus on different types of geomorphological research. Afana and del Barrio (Chapter 11) present an adaptive model of channel network development that has been designed to facilitate the future integration of additional catchment properties to enhance the depiction of landscape change. The DEM framework provides the basis for parameterisation, in contrast to the work reported by Dewez and Stewart (Chapter 12) in which the DEM *is* the landform model. Here, the results of a semi-automated analytical procedure are interpreted with respect to the local tectonic context to explain the formation of a sequence of coastal terraces. In Chapter 13, Dykes and Alcántara-Ayala use slope stability models as part of an investigation of the role of land-use changes in the incidence of damaging landslides in Mexico. The context of land-use monitoring is reviewed, and the difficulties of utilising such land-use data in stability and hazard analyses are demonstrated.

Part 2 comprises a short collection of overviews of John's contribution to geomorphological research and the application of that research to real-world problems, rounded off with a short biography, written by some of his major contemporaries, colleagues and friends. The purpose of these chapters is to provide a scientific and historical context for (i) the wide range of related contemporary research in semi-arid regions and particularly in Mediterranean countries and (ii) the interests and scientific approaches utilised by students and colleagues of John that were strongly shaped by his influence.

We believe that John would have approved of the continuing sense of scientific rigour and exploration embodied in the studies presented here. That his influence continues on to present and future generations is a fundamental reflection of the significant advances that he made both directly in the discipline of geomorphology and more broadly in developing approaches to sustainable management of the environment.

References

BBC (2006) *Torchwood* (Series 1) opening sequence. Television series created by Russell T. Davies. British Broadcasting Corporation, Cardiff.

Brandt, C. J., Thornes, J. B. (1996) *Mediterranean Desertification and Land Use*. John Wiley & Sons, Ltd, Chichester.

Brown, A. G., Tooth, S., Chiverrell, R. C., Rose, J., Thomas, D. S. G., Wainwright, J., Bullard, J. E., Thorndycraft, V., Aalto, R., Downs, P. (2013) ESEX commentary. The Anthropocene: is there a geomorphological case? *Earth Surface Processes and Landforms* **38**, 431–434.

Dykes, A. P., Thornes, J. B. (2000) Hillslope hydrology in tropical rainforest steeplands in Brunei. *Hydrological Processes* **14**, 215–235.

Embleton, C., Thornes, J. B. (eds.) (1979) *Process in Geomorphology*. Edward Arnold, London.

Francis, C. F., Thornes, J. B. (1990) Runoff hydrographs from three Mediterranean vegetation cover types. In: Thornes, J. B. (ed.) *Vegetation and Erosion*. John Wiley & Sons, Ltd, Chichester, pp. 363–384.

Geeson, N. A., Brandt, C. J., Thornes, J. B. (eds.) (2003) *Mediterranean Desertification: A Mosaic of*

Processes and Responses. John Wiley & Sons, Ltd, Chichester.

Gilman, A., Thornes, J. B. (1985) *Land Use and Prehistory in South East Spain*. George Allen & Unwin, London.

Hooke, R., Le B. (2000) On the history of humans as geomorphic agents. *Geology* **28**, 843–846.

Montgomery, D. R. (2007) Soil erosion and agricultural sustainability. *Proceedings of the National Academy of Sciences of the United States of America* **104**, 13268-13272.

Mulligan, M., Wainright, J. (2004). Modelling and model building. In: Wainwright, J., Mulligan, M. (eds.) *Environmental Modelling: Finding Simplicity in Complexity*. John Wiley & Sons, Ltd, Chichester, pp. 7–73.

Nortcliff, S., Thornes, J. B. (1988) The dynamics of a tropical floodplain environment with reference to forest ecology. *Journal of Biogeography* **15**, 49–59.

Steffen, W., Persson, A., Deutsch, L., Zalasiewicz, J., Williams, M., Richardson, K., Crumley, C., Crutzen, P., Folke, C., Gordon, L., Molina, M., Ramanathan, V., Rockström, J., Scheffer, M., Schellnhuber, H. J., Svedin, U. (2011) The Anthropocene: from global change to planetary stewardship. *Ambio* **40**, 739–761.

Thornes, J. B. (1975) Lithological controls of hillslope erosion in the Soria area, Duero Alto, Spain. *Boletin Geologico y Minero* **85**, 11–19.

Thornes, J. B. (1985) The ecology of erosion. *Geography* **70**, 222–235.

Thornes, J. B. (1990) The interaction of erosional and vegetational dynamics in land degradation: spatial outcomes. In: Thornes, J. B. (ed.) *Vegetation and Erosion*. John Wiley & Sons, Ltd, Chichester, pp. 41–53.

Thornes, J. B. (2005) Stability and instability in the management of Mediterranean desertification. In: Wainwright, J., Mulligan, M. (eds.) *Environmental Modelling: Finding Simplicity in Complexity*. John Wiley & Sons, Ltd, Chichester, pp. 303–315.

Thornes, J. B. (2007) Modelling soil erosion by grazing: recent developments and new approaches. *Geographical Research* **45**, 13–26.

Thornes, J. B., Alcántara-Ayala, I. (1998) Modelling mass failure in a Mediterranean mountain environment: climatic, geological, topographical and erosional controls. *Geomorphology* **24**, 87–100.

Thornes, J. B., Brunsden, D. (1977) *Geomorphology and Time*. Methuen, London.

Trimble, S. W., Thornes, J. B. (1990) Geomorphic effects of vegetation cover and management: some time and space considerations in prediction of erosion and sediment yield. In: Thornes, J. B. (ed.) *Vegetation and Erosion*. John Wiley & Sons, Ltd, Chichester, pp. 55–65.

Wainwright J. (2009) Earth-system science. In: Castree, N., Demeritt, D., Liverman, D., Rhoads, B. (eds.) *A Companion to Environmental Geography*. Wiley-Blackwell, Chichester, pp. 145–167.

Wainwright, J., Thornes, J. B. (2003) *Environmental Issues in the Mediterranean: Processes and Perspectives from the Past and Present*. Routledge, London.

Wilkinson, B. H. (2005) Humans as geologic agents: a deep-time perspective. *Geology* **33**, 161–164.

PART A

CHAPTER 2

Assessment of soil erosion through different experimental methods in the Region of Murcia (South-East Spain)

Asunción Romero Díaz[1] and José Damián Ruíz Sinoga[2]

[1] Department of Geography, University of Murcia, Murcia, Spain
[2] Department of Geography, University of Málaga, Málaga, Spain

Introduction

The Region of Murcia has been one of the pioneering Spanish regions in studying and assessing water erosion processes using experimental methods (García Ruiz 1999; Cerdà 2001; García Ruiz et al. 2001; Romero Díaz 2002; Solé Benet 2006; Añó Vidal et al. 2009). Studies on erosion plots started in the 1980s and were carried out almost simultaneously by two research groups. One of these research groups was coordinated by Professor López Bermúdez from the Geography Department at the University of Murcia (UMU), and the other was led by Juan Albaladejo, a researcher at the Centre for Soil Science and Applied Biology of the Segura (CEBAS)-Higher Council of Scientific Research (CSIC) who was in charge of assessing the Abanilla–Fortuna basin. Up to then, the existing values for erosion rates in the Region of Murcia had been calculated using the Fournier index (Fournier 1960) or models such as the USLE, which have been shown to overestimate erosion when compared to measured rates (Soto 1990).

This chapter presents erosion data obtained using several different techniques and methods originally initiated in the Region of Murcia by Professor John Thornes, with the aim of establishing, to the greatest extent possible, a comparison of their results. Needless to say, each method measures different processes at different scales, but they all contribute to the understanding of the causes of the wide range of values of erosion rates that are observed.

The study area: The Region of Murcia

Soil erosion is favoured in those environments where two important factors converge: rainfall erosivity and high soil erodibility. These two factors occur together in the Region of Murcia. From a climatic point of view, the Region of Murcia has a semi-arid climate, with low rainfall (300 mm a year on average) and high evapotranspiration. One of its most characteristic climatic traits is the high irregularity and variability of rainfall. Periods of long droughts are found together with heavy rainfall which, at certain times, can have catastrophic consequences due to flooding (Romero Díaz 2007). Forest-restoration policies have been, and still are, a priority in this region. On the other hand, it is important to mention the role of heavy

---

Monitoring and Modelling Dynamic Environments, First Edition. Edited by Alan P. Dykes, Mark Mulligan and John Wainwright.

rainfall periods (Albaladejo et al. 2006). Several studies based on experimental plots have suggested that a small number of heavy rainfall events have caused most of the soil loss in the Region of Murcia. López Bermúdez et al. (1986) showed in the Rambla de Gracia how 86% of soil loss was registered during three storms; Castillo et al. (1997), in a study which was also carried out on experimental plots in the Abanilla basin, pointed out that 80% of the total soil loss was caused by five storms in 5 years; and Martínez-Mena et al. (2001) illustrated how 70–80% of the total surface runoff and soil loss was caused by four extreme events in two experimental microcatchments in the northeast of the region.

The soil is poorly developed in the Region of Murcia, a fact that strongly affects its vulnerability to erosion. From a lithological point of view, there are numerous Neogene–Quaternary basins (Romero Díaz and López Bermúdez 2009) with soil parent materials dominated by marls, conglomerates and gypsums, all of which are particularly prone to erosion, forming gullies and badland landscapes (Figure 2.1). Due to the climatic conditions, the region has a sparse vegetation cover over a large part of its territory that provides only limited soil protection. Various studies can be found on the role of vegetation as a soil-protection factor. In the case of the Region of Murcia, the first works carried out, with the help of Professor John Thornes, were López Bermúdez et al. (1984, 1986), Francis et al. (1986), Fisher et al. (1987), Romero Díaz et al. (1988) and Francis and Thornes (1990). They were followed by Martínez Fernández et al. (1995), Romero Díaz et al. (1995), Castillo et al. (1997), López Bermúdez et al. (1998) and Gómez Plaza et al. (1998). In addition, Belmonte Serrato and Romero Díaz (1992, 1999) and Belmonte Serrato et al. (1999a, b) explicitly examined the role of vegetation in rainfall interception as part of this programme of research. Other studies showed how once the vegetation cover has been eradicated, its recovery is sometimes very slow and, on particular occasions,

impossible to achieve (Barberá et al. 1997; Castillo et al. 1997; Albaladejo et al. 1998).

A third factor is human actions. Human beings have favoured the acceleration of certain erosion processes by using inadequate agricultural and farming practices and, more recently, abandoning crops that were formerly terraced (Sánchez Soriano et al. 2003; Lesschen et al. 2007; Romero Díaz et al. 2007a). Sometimes, crops are found on steep slopes where high soil losses have been reported, as in the case of the western area of Lorca and the northern area of Puerto Lumbreras (Ortiz Silla et al. 1999). Ploughing in the same direction as the slope is the usual practice in the region, but it also promotes erosion. This effect has been experimentally corroborated in the field at El Ardal, where a plot that was ploughed in this direction always registered the highest erosion rate (Romero Díaz et al. 1995).

Experimental sites in the Region of Murcia

The Department of Physical Geography at the UMU has investigated erosion processes at several experimental sites including (1) Rambla de Gracia, (2) El Ardal, (3) Los Guillermos and (4) El Minglanillo:

1 The first experimental site was set up in the Rambla de Gracia (in the Mula basin) in November 1983 with the participation of Professor John Thornes (Francis et al. 1986). It was a 60 × 50 m (3000 m²) plot, placed on a hill with sparse vegetation cover. Ten Gerlach troughs were installed and distributed over the surface area. Six tanks of 972 l each were also placed at the base level of two small microcatchments within the area to collect the surface runoff. The site was eventually abandoned because of vandalism to the experimental equipment.

2 The experimental site of El Ardal was set up in 1989 and also benefitted from the presence of Professor Thornes. This area is located in the headwaters of the Mula basin, over a limestone substrate, with an average slope gradient of

Figure 2.1 Location of the Region of Murcia and erosional areas with gullies and badlands (dark grey scale) (MMA 2002).

20%, good vegetation cover with Mediterranean scrubland dominating the top part and a cereal field at the lower part (López Bermúdez et al. 1998). The cereal crop has now been replaced by an almond orchard. The experimental site houses a small microcatchment of 2 ha where a flow-measuring flume was installed with a level sampler for assessing suspended sediments. Seventeen closed plots were located, 12 of 16 m² and 5 of 20 m², with different orientations, slope gradients, density of vegetation cover and land use

(abandoned, in crop rotation, cultivated or using certain farming systems). Apart from measuring erosion and surface runoff, other experiments have been carried out on this site to assess fertility loss (Alias et al. 1997) or the role of vegetation, particularly evolution of the vegetation cover, biomass and leaf production (Martínez Fernández et al. 1993, 1995) and interception studies (Belmonte Serrato and Romero Díaz 1999). The erosion plots in El Ardal are no longer functional since they showed symptoms of soil exhaustion

(Belmonte Serrato et al. 2002). In the MEDALUS project, led by Professor Thornes in its first, second and third phases (1991–1998: see Chapter 16 by Kirkby et al., this volume), El Ardal was one of the study fields selected, and the data it provided would serve to test erosion models (Kirkby et al. 2002). Later on, the Guadalentín basin became the main experimental site, with numerous researchers from different countries working on it (Geeson et al. 2002).

3 The experimental site in Los Guillermos was set up in 1991, in the middle of the Rambla Salada, to the south of the Mula basin. The selected area is a badland landscape in a Neogene–Quaternary sedimentary basin formed by marls, conglomerates and sands (López Bermúdez et al. 1992). In the gully headcut, three stations for flow measurement with flumes were installed with different orientations in three microcatchments of 120 500 and 3 000 m²; five erosion plots of 16 m² with different slope gradients, vegetation cover and stony surfaces were also established. Several studies on soil micromorphology (Conesa García et al. 1994, 1996) and numerous rainfall simulations (Fernández Gambín et al. 1995, 1996) were carried out on this site. However, the great quantity of sediments deposited after rainfall events rendered the measuring systems useless and continuous vandalism also caused the site to be abandoned.

4 In 1995, the experimental site of El Minglanillo was set up in the basin of the Rambla Salada on a Quaternary glacis over Tertiary marly materials (López Bermúdez et al. 2000). Three plots of 20 m² were installed on the bare soil of an abandoned crop, on harvested soil and under natural vegetation. Data obtained from these plots on marls were of great interest in order to compare them with plots on limestone (Romero Díaz and Belmonte Serrato 2002). Following these studies, numerous experiments have taken place in other areas of the region, especially

in the basin of the Quípar river and in the Guadalentín valley (Figure 2.2).

Studies of erosion have been carried out at experimental stations of the CEBAS-CSIC:

5 The first station was set up in Abanilla in 1988. It was formed by two closed plots of 87 m² each, and the aim was to study the soil-degradation processes induced by the impoverishment of vegetation cover and the validation of physically based soil erosion models (Albaladejo and Stocking 1989).

6 In Santomera, two plots of 15 × 5 m were installed in 1989 (Castillo et al. 1997; Albaladejo et al. 1998; Martínez-Mena et al. 2002), one maintaining the original vegetation (open scrubland with pines) and another eradicating the vegetation cover in order to check the effects on erosion, surface runoff and soil properties.

7 In 1989, another experimental site with five plots was installed in Abanilla. The aim was to study the effects of incorporating different quantities of organic urban solid refuse (USR) into underdeveloped and degraded soils, formed from marls affected by intense erosion processes. One of the plots was left as a control, and different quantities of USR were incorporated into the other four (Albaladejo et al. 2000).

8 In 1991, the experimental microcatchments of Abanilla with 759 m² and Color with 328 m² were installed in the Chícamo river basin, on marls and Quaternary deposits, with dispersed small-sized bushes (*Stipa tenacissima* and *Rosmarinus officinalis*) and soils with low permeability and organic matter (Martínez-Mena et al. 2001).

9 In 1997, another experimental site with three basins was installed to the south of the Sierra del Picarcho (Venta del Olivo), with the aim of studying the factors which control surface runoff processes at different working scales, analysing the variability of these factors and developing and validating hydrological simulation models adapted to these environments. Two of the basins, with surface areas of 7.9

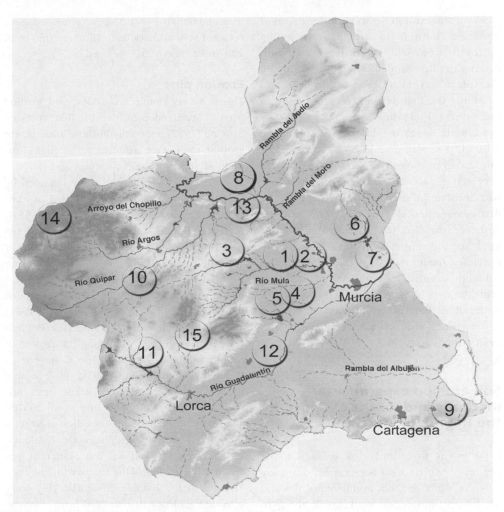

Figure 2.2 Main areas of the Region of Murcia where erosion processes have been studied. 1. Rambla de Gracia; 2. Rambla Honda; 3. El Ardal; 4. Los Guillermos; 5. El Minglanillo; 6. Abanilla; 7. Santomera; 8 Sierra del Picarcho; 9. Sierra de Cartagena-La Unión; 10. Quipar river basin; 11. Head of the Giadalentín basin; 12. Guadalentín river basin; 13. Cárcavo basin; 15. Cañada Hermosa.

and 6.4 ha, were located in a field that burned in 1994 with *S. tenacissima* in regeneration; the third one, with a larger surface area (24.3 ha), was not affected by burning and was dominated by vegetation cover with *S. tenacissima* and clear pine forest of *Pinus halepensis*. In each of the basins, a sediment sampler was installed with a battery of five bottles at different heights. Six closed plots of 30 m² (10 × 3 m) were set up in the basins, two on *S.*

tenacissima in a burned area, two on *S. tenacissima* in an unburned and well-developed area and two in a pine forest. Nine microplots of 0.24 m² were also set up for rainfall simulation experiments in areas with different soil and vegetation characteristics. The period of activity went from 1997 to 2003 (Castillo et al. 2003).

In the case of the studies carried out by the Department of Physical Geography, experimental work on soil erosion began with the arrival of Professor Thornes. He participated in setting up the first erosion plots in the Mula basin, making the first rainfall simulations, studying the role of scrubland in protecting soil and obtaining the first data on soil erosion. From there onwards, a whole series of projects was run among which the different phases of the MEDALUS programme deserve a special mention. In all of them, Professor Thornes took a very active part. His influence on this work was decisive; he participated in obtaining the first experimental data on erosion, co-authored the first publications and contributed to the enrichment of the work by providing useful methodological training and valuable scientific rigour. This chapter presents the data obtained from the many studies outlined earlier, using different techniques and methods, with the aim of comparing and evaluating the results.

Soil erosion experimental methods and results obtained in the Region of Murcia

Experiments to quantify soil erosion in the Region of Murcia have been carried out using different study methods at different scales and on different lithologies (marls, marls and gypsums, limestones, sandstones, conglomerates, schists and phyllites) and land uses (fields with crops, semi-natural vegetation, forests and abandoned fields). The methods used depend on the scale of study: (i) on a millimetric scale, erosion pins and microtopographic profiles (Sancho et al. 1991) are normally used; (ii) on a hillslope scale, the most commonly used methods are erosion plots (López Bermúdez et al. 1993), rainfall simulations (Calvo et al. 1988; Cerdà 1999), surveying and geomorphological transects (Ortigosa Izquierdo 1991; Ruiz Flaño 1993; Romero Díaz and Belmonte Serrato 2008); and (iii) on a basin scale, flow measurement with flumes and bathymetry of reservoirs are used (Avendaño and Cobo 1997; García-

Ruiz and Gallart 1997). More recently, sediment retained behind check dams (Romero Díaz et al. 2007b) has also been measured.

Erosion pins

This is a very simple and widespread method used in Spain, adopted from the first works by Scoging (1982), especially in those areas where erosion rates were assumed to be high. This method generally lacks precision and tends to provide higher values than those obtained by other methods such as erosion plots. This comparison has been discussed by Haigh (1981), Sirvent et al. (1997), Benito et al. (1992), Cerdà (2001) and Wainwright and Thornes (2004).

In the case of the Region of Murcia, erosion pins have been used for different purposes, for example, (i) calculating the volumes of eroded material in the headcut of gullies (Francis 1985) and the dynamics of piping processes (Romero Díaz et al. 2009); (ii) estimating the erosion of mining waste in the Sierra de Cartagena-La Unión (Moreno Brotons 2007); (iii) assessing the sediments retained by check dams after each period of rainfall (Romero Díaz et al. 2007c); and (iv) checking the role of different vegetation species in retaining sediments on hillslopes. This last aspect was studied using different species at the beginning of the 1980s, with Professor Thornes in the Rambla de Gracia. This work provided evidence that showed how *S. tenacissima* was, out of all the studied species (*Stipa*, *Rosmarinus* and *Thymus*), the one which retained most sediments. For this reason, *S. tenacissima* could be considered as one of the potentially most effective species for stabilising hillslopes and reducing erosion. Unfortunately, the results of these observations were not published. Sánchez (1995) reached the same conclusion in his doctoral dissertation on the architecture and dynamics of *S. tenacissima* plants.

During 3 years of sediment research (2003–2005) following the placement of numerous erosion pins upstream from 51 check dams within the Quípar river basin (Romero Díaz et al. 2007b), the erosion rate from the 3-year period of sediment research and from the total

Table 2.1 Erosion rates obtained by erosion pins in check dams, Quípar river basin.

Sub-basin	Number of controlled check dams	Rate from the 3-year period (t ha^{-1} year^{-1})	Rate from the total period (t ha^{-1} year^{-1})
1, 2, 3	11	0.11	0.71
4	11	0.90	3.39
5	5	1.49	2.64
6	6	2.41	5.29
	Total = 51	Mean = 1.37	Mean = 3.95

From Romero Díaz et al. (2007b).

period of the useful life of the check dams was calculated (Table 2.1). The erosion rate from the 3-year period was lower than that from the total period because the rainfall registered during the 3-year period was not very high and did not, therefore, generate very active erosion.

Several measurements have been carried out using microtopographic profiles, but none with significant results. In the experimental field of El Minglanillo, the evolution of a network of rills in an almond plot of 34 × 100 m (López Bermúdez et al. 2000) started to be assessed, but unfortunately, the owner of the farm ploughed the field, putting an end to the experimental work.

Erosion plots

Erosion/runoff plots are the most widely used methods to quantify the amount of soil exported, primarily by interrill runoff. The typology of the plots used is very varied depending on the objectives, the area of study or the economic resources available (López Bermúdez et al. 1993). There are two basic types of plots: (i) open and (ii) closed. Most of the data obtained in Murcia came from closed plots.

Open plots

In open plots, collectors similar to those devised by Gerlach (1967) are installed to measure the transfer of sediments along a hillside. It has the advantage that measurements can be carried out for a long time, since water flow upstream is not interrupted. On the other hand, on closed plots, sediments are depleted and must be abandoned after some time (Belmonte Serrato et al. 2002). However, open plots also have some disadvantages, such as problems with the demarcation of their slope area and the overflowing of the collectors during intense rainfall.

The open plot type of collector was installed, with Professor Thornes' collaboration, in the experimental area of the Rambla de Gracia (López Bermúdez et al. 1986). The collectors were installed in 1984 on a hillside with low-density scrubland, and the soil loss data obtained during a period of 2 years (1985–1986) ranged from 1.8 to 3.2 t ha^{-1} year^{-1} (Francis 1986; Romero Díaz et al. 1988). These values correspond with those reported by other authors who have used the same method in other parts of Spain.

From 1996 to 1999, Cammeraat (2002, 2004) carried out studies with open plots in Cañada Hermosa (at the headwaters of Guadalentín basin), obtaining very low erosion values in scrublands and areas reforested with pine on limestones (0.08 t ha^{-1} on average per event). In contrast, in marl and valley bottom areas, his measured rates exceeded 30 t ha^{-1}.

Closed plots

Closed plots also have numerous disadvantages but are, nevertheless, a widely used method and can provide very interesting data (López Bermúdez et al. 1993). In Murcia, the first erosion data from closed plots were provided by Francis (1986). They came from plots of 3 × 1 m installed in crops abandoned for 1, 2, 5 and 20 years in the Rambla Honda. The rates obtained ranged from 0.8 to 5.3 t ha⁻¹ year⁻¹. Soil loss was higher in the fields abandoned for 20 years.

From 1989 to 1999, the experiments carried out in El Ardal (López Bermúdez et al. 1996) on a limestone substrate with good vegetation cover, and under different weather conditions, obtained low erosion rates. From 1989 to 1997, the average erosion rate was less than 1 t ha⁻¹ year⁻¹ (Romero Díaz et al. 1998). Nevertheless, when different soil uses were analysed, significant differences appear. The objective of this experimental site was not to obtain high erosion rates, but rather observe the existing variations of different parameters (soil uses, orientation and hillslope). Thus, plots covered by scrubland provided the lowest erosion rates (0.06–0.22 t ha⁻¹ year⁻¹), followed by those where bush vegetation had been cut, but still kept some vegetation (0.43–0.94 t ha⁻¹ year⁻¹), and those which were abandoned (0.01–0.50 t ha⁻¹ year⁻¹). Cultivated and ploughed plots provided the highest rates (0.78–1.20 t ha⁻¹ year⁻¹). Those plots with wheat and barley crops registered high erosion rates, coinciding with periods of intense rainfall at the time when soil was least protected (from the middle of summer to winter). The highest rates were found in plots ploughed up and down (5.92 t ha⁻¹ year⁻¹) (see Figure 2.2). The experiments in El Ardal and similar ones carried out in Is Olias (Italy) and Spata (Greece) demonstrated the importance of land use and the consequences that changes in land use can have for increasing soil degradation and erosion (Romero Díaz et al. 1999).

The erosion rates obtained on marls in El Minglanillo were much higher. For the 1997–1999 period, the rates were 7.47 t ha⁻¹ year⁻¹ on crops, compared with 0.80 t ha⁻¹ year⁻¹ on scrubland and 1.12 t ha⁻¹ year⁻¹ on abandoned fields. The comparison carried out for the same rainfall events and the same period of study by Romero Díaz and Belmonte Serrato (2002) in the two experimental sites of El Ardal and Minglanillo illustrated the influence of lithology. For the same plots, López Bermúdez et al. (2000) provided values for 5 years (1996–2000) which ranged from 6 to 15 t ha⁻¹ year⁻¹

on crops, 0.7–1.4 t ha⁻¹ year⁻¹ on scrubland and 0.14–2 t ha⁻¹ year⁻¹ on abandoned fields.

The results obtained in the experiments carried out by CEBAS on the plots installed in Santomera showed the contrast between those areas covered by vegetation and those lacking it (Figure 2.3). After a follow-up period of 55 months, one of the plots deprived of vegetation

(a)

(b)

Figure 2.3 Experimental plots in Santomera: (a) adjacent plots with and without vegetation (CEBAS) and (b) an unvegetated plot ploughed up and down a hillslope after a rainfall period in El Ardal (UMU).

displayed a noteworthy loss of organic matter and a prominent decrease in the percentage of stable aggregates. Soil loss increased by 127% in the disturbed plot, which was attributed to a progressive deterioration in the physical properties of the soil. No symptom of natural recovery of the plot without vegetation was observed. Rather, the tendency in soil behaviour was towards a state of degradation (Castillo et al. 1997; Albaladejo et al. 1998). By analysing also the distribution of sediment particles, Martínez-Mena et al. (1999) observed how the vegetation cover reduced the eroding power of rain by around 50% and also reduced runoff by around 75%, especially during high-intensity rainfalls.

From October 1988 to September 1993, experiments carried out with USR in the Abanilla basin using different rates of addition (6.5, 13.0, 19.5 and 26.0 kg m^{-2}) to the soil confirmed a notable decrease in runoff and soil losses. Physical properties of soil were improved, increasing the productivity and, therefore, protecting the soil against erosion. The control plot displayed values that were much higher than those in treated plots, independently of the addition rates of USR. A rate of around 10 kg m^{-2} could be optimal for control of runoff and erosion; higher rates were not necessary and could increase the risk of soil pollution (Albaladejo et al. 2000).

Research carried out for 4 years on closed plots of 30 m^2 in the Sierra del Picarcho (Venta del Olivo) in areas of burned scrubland and unburned scrubland and pine forests showed higher rates of sediment production in the burned scrubland (0.54 t ha^{-1} year^{-1}) when compared with the unburned one (0.03 t ha^{-1} year^{-1}). In general, the erosion rates were low due to the existing vegetation cover, and the experiments demonstrated the role of high-intensity rainfall in generating runoff and erosion (Boix-Fayos et al. 2007). Notable differences were reported between burned and unburned areas. The runoff ratio was much higher (80–90%) in burned areas, which further demonstrated the important role of vegetation as a protecting factor.

On different lithologies (marls, conglomerates and schists) of the Guadalentín basin, Romero Díaz and Belmonte Serrato (2008) installed closed plots of 10×2 m during the years 2005–2006 in places with dispersed natural scrubland and near reforested areas. The aim was to compare erosion rates between forested and non-reforested areas. The rates obtained were very low on all the plots, being higher on marls (1.86 t ha^{-1} year^{-1}), followed by schists (0.11 t ha^{-1} year^{-1}) and lastly conglomerates (0.06 t ha^{-1} year^{-1}). This work illustrated differences between lithologies and the protecting role of shrubland cover, even when that cover is scarce and small in size.

Experimental catchments

Studies made on a catchment scale in the Region of Murcia are not so numerous. The first studies were carried out in small microcatchments: (i) in the Rambla de Gracia (Mula basin) on an area of 3000 m^2 from 1984 to 1986 and (ii) in the Chícamo river basin in two small microcatchments of 328 and 759 m^2 (Color and Abanilla) from 1990 to 1993. In the Rambla de Gracia, the erosion rates ranged from 0.08 to 2.36 t ha^{-1} year^{-1} (Romero Díaz et al. 1988), and in the second site, the rates ranged from 0.85 to 2.99 t ha^{-1} year^{-1} (Martínez-Mena et al. 2001).

From 1997 to 2003, there were experiments in catchments of greater surface areas (6.4, 7.9 and 24.3 ha) which took place in the Sierra del Picarcho. The erosion rates were very low compared with those in the microcatchments with mean values ranging between 0.034, 0.011 and 0.015 t ha^{-1} year^{-1} (Boix-Fayos et al. 2006). Other experiments have been carried out in different parts of the Guadalentín basin by Hooke and Mant (2000), Oostwoud Wijdenes et al. (2000) and Cammeraat (2002, 2004).

Despite the difficulties involved in the hydrological study of experimental catchments in semi-arid areas due to low rainfall, their importance has been strongly demonstrated. However, establishing the rates for erosion or sediment export is not as important (García-Ruíz and López Bermúdez 2009) as defining the factors that control runoff and erosion. Understanding the spatio-temporal variations that take place in the basins in order to predict their hydrological response is also an important factor to take into account.

Rainfall simulations

Rainfall simulations have normally been used to estimate the infiltration capacity of soils, compare the hydrological responses of different microenvironments and analyse the specific role that surface characteristics (Albaladejo et al. 2006). Sometimes, they are also used to estimate erosion rates. Nevertheless, these data are very difficult to extrapolate due to the small surface areas where simulations are carried out, the great variability in rainfall intensity and quantity and the different types of simulators (Cerdà 1999). Despite these, the values obtained using simulations allow us to compare surface areas, seasons and soil uses and identify the areas more prone to producing large quantities of sediments.

There are different types and different sizes of simulators (Calvo et al. 1988; Cerdà 1999). In the Region of Murcia, simulators of 0.25, 2 and 20 m² have been used. One of the first rainfall simulation experiments in Spain was done in Murcia in the early 1980s under the initiative of Professor Thornes (Francis 1986; Francis and Thornes 1990). In the 72 experiments carried out by Francis (1986) in the Mula basin, the rates obtained on soils with different periods of abandonment and in different hydrological conditions (wet and dry) were relatively high due to the heavy simulated rainfall that was applied. The average erosion rate on dry soil decreased from 7.7 t ha⁻¹ year⁻¹ in recently abandoned land to 1.6 t ha⁻¹ year⁻¹ in soil abandoned 20 years before. In experiments performed on wet soils, erosion rates were slightly higher.

Calvo et al. (1991) performed different simulations in the badlands located to the south and south-east of the Iberian Peninsula. One of the selected areas was the badlands formed in the Plio-Quaternary near Sucina (Campo de Cartagena).

In the Guadalentín basin, the simulations run by Cerdà (1997) on different soil types showed that cultivated soils did not generate runoff due to their high macroporosity and that soils abandoned for 3 years generated between 0.2 and 6.4 t ha⁻¹ year⁻¹. Rates decreased when the abandonment period was longer or when there was more vegetation cover. Depending on the type of soil and the level of rainfall, abandonment may increase or reduce the erosion process.

In the Abanilla basin, Martínez-Mena et al. (2001) obtained higher averages on Miocene marls than on Keuper marl outcrops using a similar simulator (0.24 m²) to those employed by the aforementioned authors and under dispersed scrubland. In Fuente Librilla, Martínez-Mena et al. (2002) used a larger simulator (2 m²) in lemon tree groves and obtained higher soil losses than on Tertiary marls when compared to limestone colluvia.

The studies performed in El Ardal with a large simulator (2 × 10 m) (Figure 2.4) must be highlighted. In this case, very little runoff was registered during the simulations, which may be due to the lithological characteristics and the existing vegetation cover. Using simulated rainfall in El Ardal, Bergkamp et al. (1996) demonstrated that the infiltration pattern was related to vegetation organisation in patches separated by bare soil. On the bare soil, surface runoff and ponding are favoured, whereas vegetation favoured infiltration.

Bathymetry of reservoirs

The use of bathymetry in reservoirs is another method of estimating erosion rates. This method has also been used in the Segura basin (in one of the areas within the Region of Murcia). It is laborious as it requires the combined use of photogrammetry, bathymetric and sedimentological techniques, as well as the calculation of parameters such as the grading, density and composition of the sediment and the exploitation regime and retention capacity of the reservoir. The calculation of sediments retained by a dam provides data on 'specific degradation',

Figure 2.4 A 2 × 10 m rainfall simulator on a plot in the El Ardal experimental field.

that is, the total amount of sediment due to erosion and transport in a fluvial basin, expressed as mass per unit surface area per unit time. Bathymetry allows us to discover changes in the topography of the reservoir bottom, thus providing the information necessary to calculate the amount of sediment accumulated. This information, when extrapolated to the whole of the surface area of the draining basin, provides estimates of erosion rates. Moreover, the sedimentation rate can also be used to estimate the useful operating lifespan of the reservoirs.

Sedimentation retention coefficients are not the same in every reservoir. Some of them, like the Cenajo reservoir, retain approximately 98% of the incoming sediment (Almorox et al. 1994). Many of the reservoirs in the Segura basin have lost more than 30% of their capacity. Some examples of these are the Valdeinfierno (Figure 2.5), Puentes and Alfonso XIII reservoirs. The main factors are the erosive activity of the basin, the frequency of heavy rainfall, the construction age of the reservoir or the basin surface area.

The Segura basin has a wide range of reservoirs and is one of the most regulated basins in Europe. The building of the first dams dates back to 1884 (Puentes) and 1897 (Valdeinfierno). The basin surface areas range from less than 100 to

Figure 2.5 Valdeinfierno reservoir, full of sediments.

1200 km², and they are spread throughout the territory. At present, there are 31 reservoirs built in the Segura basin, and they fulfil different functions (regulation, irrigation, water supply, hydroelectricity generation and flood mitigation), with a total reservoir capacity of 1256 hm³.

Bathymetry has some obvious deficiencies. These include the presence of autochthonous sediments in reservoirs (e.g. the precipitation of carbonates found in the vertical profiles of the Puentes reservoir: Cobo et al. 1997), errors in surface-area measurements (mainly in the oldest reservoirs), losses due to infiltration, partial emptying due to the opening of the bottom gates or reservoir management. Despite all of these problems, it is still a useful method in order to know the specific degradation of a basin.

The data published by different authors for the main reservoirs in the Segura basin sometimes showed different values (Table 2.2). López Bermúdez and Gutiérrez Escudero (1982) obtained an average value of 8.3 t ha⁻¹ year⁻¹ for eight reservoirs and with bathymetries performed until 1976–1977; Soto (1990) and Romero Díaz et al. (1992) obtained lower average values of 4.3 and 4.7 t ha⁻¹ year⁻¹, respectively, for 11 reservoirs and with more recent bathymetries; Almorox et al. (1994) obtained a

similar value of 4.4 t ha⁻¹ year⁻¹ for seven reservoirs; and finally, Avendaño et al. (1997), who studied a longer period of bathymetries (until 1994), reduced the average to 3.3 t ha⁻¹ year⁻¹. The data obtained by Soto (1990), Romero Díaz et al. (1992) and Almorox et al. (1994) for the same reservoirs and same study period are very similar. Notable differences are highlighted by the data reported by López Bermúdez and Gutierrez Escudero (1982). These differences may be due to a possible error in the data, expressing m³ ha⁻¹ year⁻¹ as t ha⁻¹ year⁻¹. With this change in measurement units, the data from these authors correspond with the other results.

The existing variation in retained sediments in reservoirs is usually related to the reservoir surface area and shape, the reservoir management (whether it retains or does not retain rainfall), climatic, topographic and soil use characteristics, the presence of gullies and other catchment characteristics. In general terms, the highest erosion rates were recorded in small reservoirs due to the effect of the scale, and there was an inverse relation between specific degradation and basin dimensions. Nevertheless, Verstraeten et al. (2003) stated that the basin size only explains 17% of the variability of accumulated sediment in 60 Spanish reservoirs. In the reservoirs analysed, erosion values obtained by Romero Díaz et al. (1992) and Soto (1990) were the ones which showed a relationship with the basin size ($r^2 = 0.3913$). On the contrary, no significant relationship was found between the values obtained by López Bermúdez and Gutierrez Escudero (1982) and those reported by Avendaño et al. (1997) ($r^2 = 0.0046$). Avendaño et al. (1997), in a study carried out on 60 Spanish reservoirs, found a very good relationship between the basin surface area and the annual sediment contribution, having previously grouped basins with similar characteristics.

In a study carried out on 12 reservoirs of the Segura basin, Sanz Montero et al. (1998) found that specific degradation ranged from 1.85 to 6.97 t ha⁻¹ year⁻¹. Without including the Fuensanta reservoir which had the highest

Table 2.2 Erosion rates calculated using bathymetries of reservoirs in the Segura basin by different authors.

Reservoir	López Bermúdez and Gutiérrez Escudero (1982)		Soto (1990)		Romero Díaz et al. (1992)		Almorox et al. (1994)		Avendaño et al. (1997)	
	Period	t ha⁻¹ year⁻¹	Period	t ha⁻¹ year⁻¹	Period	t ha⁻¹ year⁻¹	Period	t ha⁻¹ year⁻¹	Period	t ha⁻¹ year⁻¹
Alfonso XIII	1916–1976	10.2	1916–1976	4.2	1916–1976	4.4	1916–1985	3.4	1916–1985	2.9
Anchuricas			1957–1979	4.1	1957–1979	4.3				
Argos	1970–1976	4.8	1970–1976	1.9	1970–1976	2.1			1970–1991	2.0
Camarillas	1960–1977	3.1	1960–1983	1.1	1960–1983	2.6	1960–1983	1.3	1960–1993	3.2
Cenajo			1960–1984	10.7	1960–1984	11.0	1960–1984	10.8	1960–1992	2.2
Fuensanta	1933–1977	14.0	1933–1977	7.4	1933–1977	10.0			1933–1991	6.8
La Cierva	1929–1976	6.7	1929–1985	2.3	1929–1985	2.4	1929–1985	2.6	1929–1987	2.8
Puentes	1884–1976	4.4	1884–1976	1.9	1884–1976	2.7	1884–1985	2.1	1884–1994	2.0
Taibilla			1973–1981	3.4	1973–1981	3.7			1973–1991	3.9
Talave	1918–1976	12.0	1918–1983	5.4	1918–1983	5.8	1918–1983	5.9	1918–1993	2.5
Valdeinfierno	1897–1976	11.0	1897–1976	4.7	1897–1976	2.8	1897–1984	5.1	1897–1984	4.8
Average		8.3		4.3		4.7		4.4		3.3

specific degradation, the data ranged from 1.85 to 3.86 t ha^{-1} year^{-1}. The reasons for this variation in the spatial distribution of values were not explainable by environmental factors (e.g. rainfall, soil uses, slope, geology) but were more related to human actions (e.g. public works, quarries, deforestation). When comparing erosion values obtained using bathymetries in the Segura basin with those from other basins, the Segura basin is among those with the highest specific degradation values (Avendaño et al. 1997).

Check dams

Another method to assess erosion, recently used in the Region of Murcia, is that of measuring the sediments retained in check dams. Building check dams is quite a usual method in forestry hydrological restoration projects. In the Segura basin, there are many such structures.

Romero Díaz et al. (2007b) studied 425 check dams in the Quípar river basin (814 km^2). They calculated the erosion rate in 195 of them, which were not silted up. This method is a relatively simple and efficient way to estimate erosion rates in basins. However, it requires data to be recorded for a long enough period of time, and the check dams must be distributed over the total surface area of the basin (Romero Díaz et al. 2007c).

To evaluate the volume of the sediments and the erosion rates, the dimensions of each check dam, the depth of the sediments and the area of the sedimentation 'wedge' were measured. The surface area of the wedges was measured directly in the field by GPS. The volume of the accumulated sediments was calculated from a geometric figure which best represented the three-dimensional shape of the sediment retained by each check dam, in most cases a vertical pyramid with a trapezoidal base (Hernández Laguna et al. 2004; Romero Díaz et al. 2007b). For those check dams whose sediment wedge did not correspond very well to the above form, other geometric figures, and even a combination of several figures, were used. Sediment mass was calculated by obtaining samples and determining the density of the material in the laboratory. The erosion rates are expressed as t ha^{-1} year^{-1}, calculated from the previously obtained sediment volume (t), the area of the check dam drainage basin (ha) and the length of time that the check dam has been operational (years).

The average erosion rate obtained for the whole of the Quípar basin was 3.95 t ha^{-1} year^{-1}. The variability was very high, with values ranging between 0.03 and 72.47 t ha^{-1} year^{-1}). More than half of the check dams displayed a very low erosion rate (less than 1 t ha^{-1} year^{-1}); 30% showed a rate between 1 and 5 t ha^{-1} year^{-1}; and in 8% of the check dams, the value registered was higher than 10 t ha^{-1} year^{-1} (Table 2.3). Of

Table 2.3 Check dams grouped according to erosion rates and main topographical characteristics, Quípar river basin.

Erosion rate (t ha^{-1} year^{-1})	Check dam		Mean erosion rate	Mean surface area of the basin (ha)	Mean slope of the channel
	Number	%			
<1	104	53.3	0.37	137.1	4.00
1–5	59	30.3	2.49	50.05	4.29
5–10	16	8.2	6.78	14.29	3.80
10–20	8	4.1	14.16	17.01	2.66
>20	8	4.1	45.29	17.63	2.66
Total	195	100	3.95*	90.86*	3.96*

From Romero Díaz et al. (2007b).
* Mean value for the 195 check dams.

the 195 check dams, 146 were built in 1962 and 49 in 1996. From those in the first group, 36 were enlarged in 1996 because they were full (Figure 2.6). It is important to note that check dams enlarged or newly built in 1996 displayed higher erosion rates. The reason for this increase may be that most of the new ones were built in areas with a high predominance of marls (Table 2.4). Therefore, lithology appears to be a determining factor in relation to erosion rates (Romero Díaz 2008). It must also be noted that the average erosion rate obtained in the Quípar

Figure 2.6 Silted-up check dam in the Quípar river basin.

basin is very similar to bathymetry values, but this may be simply a coincidence.

Studies on check dams in smaller basins (Cárcavo, Rogativa and Torrecilla) have been carried out by other authors. Castillo et al. (2007) studied the effectiveness and geomorphological impact of check dams on erosion control in the Cárcavo basin (a small basin of 27.3 km² located near the Quípar basin), where 32 check dams were analysed. Conesa García and García Lorenzo (2007) focused on the role of check dams in the hydrological dynamics of the Rambla de la Torrecilla (Guadalentín basin) with a surface area of 14.7 km² and 30 check dams in its channel. Boix-Fayos et al. (2007, 2008) and Castillo et al. (2009) studied the effectiveness of forestry hydrological restoration projects on erosion control in relation to soil-use changes in the Rogativa basin (north face of the Sierra de Revolcadores). This basin has a surface area of 47 km² and 58 check dams.

The study carried out in the Rogativa basin showed how the efficiency of check dams decreased as time passed, basin vegetation cover increased and the dams filled. It also showed that part of the sediments retained in check dams may come from erosion induced downstream from check dams or from soil movement during their construction. This fact had also been corroborated by the research carried out in the Quípar river basin (Romero

Table 2.4 Erosion rates, year of construction and predominant lithology in the basin of the Quípar river.

Year of construction	Number of check dams analysed	Erosion rates (t ha⁻¹ year⁻¹)	Number of check dams >5 t ha⁻¹ year⁻¹	% of check dams >5 t ha⁻¹ year⁻¹	Number of check dams by lithology*
1962	100	1.77	5	4.5	5 (1)
Improved 1962	36	4.00	10	27.7	10 (1)
1996	49	8.80	17	34.7	14 (1), 3 (2)
Total	195		32	66.9	29 (1), 3 (2)

From Romero Díaz et al. (2007b).
* Lithology = (1) marls, marlaceous lime and clays; (2) conglomerates, alluvium and sands.

Díaz and Belmonte Serrato 2009) and in the Cárcavo basin (Castillo et al. 2007). The average erosion rate calculated using sediments contained in check dams in the Rogativa basin is 5.39 t ha^{-1} year^{-1} (Boix-Fayos et al. 2008). As in the case of the Quípar river basin, the Rogativa basin also displayed a high variability, ranging from 0.25 to 107.33 t ha^{-1} year^{-1}. When compared to the Quípar river basin, the highest average rate in the Rogativa basin may be due, among other reasons, to the fact that the rates for the check dams that were completely filled with sediments were also estimated, whereas in the Quípar river basin, only the rates from functional check dams were calculated.

In the Rambla del Cárcavo and Torrecilla basins, the average erosion rates calculated were 6.61 and 3.21 t ha^{-1} year^{-1}, respectively (Conesa García and García Lorenzo 2007). The lower rate recorded can be explained by the lithology of slates and phyllites. However, to give more accurate values of erosion rates from dams, it is necessary to calculate the trap efficiency (TE). The check dams behave as small dams, and the rainwater is retained until it evaporates, infiltrates the soil or passes through the check dam. Since none of the check dams in question was constructed to be watertight and since not all the sediments are retained, TE should be calculated. The TE is a function of the capacity–inflow ratio, type of reservoir and method of operation (Brune 1953) and also the retention time of the water in the pond. Many methods exist to estimate the TE (Verstraeten and Poesen 2000, 2001). In the Quípar river basin, the method of Brown (1943) was followed. The TE values calculated according to Brown (1943) from different TE coefficients (D) had a minimum value of 42.3% ($D=0.046$), a maximum of 83.9% ($D=1$) and a mean of 55.2% ($D=0.1$) (Romero Díaz et al. 2012). Thus, the TE method indicates higher estimated erosion rates upstream of the check dams (Romero Díaz et al. 2012).

Geomorphological transects and longitudinal profiles

The geomorphological transect method has been widely used in studies of geomorphological surveys, abandoned fields, geomorphological effects due to the construction of mountain roads and the effects of reforestation. Mediterranean environments efficiently demonstrate soil erosion processes, particularly when the erosion rates are high. Geomorphological transects allow us to gain a quick characterisation of the intensity of erosion and sedimentation processes on slopes and make it easier to compare different slopes (García-Ruiz and López Bermúdez 2009). In the Region of Murcia, these methods have been applied to assess the effects of reforestation. In transects performed perpendicular to the slope, soil erosion can be quantified by cross-sectional measurement of the rills and gullies intersecting each transect. The technique consists of placing a 20 m tape measure on the slope parallel to the contour lines. The length of the erosive form is measured along the tape measure in order to obtain the frequency of gullies, rills, crust, rock fragments and plant cover. Special attention was paid to rills and gullies. Length, width and depth were measured for each incision found along the tape measure. Three widths and three depths were measured so as to establish an average value at each rill and gully. The rill and gully survey allowed the volume of the eroded soil to be calculated (Romero Díaz et al. 2010).

In reforestation carried out with terraces during the 1970s, Chaparro Fuster and Esteve Selma (1995) and Romero Díaz and Belmonte Serrato (2008) assessed erosion rates generated as a consequence of terracing in the Guadalentín valley and in the depression located between the Sierra del Gallo and Los Villares. The value obtained by Chaparro Fuster and Esteve Selma in marls ranged from 71 to 93 t ha^{-1} year^{-1}, but Romero Díaz and Belmonte Serrato obtained an average value of 105 t ha^{-1} year^{-1}. In conglomerates, erosion rates obtained by Chaparro Fuster and Esteve Selma were lower, ranging from 17 to 40 t ha^{-1} year^{-1}, and the rate calculated by

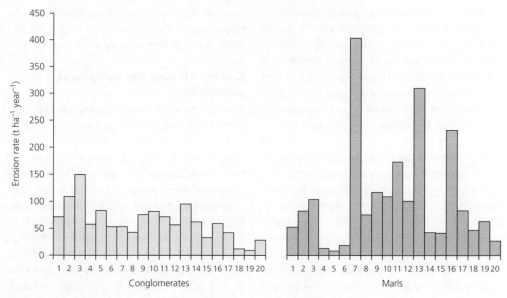

Figure 2.7 Comparison of erosion rates obtained through geomorphological transects in marls and conglomerates (Romero Díaz and Belmonte Serrato 2008).

Romero Díaz and Belmonte Serrato was 62 t ha^{-1} year^{-1} (Figure 2.7). Despite the fact that high erosion rates are obtained with this method, values are always lower than directly measured rates on the same surfaces, since only rills and gullies are measured and other types of erosion are excluded.

Erosion rates can also be calculated from transects performed in the direction of the slope (longitudinal profiles). This method has been used by several authors to determine erosion rates and slope evolution (Díaz-Fierros and Pérez-Moreira 1984; La Roca 1984; García-Ruiz et al. 1989; Ortigosa Izquierdo 1991; Chaparro Fuster and Esteve Selma 1995), by means of fixing a rope above the ground in the direction of the slope and covering at least two consecutive terraces. The rope must be tightened to form a straight line between two metal posts at the ends, and a tape measure is placed over it. Every 200 mm, the offset of the ground below the rope is measured, obtaining a very accurate profile of the topography, erosion and accumulation areas and the different modifications that the terracing

and its subsequent geomorphologic evolution has caused. Using the slope profiles, the amounts of sediment removed and deposited at each site were calculated. The slope profiles give information related to the slope length, disturbed area and excavated and accumulated topography. A comparison between the original slope reconstructed on the undisturbed sites and the current slope terraced by bulldozing was set up. Calculations were made from the comparison of the bulldozed and the natural slope to determine the volume of soil accumulated and excavated. This method allows the soil erosion or deposition rates at each profile to be determined. After calculation of the eroded and accumulated volumes at each portion of the transects, a mean soil loss (in m^3) was obtained for each transect, as well as the total soil loss (m^3 m^{-2}), the total mass of eroded material (Mg ha^{-1}) and the annual soil loss (Mg ha^{-1} year^{-1}) (Romero Díaz et al. 2010).

Romero Díaz and Belmonte Serrato (2008) used the method of longitudinal profiles in the same reforested and terraced areas. The values obtained in marls were lower than those obtained from transects (68 t ha^{-1} year^{-1}). On the contrary, they were higher in conglomerate areas (103 t ha^{-1} year^{-1}). The reason for this variation was that rills and gullies were not included when measuring longitudinal profiles in marls, but they were included when performing transects, so that the results were homogeneous in all areas. In metamorphic rocks (phyllites and schists), average values calculated in the Sierra de La Torrecilla were reduced to 29 t ha^{-1} year^{-1}. These high erosion rates in different lithologies highlighted problems with reforestation in semi-arid environments, where vegetation recolonisation was very difficult. On the other hand, the aggressiveness of the terracing technique caused the reactivation of geomorphological activity and an increase of the erosion processes, resulting in very high erosion rates.

Another type of erosion that has been assessed using longitudinal profiles is the erosion caused by tillage. In the Sierra de La Torrecilla (Guadalentín basin), Poesen et al. (1997) have carried out several studies in almond groves on slopes over a substrate of slates and phyllites. These authors showed how tillage was responsible for soil displacement, producing strong denudation on slope convexities and the filling of valley bottoms (see also Wesemael et al. 2006). They also assessed the influence of tillage type and found that up-and-down tillage produced a displacement of material higher than that produced by contour tillage. In a 50 m long field and with a 20% gradient, a downslope tillage pass moved 282 kg m^{-1}, but a cross-slope tillage pass (i.e. following the contour line) only moved 139 kg m^{-1}. The equivalent erosion if the field was ploughed between three and five times per year would be between 1.5 and 2.6 mm year^{-1} in the first case and between 3.6 and 5.9 mm year^{-1} in the second case (Poesen et al. 1997). Wesemael et al. (2006) calculated an erosion rate of 26.6 t ha^{-1} year^{-1}.

In comparison, Quine et al. (1999) estimated the differences there would be when slopes were ploughed using conventional tillage or

using a duckfoot chisel. They found that the fluxes would be 200 and 657 kg m^{-1}, respectively, for each time ploughed.

Geometric and topographical parameters

Another method for estimating erosion rate, especially in areas where gullies and piping are widespread, is by direct measurements in the field, carrying out topographic surveys and complementing the estimates with aerial photographs. Poesen et al. (2006) reported data obtained from rills and gullies in the Guadalentín basin in 2002 and calculated erosion rates of 36.6 and 37.6 t ha^{-1} year^{-1}. Oostwoud Wijdenes et al. (2000) obtained rates of 1.2 t ha^{-1} year^{-1} in other study performed between 1997 and 1998 in 458 gully headcuts in the Rambla Salada (Guadalentín basin). Vandekerckhove et al. (2003) obtained a higher gully retreat rate (17.4 m^3 year^{-1}) in the Guadalentín basin by studying 12 gullies over a 40–43-year time period using aerial photographs. This showed the importance of obtaining medium-term erosion rates, which were undoubtedly more reliable.

Romero Díaz et al. (2009) estimated the volume of soil loss in an area affected by piping processes in Campos del Río (Mula basin). Piping processes were reported for 96 out of the 122 analysed fields. The total surface area of affected plots was 166,417 m^2, with an average surface area per plot of 1,983 m^2. The average proportion of plot areas affected by piping was 35%, and some plots were found with 90% of their areas affected. The total displaced soil was 57,858 tonnes. Assuming a 40-year abandonment period (although it was shorter in some cases), the average erosion rate per plot was 287 and 120 t ha^{-1} year^{-1} in the different analysed areas. Erosion rates higher than 100 t ha^{-1} year^{-1} were found at 34% of plots (Figure 2.8). This value was higher than the rates obtained in gullies but lower than that reported by García-Ruíz and López Bermúdez (2009), who obtained a value of 550 t ha^{-1} year^{-1} in other areas with

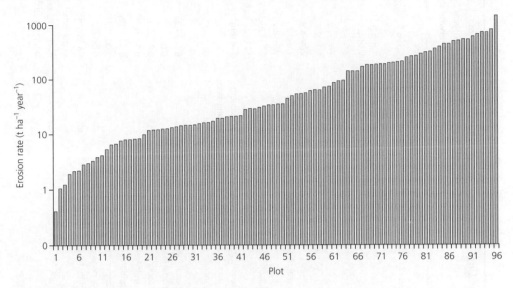

Figure 2.8 Erosion by piping in 96 analysed plots in Mula basin.

piping in Spain. There is no doubt that although estimated volumes of soil loss and erosion rates reach very high values, they are lower than true values due to the fact that the full measurement of the conduits that connect pipes at deep levels is impossible to carry out.

Discussion

After analysing the rates obtained with different methods (Table 2.5), we can make a number of general comments about the different methods and the data that they produce:

- With rainfall simulations, erosion rates ranged from 0 to $14.36\,t\,ha^{-1}\,year^{-1}$. The higher rates were obtained on marl lithologies and soils without vegetation and recently abandoned. In general, simulations displayed low erosion rates, except for the cases where high rainfall intensities were applied.
- Normally, in other researchers' studies, high rates were obtained using erosion pins. Here, however, the only experiment carried out showed a rate slightly higher than $1\,t\,ha^{-1}\,year^{-1}$.

- Data available for open plots also displayed lower rates on slopes (1.8–$3.2\,t\,ha^{-1}\,year^{-1}$) and higher rates in valley bottoms where sediments accumulated ($30\,t\,ha^{-1}\,year^{-1}$).
- The large number of studies carried out on closed plots showed a range from insignificant erosion to maximum rates of $7.5\,t\,ha^{-1}\,year^{-1}$. The highest rates were found (i) on marl lithologies with little vegetation, (ii) in the first years after abandonment, (iii) on burned surface areas, (iv) on crops and especially (v) on fields where tillage was carried out in a downslope direction.
- Microcatchments displayed higher rates than basins (0.08–$2.99\,t\,ha^{-1}\,year^{-1}$ compared with 0.011–$0.034\,t\,ha^{-1}\,year^{-1}$). This pattern is found by other authors who showed that erosion rates decreased as basin surface areas increased (Gómez Plaza et al. 1998; Boix-Fayos et al. 2008). Nevertheless, in the Region of Murcia, the lithology was another important factor to consider. Microcatchments were generally located on marls, and basins were located on limestones.
- Geomorphological transects generated higher erosion rates, averaging $105\,t\,ha^{-1}\,year^{-1}$ on

Table 2.5 Erosion rates obtained by different methods in the Region of Murcia

Reference	Size	Location	Period	Land use	Lithology	Soil loss (t ha⁻¹ year⁻¹ unless otherwise shown)

Rainfall simulations

Reference	Size	Location	Period	Land use	Lithology	Soil loss ($t\,ha^{-1}\,year^{-1}$ unless otherwise shown)
Francis (1986)	0.20 m²	Mula basin	6 simulations (25.8–100.7 mm h⁻¹)	1 year of abandonment	Marls	7.7–8.6 Wet-dry
				2 year of abandonment		4.9–6.6 Wet-dry
				(valley) 5 year of abandonment		2.5–4.6 Wet-dry
				(hill) 5 year of abandonment		5.7–8.6 Wet-dry
				20 year of abandonment		1.6–1.6 Wet-dry
Calvo et al. (1991)	0.24 m²	Sucina (Campo de Cartagena)	2 simulations (63–48 mm h⁻¹)	Badlands, bare soil	Marls	24.1–76.6 g m⁻² h⁻¹
Fernández Gambin et al. (1996)	0.25 m²	Los Guillermos, Rambla Salada	30 simulations (50 mm h⁻¹ 30 min⁻¹)	Bare soil, vegetation, stony	Conglomerates, marls and sands	12.8–27.9 g l⁻¹
Bergkamp et al. (1996)	20 m²	El Ardal (Mula basin)	2 simulations (70 mm h⁻¹ 30 min⁻¹)	Abandonment and scrub (romeral)	Limestones	1–4.99 g l⁻¹
						0.0
Cerdà (1997)	0.24 m²	Guadalentin basin	3 simulations (47 mm h⁻¹)	Espartal (bare soil)	Marls	6.1–14.36
				Espartal (*Stipa* sp.)		0.0
	0.25 m²		6 simulations	Abandonment (bare soil)		0.0
				Abandonment (3 years)		0.20–5.25
				Abandonment (10 years)		0.00–0.05
				Abandonment (*Stipa tenacissima*)		0.00–0.01
				Abandonment (*Pinus pinaster*)		0.01–0.03
Martínez-Mena et al. (2001)	0.24 m²	Abanilla (Chicamo basin)	8 simulations (50 mm h⁻¹ 30 min⁻¹)	Shrubs 10–20%	Marls	68.15 g m⁻² (average) Unpublished data
		Color (Chicamo basin)	8 simulations (50 mm h⁻¹ 30 min⁻¹)	Shrubs 10–30%	Keuper outcrops	35.40 g m⁻² (average) Unpublished data

Reference	Size	Period	Location	Land use	Lithology	Value
Martínez-Mena et al. (2002)	2 m²	8 simulations	Fuente Librilla	Lemon orchard	Quaternary calcareous colluvium	0.70 $g\,m^{-2}\,min$
		7 simulations			Tertiary marl	1.34 $g\,m^{-2}\,min$
Erosion pins						
Romero Díaz et al. (2007b)		1993–1995	Quipar river basin	51 check dam	Various sedimentary rocks	1.37 (average)
Open plots						
Francis (1986)		1985–1986 (10 Gerlach trough)	Rambla de Gracia (Mula basin)	Scrub scattered	Marls	1.8–3.2
Cammeraat (2004, 2002)		1996–1999	Cañada Hermosa (Guadalentín basin)	Scrub and pine forest (slopes)	Marls and limestones	0.08 $t\,ha^{-1}$ (average per event)
				Scrub and pine forest (valley bottoms)		30 $t\,ha^{-1}$ (average per event)
Closed plots						
Francis (1986, 1990)	3 m²	3 years	Cañada Honda (Mula basin)	Fields abandoned by 1, 2, 5 and 20 years	Marls	0.8–5.3
Romero Díaz et al. (1998)	16–20 m²	1989–1997	El Ardal (Mula basin)	Various uses	Limestones	0.02–2.82 (annual average)
						0.83 (average 9 years)
				Scrub		0.06–0.22
				Scrub cut		0.43–0.94
				Abandonment		0.01–0.50
				Cereal crop		0.78–1.20
				Contour tillage		1.15
				Up-and-down tillage		5.92
Tudela Serrano (1988)	16 m²	1987–1988	Javalí Viejo	Pine forestation	Conglomerates	0.12
Romero Díaz and Belmonte	10 × 2	1997–1999	El Minglanillo (Rambla Salada)	Crop	Marls	7.47
Serrato (2002)	20 m²			Scrub		0.80
				Abandonment		1.12

(Continued)

Table 2.5 (*Continued*)

Reference	Size	Location	Period	Land use	Lithology	Soil loss (t ha⁻¹ year⁻¹ unless otherwise shown)
López Bermúdez et al. (2000)	10 × 2 20 m²	El Minglanillo (Rambla Salada)	1996–2000	Crop Scrub Abandonment	Marls	6.0–15.0 0.7–1.4 0.14–2.0
Albaladejo et al. (1991)	15 × 5 75 m²	Abanilla basin	1988–1990	Control plot Organic amendment (13 kg m⁻²) Organic amendment (26 kg m⁻²)	Marls	0.39 0.02 0.006
Castillo et al. (1997)	15 × 5 75 m²	Santomera	1989–1993	With vegetation Without vegetation	Marls	0.13 0.3
Boix-Fayos et al. (2007)	10 × 3 30 m²	Venta del Olivo (Sierra del Picarcho)	1997–2003	Burning scrub Unburned scrub Pinewood	Limestones	0.54 0.029 0.017
Romero Díaz and Belmonte Serrato (2008)	10 × 2 20 m²	La Atalaya (Guadalentín basin)	2005–2006 (21 events)	Scattered scrub (21% coverage)	Marls	1.86
		La Hoya (Guadalentín basin)	2005–2006 (20 events)	Scattered scrub (38% coverage)	Conglomerates	0.06
		Rambla de la Torrecilla (Guadalentín basin)	2005–2006 (16 events)	Scattered scrub (30% coverage)	Slates and phyllites	0.11
Microcatchments						
Romero Díaz et al. (1988)	3000 m²	Rambla de Gracia (Mula basin)	1984 (142 mm lluvia) 1985 (288 mm lluvia) 1986 (445 mm lluvia)	Shrubland	Marls and sands	0.08 2.55 2.36
Martínez-Mena et al. (2001, 2002)	328 m²	Color (Chícamo basin)	1990–1993 (26 events in 3 years)	Shrubland	Keuper outcrops	2.99
Martínez-Mena et al. (2001, 2002)	759 m²	Abanilla (Chícamo basin)	1990–1993 (35 events in 3 years)	Shrubland	Miocene marls	0.85
Catchments						
Hooke and Mant (2000)	4110 ha	Rambla de Torrealvilla (Guadalentín basin)	September 1997	Grazing	Various lithologies	53.185 kg m⁻¹

Reference	Location	Size	Period	Land use	Lithology	Rate
Oostwood Wijdenes et al. (2000)	Guadalentín basin		April 1997 to March 1999	Shrubs, alpha grass	Quaternary fill and marls	4 m³ year⁻¹
Boix-Fayos et al. (2006)	Sierra del Picarcho	6.4 ha / 7.9 ha / 24.3 ha	1997–2003	Burned area / Burned area / *Stipa* and unburned pine	Limestones	0.034 / 0.011 / 0.015
Geomorphological transect						
Chaparro Fuster and Esteve Selma (1995)	Guadalentín basin and depression Cresta del Gallo–Sierra Los Villares	20 m (length)	Afforestation (20 years)	Terraced afforestation with *Pinus halepensis*	Marls / Conglomerates	71–93 / 14–40
Romero Díaz and Belmonte Serrato (2008)	Guadalentín basin	20 m length (40 transects)	Afforestation (30 years)	Terraced afforestation with *P. halepensis*	Marls / Conglomerates	105.3 (average) 8.6–310.2 (range) / 62.7 (average) 9.11–150.3 (range)
Longitudinal profiles						
Poesen et al. (1997)	Sierra de La Torrecilla (Guadalentín basin)	50 m length	5 sites	Almonds (up-and-down tillage) / Almonds (contour tillage)	Slates and phyllites	282 kg m⁻¹ per pass / 1.5–2.6 mm year⁻¹ (3–5 passes per year) / 54–58 t ha⁻¹ year⁻¹ / 139 kg m⁻¹ per pass / 3.6–5.9 mm year⁻¹ (3–5 passes per year) / 22–39 t ha⁻¹ year⁻¹
Quine et al. (1999)	Sierra de La Torrecilla (Guadalentín basin)			Almonds (conventional tillage) / Almonds (duckfoot chisel) / Almond groves	Slates and phyllites / Slates and phyllites	200 kg m⁻¹ per pass / 657 kg m⁻¹ per pass
Vas Wesemael (2006)	Sierra de La Torrecilla (Guadalentín basin)	21 ha				26.6
Romero Díaz and Belmonte Serrato (2008)	Guadalentín basin	12–13 m (3 profiles) / 12–17 m (3 profiles) / 15–18 m (3 profiles)	Afforestation (30 years)	Terraced afforestation with *P. halepensis*	Marls / Conglomerates / Slates and phyllites	67.8 (average) / 103.6 (average) / 29.4 (average)
Check dam						
Romero Díaz et al. (2007a,b)	Quípar river basin	816 km²	425 built 195 analysd	Various	Sedimentary rocks	0.03–72.47 (average 3.95)

(Continued)

Table 2.5 (*Continued*)

Reference	Size	Location	Period	Land use	Lithology	Soil loss ($t\,ha^{-1}\,year^{-1}$ unless otherwise shown)
Boix-Fayos et al. (2008)	47 km²	Rambla de La Rogativa	58 built / 58 analysed	Pine forest and abandoned fields	Sedimentary rocks	0.25–107.33 (average 5.39)
Conesa García and García Lorenzo (2007)	27.1 km²	Rambla del Cárcavo	32 built / 28 analysed	Scrub, pine and almond and olive trees	Sedimentary rocks	6.61
	14.7 km²	Rambla de la Torrecilla (Guadalentín basin)	30 built / 26 analysed		Slates and phyllites	3.21
Geometric and topographical parameters						
Poesen et al. (2002)		Guadalentín basin	Field mapping (December 1996)	Almond groves Gullies (51%) / Almond groves Interrill and rill	Marls and conglomerates	37.6 / 36.6
Oostwoud Wijdenes et al. (2000)	458 heads of gullies	Rambla Salada (Guadalentín basin)	1997–1998	Almond trees	Marls and conglomerates	1.2
Vandekerckhove et al. (2003)	12 gullies	Guadalentín basin	40–43 years (aerial photographs)	Almond trees	Marls and conglomerates	17.4 m³ year⁻¹
Romero Díaz et al. (2009)	96 plots	Campos del Río (Mula basin)	Field measurements 2008	40 years abandoned fields with piping	Marls	287.3
Bathymetry of reservoirs						
López Bermúdez and Gutiérrez Escudero (1982)	8 dams	Segura basin	Last bathymetry 1977	Various land uses	Various lithologies	8.3
Soto (1990)	11 dams		Last bathymetry 1985			4.3
Romero Díaz et al. (1992)	11 dams		Last bathymetry 1985			4.7
Almorox et al. (1994)	7 dams		Last bathymetry 1985			4.4
Avendaño et al. (1997)	10 dams		Last bathymetry 1994			3.3
Sanz Montero et al. (1998)	12 dams (including Fuensanta)		Last bathymetry 1994			1.85–6.97
	12 dams (excluded Fuensanta)					1.85–3.86

marls and 63 t ha^{-1} year^{-1} on conglomerates in terraced reforestation slopes. When using longitudinal profiles to assess erosion, rates were also high in the studied areas. In areas with metamorphic lithologies, the rates obtained on slopes with almond groves (26.6 t ha^{-1} year^{-1}) were very similar to those obtained on terraced areas for reforestation (29.4 t ha^{-1} year^{-1}). On sedimentary lithologies, values were very high on conglomerates (103.6 t ha^{-1} year^{-1}) and lower on marls (67.8 t ha^{-1} year^{-1}).

- When quantifying sediments retained in check dams in the Quípar and Rogativa basins, the erosion rates ranged from 3.95 to 5.39 t ha^{-1} year^{-1}. However, it is necessary to take into account that in the case of La Rogativa basin, the data included check dams completely full of sediment. In all cases, the TE has not been calculated; therefore, all erosion values are probably underestimates.

- Assessment using geometric and topographic measures provided erosion rates in gullies of up to 37.6 t ha^{-1} year^{-1}. In areas with piping, average values reached 287 t ha^{-1} year^{-1}.

- Finally, using bathymetry and depending on the period of study, erosion rates ranged from 3.3 to 4.7 t ha^{-1} year^{-1}.

Using the results obtained from different measuring techniques, it is possible to establish a gradient of mean erosion rates in relation to the method used (Figure 2.9). From lowest to highest erosion rates, the methods are (i) catchments, (ii) erosion pins, (iii) closed plots, (iv) microcatchments, (v) rainfall simulations, (vi) check dams, (vii) bathymetry, (viii) open plots, (ix) longitudinal profiles, (x) geomorphological transects and (xi) geometric and topographical parameters.

However, the erosion rate obtained is not directly related to the method used but instead

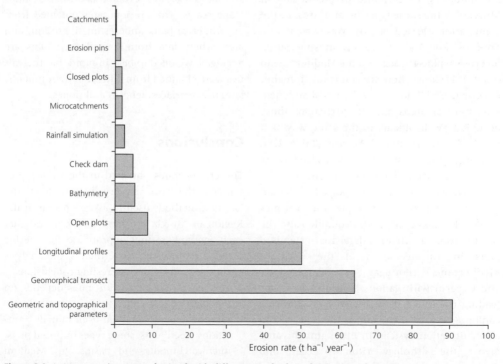

Figure 2.9 Average erosion rates obtained with different methods in the Region of Murcia.

depends on the erosion process being measured. Thus, the sheet and rill erosion measured with erosion pins, closed plots or rainfall simulations gives lower erosion rates than gully erosion or piping measured with longitudinal profiles, geomorphological transects or geometric and topographical parameters. Poesen et al. (2002) indicated how the erosion rates rise with the increment of gullies in the basin. Solé Benet (2006) reported erosion rates higher than $100\,t\,ha^{-1}\,year^{-1}$ in badland areas, and Poesen et al. (2006) found erosion rates exceeding $200\,t\,ha^{-1}\,year^{-1}$ in active badland areas of the Mediterranean. These erosion rates are consistent with those obtained in the Region of Murcia using geomorphological transects and geometric and topographical parameters in areas of gully erosion or piping. On the other hand, sheet and rill erosion rates measured using experimental plots in different parts of Europe (Cerdan et al. 2006) showed much lower mean erosion rates that are similar to those found in the Region of Murcia for the same types of land uses. In the compilation carried out by Wainwright and Thornes (2004) about erosion rates measured on erosion plots throughout the Mediterranean basin, it is shown how erosion is much higher in agricultural fields rather than in abandoned, semi-natural areas and, in particular, those occupied by shrubland, in the same way as it occurs in the region of Murcia (Table 2.5). However, the erosion rates measured show great variability in both rates obtained in the Region of Murcia and rates reported by Cerdan et al. (2006) or Wainwright and Thornes (2004). Kosmas et al. (2002) compiled also the results from an extensive database of standard plots in various areas of the northern Mediterranean (Portugal, Spain, France, Italy and Greece) with various land uses in fields, leading them to believe that the erosion rate can be seen to be a function of the type of crop and cultivation system. In other uncultivated areas, the lithology (Romero Diaz and Belmonte Serrato 2002) and the presence of

vegetation or not (Francis and Thornes 1990) are, undoubtedly, one of the most important factors to consider.

It would have been very interesting to establish relationships between the erosion rates obtained by different methods and different variables, but the only site data available for all cases described the land use and the lithology (Table 2.5). Important variables such as percentage of vegetation cover, slope, amount and intensity of rainfall and intensity of grazing are often missing. Furthermore, most of the experiments have been carried out in semi-natural areas, and there are very few data related to agricultural fields, with the exception of some experimentation plots in which cereal crops were introduced.

We have to consider that there is no perfect method to measure erosion and that each method presents some difficulties. For example, erosion plots are widely used and are very useful to understand the erosion process, but they are not as useful for evaluating erosion at landscape scales. The erosion rates obtained from experimental plots underestimate erosion, and also, when data from several such plots are averaged, we need to bear in mind that the values were obtained from different environments, land uses, surfaces, lengths and slopes.

Conclusions

The erosion rates obtained in the studies presented in this work are very different due to the fact that methods used to assess erosion in the Region of Murcia are very different and that each method is only appropriate for specific scales and processes. Nevertheless, the analysis of these data provides interesting conclusions:

1 The magnitude of erosion rates depends on the method of study. (i) In general, lower rates have been obtained from rainfall simulations, open plots, some types of closed plots, microcatchments and basins. (ii) Medium rates are the result of assessing different types

of closed plots (crops, plots without vegetation cover and on very easily erodible lithologies), check dams and bathymetry of reservoirs. (iii) Very high rates are the result of quantifying erosion through geomorphological transects, longitudinal profiles and geometric and topographic parameters.

2 None of the values obtained can be considered to be absolute erosion rates. However, they provide orders of magnitude under different environmental conditions. It is important to take into consideration that the objectives defined for each of the studies have determined the selection of method and measurement scale.

3 As stated by other authors (Boix-Fayos et al. 2006, 2007; Cammeraat 2002, 2004), it is not easy to extrapolate the results for erosion and sediment production from plots to basins due to the different types of connectivity between, for example, flows and sediments.

4 There are many different factors influencing soil loss, and the period of measurement must also be taken into account. Observation periods should be long, something that is not always possible. Therefore, the comparison of data is not always valid, even if the same method is used.

5 In general terms, erosion rates in the Region of Murcia are lower than $2\,t\,ha^{-1}\,year^{-1}$ except for the areas of badlands, piping and mismanaged or abandoned farmland. However, sometimes, the rates are high for periods of heavy rainfall. In these cases, values can be higher than $50\,t\,ha^{-1}\,year^{-1}$ as shown in Table 2.5.

6 The highest erosion rates have been recorded from simulations of heavy rainfall or from very high levels of natural rainfall, which are responsible for 70–90% of the total annual soil losses.

7 The highest erosion rates are the ones obtained from methods that are normally used to quantify areas with higher erosion, especially those with concentrated erosion such as rills, gullies or pipes.

8 The highest erosion rates, regardless of the method used, have been obtained on marls, high gradient slopes, soils not covered by vegetation and abandoned crops.

9 In semi-arid environments like the Region of Murcia, shrubland may play a protective role as efficient as, or even more efficient than, that of forest. This fact was pointed out by Professor Thornes during his studies in the 1980s (e.g. Francis et al. 1986; Francis and Thornes 1990).

10 Generally, the magnitudes of the results for the soil loss assessment depend on the scale of measurement. In different locations, some authors have shown the differences between the various scales. When a redistribution of material over the whole basin takes place, a decrease in erosion rates has been observed when there is an increase in the surface area. This effect has been demonstrated when comparing microcatchment data with results from experimental catchments. However, this is not the case when the study methods are bathymetry of reservoirs or check dams.

11 In the Region of Murcia, there are many studies of soil erosion assessment but fewer of restoration. Therefore, the studies on the incorporation of USR carried out by Albaladejo and his collaborators deserve a special mention as they confirm how runoff coefficients and erosion rates are significantly reduced, even when adding small quantities of such materials to the slopes. More recently, within the DESIRE project, different conservation strategies in the Guadalentín basin are being developed (De Vente et al. 2009) which are undoubtedly very necessary.

Acknowledgements

I would like to thank John Thornes for his teaching; Juan Albaladejo, María Martínez-Mena, Carolina Boix-Fayos, Adolfo Calvo,

Artemi Cerdà and Carmelo Conesa for providing some of the references and data included here; and the manuscript reviewers who have contributed to its improvement.

References

Albaladejo, J., Stocking, M. A. (1989) Comparative evaluation of two models in predicting storm soil loss from erosion plots in semi-arid Spain. *Catena* **16**, 227–236.

Albaladejo, J., Castillo, V., Roldan, A. (1991) Analysis, evaluation and control of soil erosion processes in semiarid environment: SE Spain. In: Sala, M., Rubio, J. L., García-Ruiz, J. M. (eds.) *Soil Erosion Studies in Spain*. Geoforma Ediciones, Logroño, pp. 9–26.

Albaladejo, J., Mártínez Mena, M., Roldan, A., Castillo, V. (1998) Soil degradation and desertification induced by vegetation removal in a semiarid environment. *Soil Use and Management* **14**, 1–5.

Albaladejo, J., Castillo, V., Díaz, E. (2000) Soil loss and runoff on semiarid land as amended with urban solid refuse. *Land Degradation and Development* **11**, 363–373.

Albaladejo, J., Boix-Fayos, C., Martínez-Mena, M. (2006) La erosión hídrica en la Región de Murcia. In: Conesa Garcia, C. (ed.) *El Medio Físico de la Región de Murcia*. Universidad de Murcia, Murcia, pp. 219–244.

Alias, L. J., López Bermúdez, F., Marín, P., Romero Díaz, A., Martínez Fernández, J. (1997) Clay minerals and soil fertility loss on Petric Calcisol under a semiarid Mediterranean environment. *Soil Technology* **10**, 9–19.

Almorox, J., De Antonio, R., Saa, A., Díaz, M. C., Gascó, J. M. (1994) *Métodos de estimación de la erosión hídrica*. Editorial Agrícola Española, Madrid.

Añó Vidal, C., Peris Mendoza, M., Sánchez Díaz, J. (2009) El estudio de la erosión hídrica en España (1980–2004). *Análisis bibliométrico. Cuaternario y Geomorfología* **23** (1–2), 141–151.

Avendaño, C., Cobo, R. (1997) Metodología para estimar la erosión de cuencas fluviales a partir de la batimetría de embalses. In: Ibáñez, J. J., Valero Garcés, B. L., Machado, C. (eds.) *El paisaje mediterráneo a través del espacio y del tiempo. Implicaciones en la desertificación*. Geoforma Ediciones, Logroño, pp. 239–258.

Avendaño, C., Cobo, R., Gómez, J. L., Sanz Montero, E. (1997) *Capacity Situation in Spanish Reservoirs*.

19ème Congrès des Grands Barrages, Florencia, pp. 849–861.

Barberá, G. G., López Bermúdez, F., Romero Díaz, A. (1997) Cambios del uso del suelo y desertificación en el mediterráneo: el caso del Sureste Ibérico. In: García-Ruiz, J. M., López García, P. (eds.) *Acción humana y desertificación en ambientes mediterráneos*. Instituto Pirenaico de Ecología, C.S.I.C, Zaragoza, pp. 9–39.

Belmonte Serrato, F., Romero Díaz, A. (1992) Evaluación de la capacidad de interceptación de la lluvia por la vegetación y su relación con la erosión de los suelos en el Sureste semiárido español. Primeros resultados. In: López Bermúdez, F., Conesa García, C., Romero Díaz, A. (eds.) *Estudios de Geomorfología en España*. SEG y Universidad de Murcia, Murcia, pp. 33–43.

Belmonte Serrato, F., Romero Díaz, A. (1999) *Interceptación en algunas especies del matorral mediterráneo*. Colección Cuadernos de Ecología y Medio Ambiente n° 7, Universidad de Murcia, Murcia.

Belmonte Serrato, F., Romero Díaz, A., López Bermúdez, F., Hernández Laguna, E. (1999a) Óptimo de cobertura vegetal en relación a las pérdidas de suelo por erosión hídrica y las pérdidas de lluvia por interceptación. *Papeles de Geografía* **30**, 5–15.

Belmonte Serrato, F., Romero Díaz, A., López Bermúdez, F. (1999b) Efectos sobre la cubierta vegetal, la escorrentía y la erosión del suelo, de la alternancia cultivo-abandono en parcelas experimentales. *Investigaciones Geográficas* **22**, 95–107.

Belmonte Serrato, F., Romero Diaz, A., Lopez Bermúdez, F., Delgado Iniesta, M. J. (2002) Changes in the physical and chemical properties of the soil in confined erosion plots (Murcia, Spain). In: Rubio, L., Morgan, R. P. C., Asins, S., Andreu, V. (eds.) *Proceedings of the Third International Congress Man and Soil at the Third Millennium*. Geoforma Ediciones, Logroño, pp. 1459–1470. Valencia, 28 March–1 April 2002.

Benito, G., Gutierrez, M., Sancho, C. (1992) Erosion rates in badlands areas of the Central Ebro basin (NE Spain). *Catena* **19**, 269–286.

Bergkamp, G., Cammeraat, E., Martínez Fernández, J. (1996) Water movement and vegetation patterns on shrublands and abandoned fields in desertification threatened areas. *Earth Surface Processes and Landforms* **21**, 1073–1090.

Boix-Fayos, C., Martínez-Mena, M., Arnau-Rosalén, E., Calvo-Cases, A., Castillo, V., Albaladejo, J. (2006) Measuring soil erosion by field plots: understanding the sources of variation. *Earth Science Reviews* **78**, 267–285.

Boix-Fayos, C., Martínez-Mena, M., Calvo Cases, A., Arnau Rosalén, E., Albaladejo, J., Castillo, V. (2007) Causes and underlying processes of measurement variability in field erosion plots in Mediterranean conditions. *Earth Surface Processes and Landforms* **32**, 85–101.

Boix-Fayos, C., De Vente, J., Martínez-Mena, M., Barberá, G. (2008) The impact of land use change and check-dams on catchment sediment yield. *Hydrological Processes* **22**, 4922–4935.

Brown, C. B. (1943) Discussion of sedimentation in reservoirs. *Proceedings of the American Society of Civil Engineers* **69**, 1493–1500.

Brune, G. M. (1953) Trap efficiency of reservoirs. *Transactions of the American Geophysical Union* **34**, 407–418.

Calvo, A., Gisbert, B., Palau, E., Romero, M. (1988) Un simulador de lluvia portátil de fácil construcción. In: Sala, M., Gallart, G. (eds.) *Métodos y técnicas para la medición de procesos geomorfológicos*. S.E.G. Monografía 1, Barcelona, pp. 6–15.

Calvo, A., Harvey, A. M., Payà, J., Alexander, R. W. (1991) Response of badland surfaces in South East Spain to simulated rainfall. *Cuaternario y Geomorfología* **5**, 3–14.

Cammeraat, L. H. (2002) A review of two strongly contrasting geomorphological systems within the context of scale. *Earth Surface Processes and Landforms* **27**, 1201–1222.

Cammeraat, L. H. (2004) Scale dependent thresholds in hydrological and erosion response of a semi-arid catchment in southeast Spain. *Agriculture, Ecosystems and Environment* **104**, 317–332.

Castillo, V. M., Martínez-Mena, M., Albadalejo, J. (1997) Runoff and soil loss response to vegetation removal in semiarid environments. *Soil Science Society of America Journal* **61**, 1116–1121.

Castillo, V. M., Gómez Plaza, A., Martínez-Mena, M. (2003) The role of antecedent soil water content in the runoff response of semiarid catchment: a simulation approach. *Journal of Hydrology* **284**, 114–130.

Castillo, V. M., Mosch, W. M., Conesa, C., Barberá, G. G., Navarro, J. A., López Bermúdez, F. (2007) Effectiveness and geomorphological impacts of check dams for soil erosion control in a semiarid Mediterranean catchment: El Cárcavo (Murcia, Spain). *Catena* **70**, 416–427.

Castillo, V. M., Boix-Fayos, C., De Vente, J., Martínez-Mena, M., Barberá, G. G. (2009) Efectividad de los proyectos de restauración hidrológico forestal para el control de la erosión en cuencas mediterráneas. In: Romero Díaz, A., Belmonte, F., Alonso, F., López

Bermúdez, F. (eds.) *Advances in Studies on Desertification*. Editum, Murcia, pp. 199–202.

Cerdà, A. (1997) Soil erosion after land abandonment in a semiarid environment of Southeastern Spain. *Arid Soil Research and Rehabilitation* **11**, 163–176.

Cerdà, A. (1999) Simuladores de lluvia y su aplicación a la Geomorfología. Estado de la cuestión. *Cuadernos de Investigación Geográfica* **XXV**, 45–84.

Cerdà, A. (2001) La erosión del suelo y sus tasas en España. *Ecosistemas* **10**, 1–16.

Cerdan, O., Poesen, J., Govers, G., Saby, N., Le Bissonnais, Y., Gobin, A., Vacca, A., Quinton, J., Auerswald, K., Klik, A., Kwaad, F., Roxo, M. J. (2006) Sheet and rill erosion. In: Boardman, J., Poesen, J. (eds.) *Soil erosion in Europe*. John Wiley & Sons, Ltd, Chichester, pp. 500–513.

Chaparro Fuster, J., Esteve Selma, M. A. (1995) Evolucion geomorfológica de laderas repobladas mediante aterrazamientos en ambientes semiáridos (Murcia, SE de España). *Cuaternario y Geomorfologia* **9**, 39–49.

Cobo, R., Sanz Montero, M. E., Gómez, J. L., Avendaño, C., Plata, A. (1997) Influence of the Puentes reservoir operation procedure on the sediment accumulation rate between 1954–1994. In: Nineteenth International Congress on Large Dams, Florence, Italy, 26–30 May, pp. 835–847.

Conesa García, C., García Lorenzo, R. (2007) *Erosión y diques de retención en la Cuenca Mediterránea*. IEA, Murcia.

Conesa García, C., López Bermúdez, F., Alonso Sarria, F. (1994) Morfometría de grietas de retracción en un badland del sureste semiárido peninsular. In: Arnáez Vadillo, J., García Ruiz, J. M., Gómez Villar, A. (eds.) *Geomorfología en España, T.II.* S.E.G., Logroño, pp. 41–54.

Conesa Garcia, C., López Bermúdez, F., Romero Díaz, A. (1996) Scale and morphometry interaction in the drainage network of badlands areas. *Zeitschrift für Geomorphologie, Supplementband* **106**, 107–124.

De Vente, J., Poesen, J., Govers, G., Boix-Fayos, C. (2009) The implications of data selection for regional erosion and sediment yield modelling. *Earth Surface Processes and Landforms* **34**, 1994–2007.

Díaz-Fierros, F., Pérez Moreira, R. (1984) Valoración de los diferentes métodos empleados en Galicia para la medida de la erosión de los suelos, con especial referencia a los suelos afectados por incendios forestales. *Cuadernos de Investigación Geográfica* **10**, 29–42.

Fernández Gambín, I., Alonso Sarria, F., López Bermúdez, F. (1995) Hydrological response to rainfall simulation in badland areas. Spatial and seasonal changes. In: Erosion and Land Degradation in the Mediterranean. Proceedings of the International Geographical Union. European Commission, DG-XII. The University of Aveiro, Aveiro, Portugal, 14–18 June, pp. 47–58.

Fernández Gambín, I., López Bermúdez, F. Alonso Sarria, F., Le Goue, P. (1996) Comportamiento hídrico, modificación micromorfológica y erosión del suelo en los badlands de rambla Salada (Murcia), bajo la acción de lluvias simuladas. *Papeles de Geografía* **23–24**, 127–145.

Fisher, G., Romero Díaz, A., López Bermúdez, F., Thornes, J. B., Francis, C. (1987) La producción de biomasa y sus efectos en los procesos erosivos en un ecosistema mediterráneo semiárido del Sureste de España. *Anales de Biología* **12**, 91–102.

Fournier, F. (1960) *Climat et Erosion*. Presses Universitaires de France, Paris.

Francis, C. (1985) Hydrological investigation of soils in relation to gully head development in South East Spain. *Cuadernos de Investigación Geográfica* **X**, 55–63.

Francis, C. (1986) Soil erosion on fallow fields: an example from Murcia. *Papeles de Geografía Física* **11**, 21–28.

Francis, C. (1990) Soil erosion and organic matter losses on fallow land: a case study from south-east Spain. In: Boardman, J., Foster, I. D. L., Dearing, J. A. (eds.) *Soil Erosion on Agricultural Land*. John Wiley & Sons, Ltd, Chichester, pp. 331–338.

Francis, C., Thornes, J. B. (1990) Runoff hydrographs from three Mediterranean vegetation cover types. In: Thornes, J. B. (ed.) *Vegetation and Erosion: Processes and Environments*. John Wiley & Sons, Ltd, Chichester, pp. 363–384.

Francis, C., Thornes, J. B., Romero Díaz, A., López Bermúdez, F., Fisher, G. C. (1986) Topographic control of soil moisture, vegetation cover and degradation in a moisture-stressed Mediterranean environment. *Catena* **13**, 211–255.

García-Ruiz, J. M. (1999) *La producción científica de la Geomorfología española y su impacto, a través de las publicaciones periódicas*. CSIC, IPE, SEG, Zaragoza.

García-Ruíz, J. M., Gallart, F. (1997) Las cuencas experimentales como base para el estudio de la erosión y la desertificación. In: Ibáñez, J. J., Valero Garcés, B. L., Machado, C. (eds.) *El paisaje mediterráneo a través del espacio y del tiempo. Implicaciones para la desertificación*. Geoderma Ediciones, Logroño, pp. 221–238.

García-Ruiz, J. M., López Bermúdez, F. (2009) *La erosión del suelo en España*. Sociedad Española de Geomorfología, Zaragoza.

García-Ruiz, J. M., Ortigosa, L., Martínez Castroviejo, R. (1989) Notas sobre la geomorfología de ambientes degradados del Pirineo aragonés. *Monografía del Instituto pirenaico de Ecología* **4**, 983–991.

García-Ruiz, J. M., López Bermúdez, F., Romero Díaz, A. (2001) Geomorfología de vertientes y procesos de erosión. In: Gómez-Ortiz, A., Pérez-González, A. (eds.) *Evolución reciente de la Geomorfología Española (1980–2000*. SEG-UB-Geoforma Ediciones, Logroño, pp. 223–254.

Geeson, N. A., Brandt, C. J., Thornes, J. B. (eds.) (2002) *Mediterranean Desertification: A Mosaic of Processes and Responses*. John Wiley & Sons, Ltd, Chichester.

Gerlach, T. (1967) Hillslope troughs for measuring sediment movement. *Revue de Géomorphologie Dynamique* **17**, 173–174.

Gómez Plaza, A., Castillo, V. M., Albaladejo, J. (1998) Estudio de procesos hidrológicos a diferentes escalas (marco teórico y propuesta metodológica). *NORBA Revista de Geografía* **X**, 81–93.

Haigh, M. J. (1981) The use of erosion pins in the study of slope evolution. *British Geomorphological Research Group, Technical Bulletin* **18**, 31–49.

Hernández Laguna, E., Martínez Lloris, M., Romero Díaz, A. (2004) Método de determinación del volumen de sedimentos retenidos en diques de corrección hidrológica. In: Benito, G., Diez Herrero, A. (eds.) *Riesgos naturales y antrópicos en Geomorfología*. SEG and CSIC, Madrid, pp. 201–210.

Hooke, J. M., Mant, J. M. (2000) Geomorphological impact of a flood event on ephemeral channels in SE Spain. *Geomorphology* **34**, 163–180.

Kirkby, M. J., Abrahart, R. J., Bathurst, J. C., Kilsby, C. G., McMahon, M. L., Osborne, C. P., Thornes, J. B., Woodward, F. I. (2002) MEDRUSH: a basin-scale physically based model for forecasting runoff and sediment yield. In: Geeson, N. A., Brandt, C. J., Thornes, J. B. (eds.) *Mediterranean Desertification: A Mosaic of Processes and Responses*. John Wiley & Sons, Ltd, Chichester, pp. 203–228.

Kosmas, C., Danalatos, N., López Bermúdez, F., Romero Díaz, A. (2002). The effect of land use on soil erosion and land degradation under Mediterranean conditions. In: Geeson, N. A., Brandt, C. J., Thornes, J. B. (eds.) *Mediterranean Desertification: A Mosaic of Processes and Responses*. John Wiley & Sons, Ltd, Chichester, pp. 57–70.

La Roca, N. (1984) La erosión por arroyada en una estación experimental (Requena, Valencia). *Cuadernos de Investigación Geográfica* **10**, 85–98.

Lesschen, J. P., Kok, K., Verburg, P. H., Cammeraat, L. H. (2007) Identification of vulnerable areas for gully erosion under different scenarios of land abandonment in Southeast Spain. *Catena* **71**, 110–121.

López Bermúdez, F., Gutiérrez Escudero, D. (1982) Estimación de la erosión y aterramientos de embalses en la Cuenca hidrográfica del río Segura. *Cuadernos de Investigación Geográfica* **VIII** (1), 3–18.

López Bermúdez, F., Thornes, J. B., Romero Díaz, A., Fisher, G., Francis, C. (1984) Erosión y Ecología en la España semiárida (Cuenca de Mula. Murcia). *Cuadernos de Investigación Geográfica* **X** (1 y 2), 113–126.

López Bermúdez, F., Thornes, J. B., Romero Díaz, A., Francis, C., Fisher, G. (1986) Vegetation–erosion relationships: Cuenca de Mula, Murcia. Spain. In: López Bermúdez, F., Thornes, J. B. (eds.) *Estudios sobre Geomorfología del sur de España*. Universidad de Murcia, Murcia, pp. 101–104.

López Bermúdez, F., Alonso Sarria, F., Romero Díaz, A., Conesa García, C., Martínez Fernández, J., Martínez Fernández, J. (1992) Caracterización y diseño del Campo experimental de "Los Guillermos" (Murcia) para el estudio de los procesos de erosión y desertificación en litologías blandas. In: López Bermúdez, F., Conesa García, C., Romero Díaz, A. (eds.) *Estudios de Geomorfología en España*. Sociedad Española de Geomorfología, Murcia, pp. 151–160.

López Bermúdez, F., García Ruiz, J. M., Romero Díaz, A., Ruiz Flaño, P., Martínez Fernández, J., Lasanta, T. (1993) Medidas de flujos de agua y sedimentos en parcelas experimentales. *Cuadernos Técnicos de la SEG, n° 6*, Geoforma Ediciones, Logroño.

López Bermúdez, F., Romero Diaz, A., Martínez Fernández, J., Martínez Fernández, J. (1996) The Ardal field site: soil and vegetation cover. In: Brandt, J., Thornes, J. (eds.) *Mediterranean Desertification and Land Use (MEDALUS)*. John Wiley & Sons, Ltd, Chichester, pp. 169–188.

López Bermúdez, F., Romero Diaz, A., Martínez Fernández, J., Martínez Fernández, J. (1998) Vegetation and soil erosion under semi-arid Mediterranean climate: a case study from Murcia (Spain). *Geomorphology* **24**, 51–58.

López Bermúdez, F., Conesa García, C., Alonso Sarría, F., Belmonte Serrato, F. (2000) La cuenca experimental de Rambla Salada (Murcia). Investigaciones hidrogeomorfológicas. *Cuadernos de Investigación Geográfica* **26**, 95–112.

Martínez Fernández, J., Romero Díaz, A., López Bermúdez, F., Martínez Fernández, J. (1993) Parámetros estructurales y funcionales de *Rosmarinus officinalis* en ecosistemas mediterráneos semiáridos. *Studia Oecologica* **X–XI**, 289–296.

Martínez Fernández, J., López Bermúdez, F., Martínez Fernández, J., Romero Díaz, A. (1995) Land use and soil–vegetation relationships in a Mediterranean ecosystem: El Ardal, Murcia, Spain. In: Poesen, J., Govers, G., Goossens, D. (eds.) *Special Issue: Experimental Geomorphology and Landscape Ecosystem Changes*. Catena, Vol. **25**, Elsevier, Amsterdam/New York, pp. 153–167.

Martínez-Mena, M., Alvarez Rogel, J., Albaladejo, J., Castillo, V. (1999) Influence of vegetal cover on sediment particle size distribution in natural rainfall conditions in a semiarid environment. *Catena* **38**, 175–190.

Martínez-Mena, M., Castillo, V., Albaladejo, J. (2001) Hydrological and erosional response to natural rainfall in a semi-arid area of SE Spain. *Hydrological Processes* **15**, 557–571.

Martínez-Mena, M., Castillo, V., Albaladejo, J. (2002) Relations between interrill erosion processes and sediment particle size distribution in a semiarid Mediterranean area of SE of Spain. *Geomorphology* **45**, 261–275.

MMA. (2002) *Inventario nacional de erosión de suelos 2002–2012, Región de Murcia*. Dirección General de Conservación de la Naturaleza, MMA, Madrid.

Moreno Brotons, J. (2007) *Erosión eólica e hídrica en estériles de minería en el campo de Cartagena*. PFC Ciencias Ambientales, Murcia.

Oostwoud Wijdenes, D. J., Poesen, J., Vandekerckhove, L., Ghesquiere, M. (2000) Spatial distribution of gully head activity and sediment supply along an ephemeral channel in a Mediterranean environment. *Catena* **39**, 147–167.

Ortigosa Izquierdo, L. M. (1991) *Las repoblaciones forestales en la Rioja: resultados y efectos geomorfológicos*. Geoforma Ediciones, Logroño.

Ortiz Silla, R., Albaladejo, J., Martínez-Mena, M., Guillen, F., Álvarez, J. (1999) Erosión hídrica en zonas agrícolas. In: *Atlas del medio Natural de la Región de Murcia*. ITGE y Consejería de Política Territorial y Obras Públicas, CARM. ITGE, Madrid, pp. 53–59.

Poesen, J., Van Wesemael, B., Govers, G., Martínez Fernández, J., Desmet, P., Vandaele, K., Quine, T., Degraer, G. (1997) Patterns of rock fragment cover generated by tillage erosion. *Geomorphology* **18**, 183–197.

Poesen, J., Vandekerckhove, L., Nachtergaele, J., Oostwoud Wijdenes, D., Verstraeten, G., Van Wesemael, B. (2002) Gully erosion in dryland environments. In: Bull, L. J., Kirkby, M. J. (eds.) *Hydrology and Geomorphology of Semi-arid Channels*. John Wiley & Sons, Ltd, Chichester, pp. 229–262.

Poesen, J., Vanwalleghem, T., De Vente, J., Knapen, A., Verstraeten, G., Martínez Casasnovas, A. (2006) Gully erosion in Europe. In: Boardman, J., Poesen, J. (eds.) *Soil Erosion in Europe*. John Wiley & Sons, Ltd, Chichester, pp. 516–536.

Quine, T., Govers, G., Poesen, J., Walling, D., Van Wesemael, B., Martínez Fernández, J. (1999) Fine-earth translocation by tillage in stony soils in the Guadalentín, south-east Spain: an investigation using caesium-134. *Soil and Tillage Research* **51**, 279–301.

Romero Díaz, A. (2002) *La erosión en la Región de Murcia*. Publicaciones de la Universidad de Murcia, Murcia.

Romero Díaz, A. (2007) Las inundaciones. In: Romero Díaz, A., Alonso Sarría, F. (coords.) *Atlas Global de la Región de Murcia*. La Verdad – CMM S.A., Murcia, pp. 250–259.

Romero Díaz, A. (2008) Los diques de corrección hidrológica como instrumentos de cuantificación de la erosión. *Cuadernos de Investigación Geográfica* **34**, 83–99.

Romero Díaz, A., Belmonte Serrato, F. (2002) Erosión del suelo en ambiente semiárido extremo bajo diferentes tipos de litologías y suelos. In: Pérez González, A., Vagas, J., Machado, M. J. (eds.) *Aportaciones a la Geomorfología de España en el inicio del tercer milenio*. Serie Geología N° I. Instituto Geológico y Minero de España, Madrid, pp. 315–322.

Romero Díaz, A., Belmonte Serrato, F. (2008) *Erosión en forestaciones aterrazadas en medios semiáridos: Región de Murcia*. Editum, Universidad de Murcia, Murcia.

Romero Diaz, A., Belmonte Serrato, F. (2009) Erosive and environmental impacts of hydrological correction check dam. In: Hayes, W. P., Bornes, M. C. (eds.) *Dams: Impacts, Stability and Design*. Nova Science Publishers, Hauppauge, pp. 161–178.

Romero Díaz, A., López Bermúdez, F. (2009) *Soil Erosion and Desertification in Neogene-Quaternary Basins of the Region of Murcia*. Fundación Instituto Euromediterráneo del Agua (IEA), Murcia.

Romero Díaz, A., López Bermúdez, F., Thornes, J. B., Francis, C., Fisher, G. C. (1988) Variability of overland flow erosion rates in a semi-arid Mediterranean environment under matorral cover. Murcia, Spain. *Catena Supplement* **13**, 1–11.

Romero Díaz, M. A., Cabezas, F., López Bermúdez, F. (1992) Erosion and fluvial sedimentation in the River Segura basin (Spain). *Catena* **19**, 379–392.

Romero Díaz, A., Barberá, G. G., López Bermúdez, F. (1995) Relaciones entre erosión del suelo, precipitación y cubierta vegetal en un medio semiárido del sureste de la península Ibérica. *Lurralde* **18**, 229–242.

Romero Díaz, A., López Bermúdez, F., Belmonte Serrato, F. (1998) Erosión y escorrentía en el campo experimental de 'El Ardal' (Murcia). Nueve años de experiencias. *Papeles de Geografía* **27**, 115–130.

Romero Díaz, A., Cammeraat, L. H., Vacca, A., Kosmas, C. (1999) Soil erosion at experimental sites in three Mediterranean countries: Italy, Greece and Spain. *Earth Surface Processes and Landforms* **24**, 1243–1256.

Romero Díaz, A., Marín Sanleandro, P., Sánchez Soriano, A., Belmonte Serrato, F., Faulkner, H. (2007a) The causes of piping in a set of abandoned agricultural terraces in southeast Spain. *Catena* **69**, 282–293.

Romero Díaz, A., Martínez Lloris, M., Alonso Sarriá, F., Belmonte Serrato, F., Marín Sanleandro, P., Ortiz Silla, R., Rodríguez Estrella, T., Sánchez Toribio, M. I. (2007b) *Los diques de corrección hidrológica. Cuenca del Río Quipar (Sureste de España)*. Editum, Universidad de Murcia, Murcia.

Romero Díaz, A., Alonso Sarriá, F., Martínez Lloris, M. (2007c) Erosion rates obtained from check-dam sedimentation (SE Spain). A multi-method comparison. *Catena* **71**, 172–178.

Romero Díaz, A., Belmonte Serrato, F., Plaza Martínez, J. F., Sánchez Soriano, A., Ruíz Sinoga, J. D. (2009) Estimated volume of soil lost by erosion processes by piping. Southeastern Spain. In: Romero Díaz, A., Belmonte, F., Alonso, F., López Bermúdez, F. (eds.) *Advances in Studies on Desertification*. Editum, Murcia, pp. 403–407.

Romero Díaz, A., Belmonte Serrato, F., Ruíz Sinoga, J. D. (2010) The geomorphic impact of afforestations on soil erosion in Southeast Spain. *Land Degradation and Development* **21**, 188–195.

Romero Díaz, A., Marín Sanleandro, P., Ortiz Silla, R. (2012) Loss of soil fertility estimated from sediment trapped in check dams. South-eastern Spain. *Catena* **99**, 42–53.

Ruiz Flaño, P. (1993) *Procesos de erosión en campos abandonados del Pirineo. El ejemplo del valle de Aísa*. Monografías Científicas 4, Geoforma Ediciones, Logroño.

Sánchez, G. (1995) Arquitectura y dinámica de las matas de esparto (Stipa tenacissima L.) efectos en el medio e interacciones con la erosión. PhD Thesis, Universidad Complutense de Madrid, Madrid.

Sánchez Soriano, A., Romero Díaz, A., Marín Sanleandro, P. (2003) Procesos de 'piping' en campos de cultivo abandonados (Campos del Río, Murcia). In: Bienes, R., Marques, M. J. (eds.) *Control de la erosión y degradación del suelo*. IMIA, Madrid, pp. 625–629.

Sancho, C., Benito, G., Gutiérrez, M. (1991) *Agujas de erosión y perfiladores microtopográficos*. Cuadernos Técnicos de la S.E.G. n°2, SEG, Geoforma Ediciones, Logroño.

Sanz Montero, E., Avendaño, C., Cobo, R., Gómez, J. L. (1998) Determinación de la erosión en la Cuenca del Segura a partir de los sedimentos acumulados en sus embalses. *Geogaceta* **23**, 135–138.

Scoging, H. (1982) Spatial variations in infiltration, runoff and erosion on hillslopes in semi-arid Spain. In: Bryan, R., Yair, A. (eds.) *Badland Geomorphology and Piping*. Geobooks, Norwich, pp. 89–112.

Sirvent, J., Desir, G., Gutierrez, M., Sancho, C. (1997) Erosion rates in badland areas recorded by collectors, erosion pins and profilometer techniques (Ebro Basin, NE-Spain). *Geomorphology* **18**, 61–75.

Solé Benet, A. (2006) Spain. In: Boardman, J., Poesen, J. (eds.) *Soil Erosion in Europe*. John Wiley & Sons, Ltd, Chichester, pp. 311–346.

Soto, D. (1990) Aproximación a la medida de la erosión y medios para reducir esta en la España peninsular. *Ecología, Fuera de Serie n° 1*, 169–196.

Tudela Serrano, M. L. (1988) *Erosion hídrica en conglomerados de ladera: análisis en parcelas experimentales (Murcia)*. Tesis de Licenciatura. Departamento de Geografía, Universidad de Murcia, Murcia, 329 pp.

Vandekerckhove, L., Poesen, J., Govers, G. (2003) Medium term gully headcut retreat rates in Southeast Spain determined from aerial photographs and ground measurement. *Catena* **50**, 329–352.

Verstraeten, G., Poesen, J. (2000) Estimating trap efficiency of small reservoirs and ponds: methods and implications for the assessment of sediment yield. *Progress in Physical Geography* **24**, 219–251.

Verstraeten, G., Poesen, J. (2001) Modelling the long-term sediment trap efficiency of small ponds. *Hydrological Processes* **15**, 2797–2819.

Verstraeten, G., Poesen, J, De Vente, J., Koninck, X. (2003) Sediment yield variability in Spain: a quantitative and semiqualitative analysis using reservoir sedimentation rates. *Geomorphology* **50**, 327–348.

Wainwright, J., Thornes, J. B. (2004) *Environmental Issues in the Mediterranean: Processes and Perspectives from the Past and Present*. Routledge, London.

van Wesemael, B., Rambaud, X., Poesen, J., Muligan, M., Cammeraat, E., Stevens, A. (2006) Spatial patterns of land degradation and their impacts on the water balance of rainfed treecrops: a case study in South East Spain. *Geoderma* **133**, 43–56.

CHAPTER 3

Shrubland as a soil and water conservation agent in Mediterranean-type ecosystems: The Sierra de Enguera study site contribution

Artemi Cerdà[1], Antonio Giménez-Morera[2], Antonio Jordán[3], Paulo Pereira[4], Agata Novara[5], Saskia Keesstra[6], Jorge Mataix-Solera[7] and José Damián Ruiz Sinoga[8]

[1] SEDER – Soil Erosion and Degradation Research Group, Department of Geography, University of Valencia, Valencia, Spain
[2] Departamento de Economía y Ciencias Sociales, Escuela politécnica superior de Alcoy, Universidad Politécnica de Valencia, Alicante, Spain
[3] MED_Soil Research Group, Department of Crystallography, Mineralogy and Agricultural Chemistry, University of Seville, Seville, Spain
[4] Department of Environmental Policy, Mykolas Romeris University, Vilnius, Lithuania
[5] Dipartimento dei Sistemi Agro-ambientali, University of Palermo, Palermo, Italy
[6] Soil Physics and Land Management Group, Wageningen University, Wageningen, the Netherlands
[7] Environmental Soil Science Group, Department of Agrochemistry and Environment, Miguel Hernández University, Alicante, Spain
[8] Department of Geography, University of Málaga, Málaga, Spain

Introduction

The control vegetation exerts on soil erosion and soil hydrology processes was studied by Professor John Thornes, who encouraged other scientists to be involved by means of fieldworks, lectures, international scientific exchanges, congresses and publications (Thornes 1990). Within the topic of the effect of vegetation on geomorphological processes, Professor Thornes was especially interested in Mediterranean shrublands which he visited many times. He noted that the scientific literature paid more attention to grasslands and forest as they were more economically productive landscapes. Timber and pastures are resources that are important for 21st-century societies; however,

shrubs produce low-quality fuel and timber and most of the shrubland in the north of the Mediterranean Basin was abandoned as urbanisation and globalisation progressed. Until the 1960s, shrubs were important for fuel, forage, aromatic plants and medicinal compounds, as well as to produce ash to fertilise agricultural soils.

Shrubland, also called scrubland, or matorral in Spain and chaparral in California, is a diverse assortment of plants that are sharing common properties such as the dominance of shrubs. Shrubs are woody plants of less than 5 m in height and are very abundant in the Mediterranean regions as a natural vegetation or due to the recovery of the degraded vegetation. The traditional intensive use of the shrubland

Monitoring and Modelling Dynamic Environments, First Edition. Edited by Alan P. Dykes, Mark Mulligan and John Wainwright.

contributes to its degradation and soil erosion as a consequence of the loss of vegetation cover due to the effects of grazing, charcoal and wood production. Afforestation has taken place on many shrublands mainly through planting Aleppo pine (*Pinus halepensis* Miller) in order to make the land more economically productive. On the other hand, shrubland area has increased due to land abandonment and subsequent encroachment of shrubland vegetation. When vegetation is scarce, water infiltration rates are low and surface wash contributes to high erosion rates (Cerdà 1999). For many Mediterranean governments, shrublands were seen as degraded land ready to be afforested or put to some other economic use. Those views contributed to large-scale afforestation plans for many semi-arid lands. However, the scientific research carried out in this shrubland pioneered by John Thornes gives a different picture of the role of these systems. John saw shrubland as a protective cover for soil and water resources (Thornes 1985). This view of the shrub as the protector of the land was formed over many years of study (Thornes 1976). Thus, his passion to know the role of shrubland led him to work in the process of erosion closely with Spanish researchers (e.g. López Bermúdez et al. 1986; Francis and Thornes 1990).

The relationship between vegetation and erosion was always a key part of the research of John Thornes, but it was in the Mediterranean shrublands where his study was most intensive and productive. In Spain, there is now much research inspired by John Thornes' early investigations related to shrubland processes and dynamics. Schnabel et al. (1996) studied the role of vegetation on soil erosion processes in the dehesa silvopastoral ecosystems, and Puigdefábregas et al. (1992) the tussock distribution of *Stipa tenacissima*. Rodríguez et al. (1991); Cerdà (1999) and Boix (2000) in Valencia, González Hidalgo (1996–1997) in Aragón and Romero-Diaz et al. (1995) and Belmonte Serrato et al. (1999) in Murcia all confirmed the importance of vegetation on soil erosion control. Bergkamp (1996) and Bochet

et al. (1998) are two examples of the leadership of John Thornes that brought contributions from other countries to the knowledge of the impact of shrubland on soil and water resources in the Mediterranean. Thus, due to John Thornes, studies of the effect of shrublands became a key topic of researchers in Spain and other Mediterranean countries where the research contribution of Professor Thornes stimulated and inspired many environmental scientists (Cerdà 2008).

This chapter presents the research developed to understand the control that shrubland exerts on soil erosion rates and processes. The experiments and measurements were conducted since 2003 by the Soil Erosion and Degradation Research Group (SEDER) of the Universitat de València and move forwards in the work carried out by Professor Thornes. We review the state of the art of the interaction between shrubland and its soil erosion processes and water dynamics in Mediterranean-type ecosystems. Original data from research inspired by John Thornes are shown. Data collected at the Sierra de Enguera, Eastern Spain, demonstrate the interaction between societies, vegetation and soil erosion. The soil and water losses in the shrubland of Sierra de Enguera within the province of Valencia were measured by means of rainfall simulation experiments and natural rainfall measurements to quantify the contribution of shrubs as a water and soil conservation vegetation cover strategy.

Methods

In the Sierra de Enguera, Eastern Spain, shrubland is composed primarily of *Pistacia lentiscus* L., *Quercus coccifera* L. and *Juniperus oxycedrus* L. (Figure 3.1). This type of scrubland is found as a first succession stage after the removal of oak (*Quercus ilex* L.) woodland. However, the impact of recurrent forest fires also contributed to the development of other shrubs characterised by a lower biomass. The main plants on those fire-developed shrublands are *Cistus*

(a)

(b)

Figure 3.1 View of matorral (shrubland) in the Sierra de Enguera. Since 1970, a forest fire has affected this land more than once: (a) shows an area burned 2 years before; and (b) shows an area burned 8 years before.

albidus L., *Ulex parviflorus* Pourr., *Erica multiflora* L. and *Rosmarinus officinalis* L. At the Sierra de Enguera, shrubland was used for decades as pastoral areas of goats and sheep produced high-quality meat from terrains that are unsuitable for growing grain, olive and wine. When the shrublands did not provide sufficient production, they were burned to allow regeneration of shoots and grasses. Two of the most widespread land uses were the use of bush biomass for firing the local pottery in lime kilns and use in kitchens as small pieces of wood or charcoal for cooking and heating. This was not just a local phenomenon, as the office of 'gabillero' or 'carbonero' was very common in the first half of the 20th century. 'Gabillero' was the name for workers that collected wood from shrubs to be transported to the factories where it was used as fuel. Moreover, 'carbonero' is the name for the workers that transformed the

wood into charcoal to use as fuel. Also, the dwarf shrubs *C. albidus* L., *U. parviflorus* Pourr. and *E. multiflora* L. were used for agricultural fertiliser. Those shrubs were burned, and their ashes mixed with the soil. It was a kind of slash agriculture called 'artigueo' in the Pyrenees (García Ruiz 1997) and known in Eastern Spain as 'hormiguero'. Besides its role as a nutrient source for agricultural soils, forage for livestock and energy for cooking, shrubland has proved to be a great protector of the soil.

Measurements of the effects of shrub on soil erosion and soil hydrology have been taken by the SEDER since 1990. During the 1990s, most of the measurements were collected by means of rainfall simulation experiments conducted at different rainfall intensities (Cerdà 1996). During the 2000s, larger plots were installed (subplots of 1, 2, 4 and 16 m^2) and monitored under natural rainfall at the 'El Teularet' Experimental Station (Bodí and Cerdà 2008; Cerdà and Bodí 2008), two of which were located in areas of shrub cover (Figures 3.1 and 3.2). These plots quantify the loss of soil and water after every rainfall event. Precipitation was measured by automatic rain gauges of 0.2 mm accuracy and runoff collectors allowed the measurement of the sediment and water yield. After each rainfall event, we quantified the cumulative runoff (volume) and sediment concentration (after desiccation). Measurements of soil temperature, soil water repellency (water drop penetration time (WDPT): see Cerdà and Doerr 2008), soil moisture, soil organic matter and soil bulk density were obtained in order to understand the effect of shrubland on soil properties and then on soil erosion and soil hydrology.

At the El Teularet–Sierra de Enguera research station, plots were constructed in 2002. The mean annual rainfall is 578 mm, the parent material is composed of marls and limestone, and the altitude ranges from 300 to 800 m.a.s.l. A plot (Shrubland-1) was selected in an abandoned field (abandoned for 30 years) with a cover of *C. albidus* and *U. parviflorus*. The Shrubland-2 plot was located on a slope with *Q. coccifera*, *P. lentiscus* and *J. oxycedrus* vegetation

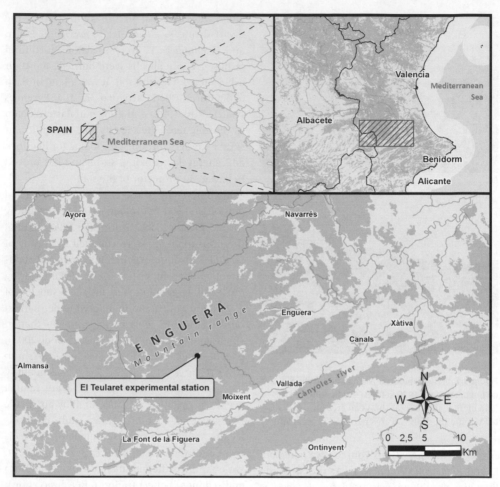

Figure 3.2 Location map of the Sierra de Enguera.

cover and showed the traditional Mediterranean maquia vegetation cover. The Shrubland-1 plot is covered mainly by dwarf shrubs; meanwhile, the Shrubland-2 plot was covered by a typical *maquia* shrub cover. Each of these plots had four subplots with the four plot sizes outlined earlier to quantify soil loss under different plot sizes (Figure 3.3). Data from the shrub-covered plots were compared with agricultural land in order to determine the contribution of the shrubs to controlling soil and water losses. A paired plot was set up on a rainfed olive orchard and one in a drip-irrigated citrus orchard with similar slope gradient (6–10°), mean annual rainfall (500 mm) and mean annual temperature (16°C). Rainfall simulation experiments were carried out in order to determine the control of shrubs on soil erosion. The characteristics of the rainfall simulation can be found in Cerdà et al. (2009) and Cerdà and Jurgensen (2011). The effect of shrub cover (*Q. coccifera*) on soil temperature was measured by means of two sensors located under the shrubs and under a bare surface, respectively (patch against inter-patch areas). The sensor depth was 1 cm below the soil surface, and the measurements were taken every 10 s and an average was recorded every 5 min.

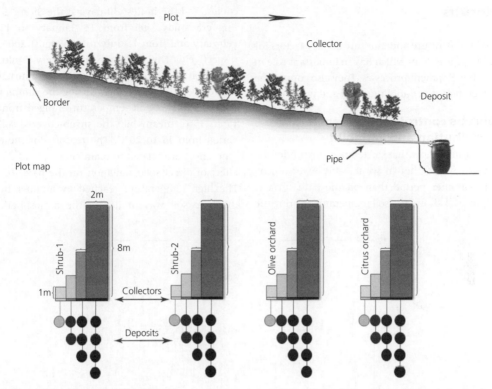

Figure 3.3 Diagram of the soil erosion plots at the El Teularet Soil Erosion and Degradation Research Station in the Sierra de Enguera, Eastern Spain.

In order to determine the effects of the patchy distribution of the shrubland on soil properties, twenty paired samples on the centre of the shrub and in the inter-patch bare area were collected. The soil organic matter content was determined by the Walkley and Black (1934) method, and the soil bulk density by the ring method by means of rings of 3 (height) × 3.1 (diameter) cm. The WDPT method (Cerdà and Doerr 2007) allows measuring the degree of water repellency by placing drops on the soil surface and measuring the time to complete infiltration of the drops. Ten drops were placed on the soil surface under shrub cover and ten more on the inter-shrub soil surface, which acted as a control. Those measurements were carried out once every month during 2004.

Simulated rainfall was performed at $55\,mm\,h^{-1}$ rain intensity during 1 h in order to determine the efficiency of shrubs to control the soil and water losses on circular plots of $0.25\,m^2$ in order to distinguish between the plots covered with shrubs and with inter-patch vegetation. The simulated rainfall lasted 60 min, which is a thunderstorm with a 10-year return period in this region. Those rainfall events trigger runoff discharges that contribute to runoff at the catchment scale (Cerdà 1997a).

A further experiment was set up at the El Teularet–Sierra de Enguera in 2005 to determine the soil moisture content measured with four Echo-20 probes connected to a data logger with a recording interval of 5 min, at 20 cm depth. The four sensors were located underneath the plant cover (two sensors) and in the inter-patch area (two sensors). The volumetric soil moisture content was determined after calibration at the laboratory.

Results

Soil temperature and moisture are key factors for biological activity which has an important role in soil development processes. They also control the vegetation cover and, thus, soil and water losses.

Shrubs control the soil temperature distribution and variability

The results show that soil temperature under the shrubs was higher in winter and lower during the summer period than on the bare surface. Figure 3.4 shows the soil temperature during the coolest and the hottest 30 days at the Sierra de Enguera study site: from 15 January to 15 February and from 15 July to 15 August 2005, respectively. The data, recorded every 5 min, show that the shrubland soils had lower diurnal soil temperature variability. During the summer season, the bare soil temperature ranged from 17 to 35°C, meanwhile the shrub-covered soil varied from 17 to 20°C. The reasons for these differences are related to plant shading and the effect of direct solar radiation on the bare soil. The high temperature reached in summer by the bare soil was similar to the atmospheric

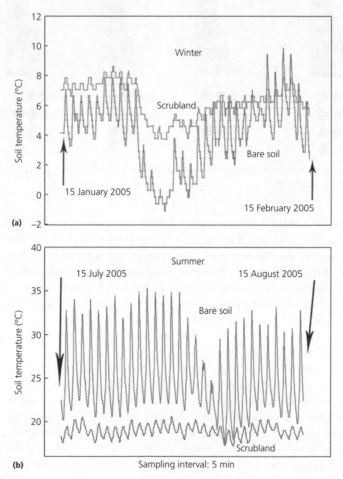

Figure 3.4 Soil temperature changes: (a) during the coolest season (15 January to 15 February) and (b) during the hottest season (15 July to 15 August) at 1 cm depth on shrub-covered and bare surfaces at the El Teularet–Sierra de Enguera experimental station.

temperature. During the winter, the daily soil temperature variability was low: 4°C for the bare soils and 1°C for the shrub-covered soils (see Figure 3.4). The shrub-covered surface was warmer in winter and cooler in summer compared with bare surfaces. Those differences were even clearer in summer when the lowest temperatures reached at night on bare surfaces were higher than the highest temperatures reached during the day on shrub-covered surfaces.

Shrubs contribute to a more stable and less variable soil temperature regime both in the summer and the winter season. Moreover, shrubs ensure that the soil temperature is less variable at both diurnal and annual scales. In fact, bare soils ranged between −1 and +36°C in 2005; the shrub-covered soil showed temperatures ranging from 4 to 20°C. During the first days of August 2005, a cold front resulted in extremely cold weather (the coolest for the last 50 years) that did not affect the soil temperatures reached under the shrubs though a clear impact was recorded on the bare soils. The data show that shrubs maintained a stable soil temperature, which is a key factor for the biota and, thus, organic matter decomposition.

Soil moisture is lower under shrubs during drought periods

The results show that the mean annual volumetric soil moisture, measured every 5 min, was 15.6% for the bare soil and 12.8% for the shrub-covered soils. Seasonal changes of rainfall are the key control on soil moisture content. Winter was the wetter season with 19.8% (shrub-covered) and 20.5% (bare) soil moisture and 71 mm of rain (autumn 2004 was extremely humid with 298 mm). Spring was dry (63 mm of rain) with soil moisture contents still wet at 20 cm depth (14.3 and 17.3% for the shrub and bare soils, respectively) due to the low temperatures and solar radiation. Summer was extremely dry (48 mm from July to September 2005) and resulted in very low soil water availability (11.6 and 7.5%). Autumn saw 64 mm of rainfall and still a low soil moisture (11.6 and 17.0% for the shrub and bare soils). The soils covered by shrubs show lower moisture contents due to either the interception or the transpiration rates of the shrubs. This is particularly clear during a very dry year such as 2005 when the soil water resources were limited (Figure 3.5).

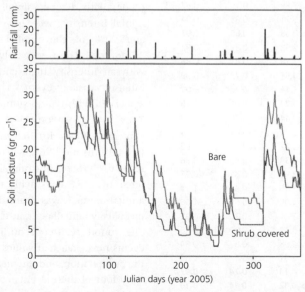

Figure 3.5 Soil moisture measured every 5 min at 20 cm depth under a canopy of *Quercus coccifera* and bare soils at the El Teularet–Sierra de Enguera research station. Rainfall shows daily measurements.

The shrubs change soil properties

The results show that the organic matter content was 8.15% on the shrub-covered soils and 4.20% on the bare patches (Table 3.1). The bulk density (0.8 g cm^{-3}) was also lower on the shrub-covered soils than on the bare ones (1.0 g cm^{-3}). Moreover, on the shrub-covered soil, the organic matter ranged from 6.0 to 10.1% and on the bare one from 3.0 to 5.5%. The soil bulk density was also different under shrubs (0.6–0.9 g cm^{-3}) compared with bare soils (1.2–0.8 g cm^{-3}).

Soil 'wettability' is enhanced under shrubs and during the dry periods

During the last decade, soil water repellency has become a key research topic in Mediterranean hydrology. Hydrophobicity can be defined as a

Table 3.1 Organic matter and soil bulk density on paired plots distinguishing between the shrub-covered and bare inter-shrub soils.

	Organic matter (%)		Bulk density (g cm^{-3})	
	Shrub	Bare	Shrub	Bare
1	7.04	3.45	0.77	1.06
2	7.26	3.89	0.75	0.87
3	9.47	5.21	0.85	0.94
4	8.36	4.58	0.88	0.91
5	6.45	2.98	0.89	0.92
6	8.45	3.78	0.81	0.86
7	9.63	4.65	0.72	0.99
8	10.05	5.14	0.76	1.00
9	6.78	3.78	0.61	1.15
10	6.66	3.44	0.75	1.03
11	7.18	3.33	0.67	1.14
12	9.78	5.02	0.78	0.86
13	7.54	3.33	0.89	0.98
14	8.88	3.98	0.81	0.97
15	6.02	3.45	0.59	1.14
16	7.58	4.05	0.56	1.06
17	8.96	5.04	0.70	0.92
18	9.9	5.47	0.90	0.86
19	8.25	3.58	0.87	0.84
20	6.74	2.97	0.61	1.00
Mean	8.15	4.10	0.74	0.97
St. dev	1.32	0.82	0.14	0.10
cv (%)	16.23	20.08	19.25	10.17

key soil property that controls the 'wettability' of the soil and, thus, plant and soil development. The research conducted at the Sierra de Enguera shows that shrubs contribute to a soil that is richer in organic matter but also to a more water-repellent soil. The results show no water repellency in the inter-shrub soils (Figure 3.6), although during summer some drops reached values higher than 5 s, which is the threshold to determine if a soil is water repellent or not (Cerdà and Doerr 2007; Mataix-Solera et al. 2007). However, on average, the inter-shrub soil surfaces should be understood as hydrophilic, while the under-shrub sites are hydrophilic in all months except June to September where they reach WDPT values higher than 5 s (hydrophobic), due to a combination of the dryness of the soil surface and the high organic matter content. The largest WDPT values were reached in August (Figure 3.6).

Soil and water losses under natural rainfall

The measurements of the soil and water losses shown here were carried in a very wet year (2004) with a total rainfall of 748 mm. Daily rainfall intensities were lower than 60 mm day^{-1} and 90% of the rainfall fell at less than 30 mm day^{-1}. There were only two rainfall events with an intensity high enough to trigger an intense erosion event. Those events were associated with typical summer thunderstorms. They took place on 31 August 2004 and 2 September 2004 with 26.8 mm falling in 165 min ($I_5 = 4.8$ mm) and 32.6 mm in 68 min ($I_5 = 6.2$ mm), respectively. Sixty-one rainfall events were registered over 81 days with rainfall. The mean annual rainfall was 9 mm day^{-1} although the mean daily rainfall per rain-day was 13 mm day^{-1}. The runoff registered during the 61 rainfall events never reached values higher than 1% of the precipitation, and the mean value was 0.2%. Only four of these events contributed to runoff at the outlet. Two of these were due to the intense summer thunderstorms mentioned

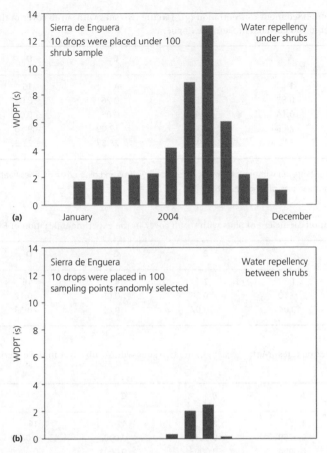

Figure 3.6 Water drop penetration time (s) measurements carried out at the Sierra de Enguera each month in 2004: (a) under shrubs and (b) in the inter-shrubs position. A hundred shrubs were characterised by 10 drops in the inter-shrub and shrub position (n=2000 samples per month).

earlier promoting infiltration excess overland flow. The other two were related to a long period of rainfall (141 mm from 3 December to 12 December 2004 and 38 mm from 9 May to 30 May 2004) promoting saturation excess overland flow.

It should be highlighted that the higher runoff rates were registered on the small plots, while the largest plots showed the lowest runoff coefficient. Runoff was negligible on the latter plots, and the sediment concentration was also extremely low (0.044 g l^{-1} on average) with no values over 0.1 g l^{-1}. No relationship was found between the sediment concentration and the size of the plot. The sediment concentration

during the two extreme events was also very low, probably due to the high vegetation cover.

Measurements on plots of different sizes (1, 2, 4, 16 m^2) during 2004 showed that the soil erosion rates on the shrub-covered soil were low. Runoff rates per rainfall event were negligible (<1%) and average values were below 0.5%. Lower values were found in runoff plots with larger sizes (Table 3.2). The sediment concentration showed no clear pattern in relation to the area of plots and in all cases was very low (<0.1 g l^{-1}). In the case of runoff, the Shrubland-1 plot had higher flows (by a factor of four), but in the case of the sediment concentration, the differences between plots were

Table 3.2 Average runoff sediment concentration $(g\,l^{-1})$ in the two sites with shrub cover in the experimental station of El Teularet in 2004 located in Sierra de Enguera.

Plot	A	B	C	D	Mean
Surface (m²)	1	2	4	16	
Matorral-1	0.82	0.61	0.26	0.04	0.43
Matorral-2	0.14	0.12	0.08	0.03	0.09
Olive orchard	24.54	15.23	12.09	9.56	0.09
Citrus orchard	34.34	29.32	28.87	23.32	0.09

The shrubland plots are compared with the data collected on nearby citrus and olive orchards. These data correspond to the average concentration of samples of runoff events.

Table 3.3 Annual runoff coefficient of plots with shrub cover in the experimental station of El Teularet in 2004.

Plot	A	B	C	D	Mean
Surface (m²)	1	2	4	16	
Matorral-1	0.10	0.06	0.05	0.12	0.08
Matorral-2	0.04	0.07	0.05	0.03	0.05

Table 3.4 Annual erosion rates $(Mg\,ha^{-1}\,year^{-1})$ in the two sites with shrub cover in the experimental station of El Teularet in 2004.

Plot	A	B	C	D	Mean
Surface (m²)	1	2	4	16	
Matorral-1	0.0059	0.0013	0.0002	0.0000	0.0019
Matorral-2	0.0004	0.0003	0.0001	0.0000	0.0002

negligible (Table 3.3). The mean soil loss was 0.0019 and 0.0002 $Mg\,ha^{-1}\,year^{-1}$ for plots 1 and 2, respectively (Table 3.4).

Soil and water losses under simulated rainfall

The study developed by means of soil erosion plots presented earlier showed that soil erosion is low on shrubland-covered soils. However, there are two as yet unsolved research questions: (i) as soil erosion is determined by the rainfall intensity, soil losses will be extremely high during low-frequency–high-magnitude rainfall events; and (ii) due to the influence of the shrub patch on the soil erosion processes, measurements should be done in the inter-patch and patch surfaces in order to determine if sediments are removed from the bare patch and collected by the vegetated patch. The use of rainfall simulation experiments can shed light on both of these questions.

Table 3.5 shows the results of 24 paired plots that compare the hydrological response of shrub and inter-shrub patches. On the shrub-covered areas, most of the plots contributed to negligible runoff discharge, but the one with the greatest runoff still showed a runoff rates four times lower than the one with the lowest runoff rate measured on the bare patches. A similar response was found when comparing the sediment concentration, soil erosion and sediment yield from the same plots (Table 3.6). The data

Table 3.5 Runoff (%, l and mm) on 0.25 m² paired plots: Shrub-covered versus bare inter-shrub areas in the Sierra de Enguera under simulated rainfall.

Runoff	Shrub	Inter-shrub	Shrub	Inter-shrub	Shrub	Inter-shrub
Plot	%	%	l	l	mm	mm
1	0.2	15.0	0.0	2.5	0.1	9.0
2	0.5	14.2	0.1	2.3	0.3	8.5
3	0.6	16.3	0.1	2.7	0.4	9.8
4	1.2	18.6	0.2	3.1	0.7	11.2
5	1.8	17.1	0.3	2.8	1.1	10.2
6	1.9	18.9	0.3	3.1	1.1	11.3
7	2.4	19.4	0.4	3.2	1.4	11.6
8	2.5	17.0	0.4	2.8	1.5	10.2
9	2.6	20.5	0.4	3.4	1.6	12.3
10	3.5	21.0	0.6	3.5	2.1	12.6
11	3.8	30.5	0.6	5.0	2.3	18.3
12	4.0	27.3	0.7	4.5	2.4	16.4
Mean	2.08	19.64	0.34	3.24	1.25	11.78
St. dev.	1.29	4.82	0.21	0.80	0.77	2.89
cv (%)	61.94	24.54	61.94	24.54	61.94	24.54

Table 3.6 Sediment production (g l⁻¹, g and g m⁻² h⁻¹) on 0.25 m² paired plots: Shrub-covered versus bare inter-shrub areas in the Sierra de Enguera under simulated rainfall.

Sediments	Shrub	Inter-shrub	Shrub	Inter-shrub	Shrub	Inter-shrub
Plot	g l⁻¹	g l⁻¹	g	g	g m⁻² h⁻¹	Mg ha h⁻¹
1	3.5	1.0	0.1	2.5	0.005	0.101
2	4.7	1.1	0.4	2.5	0.015	0.098
3	3.3	1.0	0.3	2.7	0.013	0.110
4	3.9	1.0	0.8	3.1	0.031	0.125
5	3.1	1.7	0.9	4.8	0.037	0.191
6	4.2	1.9	1.3	5.8	0.053	0.231
7	3.1	1.0	1.2	3.1	0.048	0.126
8	4.7	1.4	1.9	4.0	0.077	0.159
9	3.7	1.6	1.6	5.3	0.063	0.214
10	3.6	1.4	2.1	4.7	0.083	0.187
11	3.7	1.6	2.3	7.9	0.094	0.316
12	4.1	1.6	2.7	7.3	0.107	0.293
Average	3.79	1.35	1.31	4.48	0.052	0.179
St. dev.	0.53	0.32	0.84	1.84	0.033	0.074
cv (%)	14.10	23.49	64.11	41.00	64.110	41.000

show that the suspended sediment concentration of the runoff was three times higher on the bare sites (from 1.35 to 3.79 g l⁻¹) and the sediment yield and soil erosion was close to four times higher. Even under extremely high-intensity rainfall events, the shrubland contributed a very low sediment yield (1.31 g during 1 h), while the bare sites reached 4.48 g. These results also show that the shrublands are very stable from the erosional point of view as the soil erosion rate was 0.05 Mg ha h^{-1} under the shrubs and 0.18 Mg ha h^{-1} on the inter-patch areas.

Discussion

Shrubs reduce the direct solar radiation and protect the soil against wind, which are both key factors for evaporation. Moreover, shrub cover also develops a litter layer that reduces direct evaporation. However, there are other factors that determine the soil moisture. Shrubs transpire soil water, and during dry periods, they can also exhaust those resources. However, the 'islands' of shrubs on Mediterranean hillsides can also contribute to water infiltration from runoff coming from the surrounding bare areas (Cerdà 1997b). Shrubs contribute to reducing soil moisture, which is especially relevant during drought periods where the soil is dry. The presence of shrubs and the resulting low soil moisture and high soil infiltration rates ensure that water flowing from nearby areas will (during the rainy season) feed the patch where the shrubs are located with water, sediments, nutrients and seeds (Cammeraat et al. 2010). This will contribute to a higher biological activity that will, in turn, contribute to changes in the soil properties (as demonstrated by García-Orenes et al. 2012).

The measurements and the experiments carried out in this research show that shrubland contributes to reducing soil and water losses from slopes affected by a millennial-old tradition of exploitation and degradation. The growth of matorral shrublands after the land abandonment from the 1950s is reducing sediment yields due to a lower soil erodibility and the protection offered by these shrubs. Moreover, infiltration rates are higher under these shrublands. These findings were tested under natural and simulated rainfall of 10-year return periods and confirmed that shrubland contributes to a reduction in discharge and

sediment yield. These results were also found by other researchers such as Guerrero Campo (1996–1997) in Aragón, Úbeda (1994) in Cataluña, Soto and Díaz-Fierros (1998) in Galicia and Sánchez (1995) and Ruiz Sinoga et al. (2010) in Andalucía, among others. The research shown here demonstrates that shrublands control the soil and water losses at different scales within these slopes. The measurements by the WDPT at millimetre scale, the simulated rainfall at centimetre scale and the plot measurements at metre scale confirm that shrubland reduces the soil and water losses and develops more organic matter-rich soils, with lower soil bulk densities. This also contributes to triggering water repellency and preferential flow (Jordán et al. 2008). Studies of this interaction of water and soils are now being extended to agriculture land (e.g. González-Peñaloza et al. 2012).

The results shown here also demonstrate that runoff mechanisms in Mediterranean-type ecosystems are not Hortonian. Subsurface flow is important in Mediterranean slopes and the vegetation cover controls this. More research is needed to determine the effects of plant roots on soil erosion and soil hydrology. Likewise, very little is known about shrubland rainfall interception and stemflow. The study of effects of biota on the hydrology of these slopes should be extended to examining the fauna. Soil properties change with time under shrubs (Cerdà 1997b), and these also need to be quantified and will contribute to a better understanding of the work of Professor John Thornes. The pioneering ideas of John Thornes are now extending to the related topic of the relationship between vegetation and erosion through studies such as those on seed behaviour on slopes (García Fayos et al. 2010), biological changes due to vegetation and land treatments (Morugán et al. 2008) and the effects of ash on shrubland soils after forest fire (Bodí et al. 2011; Pereira et al. 2013). Organic matter redistribution after forest fires also shows the interaction between vegetation and erosion (Novara et al. 2011). The response of water repellency on soil covered by shrubs has been identified as a key mechanism in infiltration (Jordán et al. 2010) as was found on typical semi-arid shrublands in southeastern Spain by Cammeraat et al. (2010). The research inspired by John Thornes' visits and research in Spain has also expanded to include examination of the effects of fauna (Cerdà et al. 2010; Hallett et al. 2010). Some examples are the research related to the ants (Cerdà and Doerr 2010) and microbial activity (García-Orenes et al. 2010) which may help to explain why soil properties are different on vegetation-covered soils and on bare soils (García-Orenes et al. 2009, 2010) because under shrub cover there is a higher faunal activity (Cammeraat et al. 2010) which is related to the higher organic matter and porosity of the plant-covered soils.

Conclusions

Since the 1960s, John Thornes' insistence on the importance of shrubs as a protector of the soil was repeatedly ignored by the policymakers. The work presented here confirms the idea that the shrubland should be regarded as an excellent protector of the soil and water resources just as John Thornes suggested. The patchy distribution of the shrubs contributes to uneven distributions of the plants, water, nutrients, fauna and organic matter and to the negligible soil and water losses.

Acknowledgements

To Professor John Thornes for his teaching and encouragement. The authors wish to thank Vicent García, Carlos Jovani, Maria Burguet, Miguel Segura, Adrian Revert, Diego García, Fermin Poquet and Javier García López for their collaboration in data collection. The research was partially funded by CGL2008-02879/BT and LEDDRA 243857 CONSORTIUM AGREEMENT ENV.2009 243857.

References

Belmonte Serrato, F., Romero-Díaz, A., López-Bermúdez, F., Hernández Laguna, E. (1999) Óptimo de cobertura vegetal en relación a las pérdidas de suelo por erosión hídrica y las pérdidas de lluvia por interceptación. *Papeles de Geografía* **30**, 5–15.

Bergkamp, G. (1996) Mediterranean geoecosystems. Hierarchical organisation and degradation. PhD thesis. University of Amsterdam, Amsterdam.

Bochet, E., Rubio, J. L., Poesen, J. (1998) Relative efficiency of three representative matorral species in reducing water erosion at the microscale in a semi-arid climate (Valencia, Spain). *Geomorphology* **23**, 139–150.

Bodí, M., Cerdà, A. (2008) La Estación Experimental para el Estudio de la Erosión y Degradación de los Suelos de El Teularet-Sierra de Enguera. In: Cerdà, A. (ed.) *Erosión y degradación del suelo agrícola en España*. Universitat de València Estudi General, Valencia, pp. 209–238.

Bodí, M. B., Mataix-Solera, J., Doerr, S. H., Cerdà, A. (2011) The wettability of ash from burned vegetation and its relationship to Mediterranean plant species type, burn severity and total organic carbon content. *Geoderma* **160**, 599–607

Boix, C. (2000) Procesos geomórficos en diferentes condiciones ambientales mediterráneas: el estudio de la agregación y la hidrología de los suelos. PhD thesis. Faculty of Geography and History, University of Valencia, Valencia.

Cammeraat, E., Cerdà, A., Imeson, A. C. (2010) Ecohydrological adaptation of soils following land abandonment in a semiarid environment. *Ecohydrology* **3**, 421–430.

Cerdà, A. (1996) Seasonal variability of infiltration rates under contrasting slope conditions in Southeast Spain. *Geoderma* **69**, 217–232.

Cerdà, A. (1997a) Seasonal changes of the infiltration rates in a typical Mediterranean scrubland on limestone. *Journal of Hydrology* **198**, 209–225.

Cerdà, A. (1997b) The effect of patchy distribution of *Stipa tenacissima* L. on runoff and erosion. *Journal of Arid Environments* **36**, 37–51.

Cerdà, A. (1999) Parent material and vegetation affect soil erosion in eastern Spain. *Soil Science Society of America Journal* **63**, 362–368.

Cerdà, A. (2008) *Erosión y degradación del suelo agrícola en España*. Universitat de València Estudi General, Valencia.

Cerdà, A., Bodí, M. (2008) Erosión hídrica del suelo en el territorio valenciano. In Cerdà, A. (ed.) *Erosión y degradación del suelo agrícola en España.* Universitat de València Estudi General, Valencia, pp. 51–82.

Cerdà, A., Doerr, S. H. (2007) Soil wettability, runoff and erodibility of major dry-Mediterranean land use types on calcareous soils. *Hydrological Processes* **21**, 2325–2336.

Cerdà, A., Doerr, S. H. (2008) The effect of ash and needle cover on surface runoff and erosion in the immediate post-fire period. *Catena* **74**, 256–263.

Cerdà, A., Doerr, S. H. (2010) The effect of ant mounds on overland flow and soil erodibility following a wildfire in eastern Spain. *Ecohydrology* **3**, 392–401.

Cerdà, A., Jurgensen, M. F. (2011) Ant mounds as a source of sediment on citrus orchard plantations in eastern Spain. A three-scale rainfall simulation approach. *Catena* **85**, 231–236.

Cerdà, A., Giménez-Morera, A., Bodí, M. B. (2009) Soil and water losses from new citrus orchards growing on sloped soils in the western Mediterranean basin. *Earth Surface Processes and Landforms* **34**, 1822–1830.

Cerdà, A., Hooke, J., Romero-Diaz, A., Montanarella, L., Lavee, H. (eds.) (2010) Special issue on soil erosion and degradation on mediterranean type-ecosystems. *Land Degradation and Development* **21**, 71–217.

Francis, C., Thornes, J. B. (1990) Matorral: erosion and reclamation. In Albaladejo, J., Stocking, M. A., Díaz, E. (eds.) *Degradación y regeneración del suelo en condiciones ambientales mediterráneas.* CSIC, Murcia, pp. 87–112.

García Ruiz, J. M. (1997) La agricultura tradicional de montaña y sus efectos sobre la dinámica hidromorfológica de laderas y cuencas. In García Ruiz, J. M., López García P. (eds.) *Acción humana y desertificación en ambientes mediterráneos.Instituto Pirenaico de Ecología,* Zaragoza, 119–144.

García-Fayos, P., Bochet, E., Cerdà, A. (2010) Seed removal susceptibility through soil erosion shapes vegetation composition. *Plant Soil* **334**, 289–297.

García-Orenes, F., Cerdà, A., Mataix-Solera, J., Guerrero, C., Bodí, M. B., Arcenegui, V., Zornoza, R., Sempere, J. G. (2009) Effects of agricultural management on surface soil properties and soil-water losses in eastern Spain. *Soil and Tillage Research* **106**, 117–123.

García-Orenes, F., Guerrero, C., Roldán, A., Mataix-Solera, J., Cerdà, A., Campoy, M., Zornoza, R., Bárcenas, G., Caravaca. F. (2010) Soil microbial biomass and activity under different agricultural management systems in a semiarid Mediterranean agroecosystem. *Soil and Tillage Research* **109**, 110–115.

García-Orenes, F., Roldán, A., Mataix-Solera, J., Cerdà, A., Campoy, M., Arcenegui, V., Caravaca, F. (2012) Soil structural stability and erosion rates influenced by agricultural management practices in a semi-arid Mediterranean agro-ecosystem. *Soil Use and Management* **28**, 571–579.

González Hidalgo, J. C. (1996–1997) Efecto de la vegetación y orientación de ladera en perfiles de humedad en el suelo de un ambiente semiárido del interior de España. *Cuadernos de Investigación Geográfica* **22–23**, 81–96.

González-Peñaloza, F. A., Cerdà, A., Zavala, L. M., Jordán, A. (2012) Do conservative agriculture practices increase soil water repellency? A case study in citrus-cropped soils. *Soil and Tillage Research* **124**, 233–239.

Guerrero Campo, J. (1996–1997) Procesos erosivos intensos en las áreas marginales de la depresión del Ebro y del Pirineo. *Interpretación de los patrones de vegetación. Cuadernos de Investigación Geográficas* **22–23**, 57–79.

Hallett, P. D., Lichner, L., Cerdà, A. (2010) Preface. Biohydrology: coupling biology and soil hydrology from pores to landscapes. *Ecohydrology* **3**, 379–381.

Jordán, A., Martínez-Zavala, L., Bellinfante, N. (2008) Heterogeneity in soil hydrological response from different land cover types in southern Spain. *Catena* **74**, 137–143.

Jordán, A., González, F. A., Zavala, L. M. (2010) Re-establishment of a soil water repellency after the destruction by intense burning in a Mediterranean heathland (SW Spain). *Hydrological Processes* **24**, 736–748.

López Bermúdez, F., Thornes, J. B., Romero, A., Francis, C. F., Fisher, G. C. (1986) Vegetation–erosion relationships; Cuenca de Mula, Murcia, Spain. In López Bermúdez, F., Thornes, J. B. (eds.) *Estudios sobre Geomorfología del Sur de España.* Universidad de Murcia, Murcia, pp. 101–104.

Mataix-Solera, J., Arcenegui, V., Guerrero, C., Mayoral, A. M., Morales, J., González, J., García-Orenes, F., Gómez, I. (2007) Water repellency under different plant species in a calcareous forest soil in a semiarid Mediterranean environment. *Hydrological Processes* **21**, 2300–2309.

Morugán, A., García-Orenes, F., Mataix-Solera, J., Gómez, I., Guerrero, C. (2008) Short term effects of treated waste water irrigation on soil. Influence on chemical soil properties. In Alonso, D., Iglesias, H. J. (eds.) *Agricultural Irrigation Research Progress.* Nova Science Publishers, Hauppauge, NY, pp. 1–10.

Novara, A., Gristina, L., Bodì, M. B., Cerdà, A. (2011) The impact of fire on redistribution of soil organic matter on a Mediterranean hillslope under maquia vegetation type. *Land Degradation and Development* **22**, 530–536.

Pereira, P., Cerdà, A., Úbeda, X., Mataix-Solera, J., Jordan, A., Burguet, M. (2013) Spatial models for monitoring the spatio-temporal evolution of ashes after fire – a case study of a burnt grassland in Lithuania. *Solid Earth* **4**, 153–165.

Puigdefábregas, J., Solé-Benet, A., Lázaro, R., Nicolau, J.M. (1992) Factores que controlan la escorrentía en una zona semiárida sobre micaesquistos. In: López, F., Conesa, H., Romero, A. (eds) *En Estudios de Geomorfología en España*, Vol. **1**. Sociedad Española de Geomorfología, Murcia, 117–127.

Rodríguez, J., Pérez, R., Cerdà, A. (1991) Colonización vegetal y producción de escorrentía en bancales abandonados: Vall de gallinera, Alacant. *Cuaternario y Geomorfología* **5**, 119–129.

Romero Díaz, A., Barberá, G. G., López Bermúdez, F. (1995) Relaciones entre erosión del suelo, precipitación y cubierta vegetal en un medio semiárido del sureste de la península ibérica. *Lurralde* **18**, 229–243.

Ruiz Sinoga, J. D., Romero Díaz, A., Ferre Bueno, E., Martínez Murillo, J. F. (2010) The role of soil surface conditions in regulating runoff and erosion processes on a metamorphic hillslope (Southern Spain). Soil surface conditions, runoff and erosion in Southern Spain. *Catena* **80**, 131–139.

Sánchez, G. (1995) Arquitectura y dinámica de las matas de esparto (*Stipa tenacissima* L.) efectos en el medio e interacciones con la erosión. PhD thesis. Complutense University of Madrid, Madrid.

Schnabel, S., Gómez Amelia, D., Bernet, R. (1996) La pérdida de suelo y su relación con la cubierta vegetal en una zona de dehesa. VII Coloquio Ibérico de Geografía, Cáceres, pp. 195–206.

Soto, B., Díaz-Fierros, F. (1998) Runoff and soil erosion from areas of burnt scrub: comparison of experimental results with those predicted by the WEPP model. *Catena* **31**, 257–270.

Thornes, J. B. (1976) Semi-arid erosional system, case studies from Spain. LSE Geographical Papers No. 7. London School of Economics, London.

Thornes, J.B. (1985) The Ecology of Erosion. *Geography* **70**, 222–236.

Thornes, J. B. (1990) *Vegetation and Erosion*. John Wiley & Sons, Ltd, Chichester.

Úbeda, X. (1994) Caracterització del sòl i quantificació del transport de sediment en un bosc mediterrani (les Gavarres, Massís Litoral Càtala). *Notes de Geografia Física* **23**, 31–38.

Walkley, A., Black, I. A. (1934) An examination of the Degtjareff method for determining soil organic matter, and proposed modification of the chromic acid titration method. *Soil Science* **37**, 29–38.

CHAPTER 4

Morphological and vegetation variations in response to flow events in rambla channels of SE Spain

Janet Hooke[1] and Jenny Mant[2]

[1]Department of Geography and Planning, School of Environmental Sciences, University of Liverpool, Liverpool, UK
[2]River Restoration Centre, Cranfield University Campus, Bedford, UK

Introduction

John Thornes believed in and made major contributions to combining detailed field data with conceptual and numerical modelling. This chapter synthesises some research undertaken to measure and analyse impacts of effects of flow events and interactions with vegetation in channels in SE Spain, in order to provide validation for a simulation model of hydrological effects on river channels and so increase understanding of the possible effects of desertification. This work was initiated under the scope of MEDALUS III and John's guidance in 1996 and continued by field monitoring and various analyses since then. It focuses on the ephemeral channels of that semi-arid region.

The research was established through concern over the impacts of climate and land use disruption and to investigate whether these impacts could be modelled and likely future responses predicted. Overall, this would allow assessment of the vulnerability of these river valleys to hydrological changes and recognition of thresholds of response and sensitivity to impacts. The major impetus for the work was

the aim to construct a simulation model of channel change which would enable the effects of differing flows and hydrological regime resulting from climate and/or land use change to be simulated. A novel aspect of this model was that it included channel vegetation and its feedback effects on processes. The model, CHANGISM, was successfully constructed and was able to simulate morphological and vegetation changes in channel reaches (Hooke et al. 2005). However, it emerged early in the research programme that few data were available on morphological changes and responses in such channels and especially on the properties and responses of Mediterranean plants with which to calibrate and validate the model. The key components of vegetation change that needed to be modelled are identified in Brookes et al. (2000), and the way in which each was dealt with, given a dearth of information, was discussed. The lack of data and understanding led to the research undertaken by Mant for her PhD (Mant 2002) to try to fill gaps. The potential for use of vegetation in management of soil erosion and desertification impacts and the need for still further research were pursued through the EU

Monitoring and Modelling Dynamic Environments, First Edition. Edited by Alan P. Dykes, Mark Mulligan and John Wainwright.

RECONDES project 2004–2007.[1] Given the infrequency and variability of flow events in semi-arid regions, the field measurements were extended and the results of a much longer period are analysed here.

The aim of this paper is to analyse and summarise the evidence of morphological and vegetation dynamics and interactions with flow and hydrological conditions in the semi-arid environment of SE Spain cumulated over a 14-year monitoring period, 1996–2010. In particular, evidence will be examined to address the questions of:

• The incidence and effects of flow events and whether trends or patterns can be detected on this timescale, including any signal of desertification
• The impacts of individual events
• The critical flow conditions and threshold events for significant morphological change and for effects on vegetation of various types
• The effects of drought on vegetation
• The spatial variability in response and sensitivity of sites

Research context

This work was initiated under the impetus for increased research on the possible effects of increased desertification in the southern Mediterranean. John and colleagues undertook much of that research and progressed understanding of desertification trends and causes as well as processes and responses of the physical and biotic systems. These wider issues are not reviewed here but provide the context to our work. Research on runoff generation, especially in the SE Spain region, has given new insights into the spatial distribution and characteristics of the major generating areas (e.g. Bracken and Kirkby 2005; Cammeraat 2004; Kirkby et al. 2005; Reaney et al. 2007) and on the relationship with rainfall and storm properties (e.g. Bracken et al. 2008), showing the importance of bare rock

and low infiltration areas on runoff generation and the complex relationship of flow discharge to rainfall amounts and intensity. Short bursts of high-intensity rainfall tend to produce the very high flows, especially if these occur after a more prolonged wet period (Bull et al. 1999).

The dearth of data on morphological change and responses of these ephemeral channels was a major stimulus to the establishment of the research programme analysed here, but such data are still limited. Thornes' early work on spatial variability of channel morphology (Thornes 1977, 1980) and his work on the impact of the 1973 flood (Mairota et al. 1998) still remain seminal. Conesa-Garcia's (1995) analysis of the effects of different size flow events on these channels provides a useful classification of events and their impacts (Table 4.1). The morphodynamics of ephemeral streams were reviewed by Hooke and Mant (2002), and the overall impacts of human activities, direct and indirect, on fluvial systems in the Mediterranean have been examined by Hooke (2006). A major flow event in 1997, during the MEDALUS III project, produced much change which was analysed by Hooke and Mant (2000). Many of these channels are highly modified, particularly by multiple check dams constructed along the course. More recently, Boix-Fayos et al. (2008) and Castillo et al. (2007) have shown the profound morphological effects of these structures and have calculated amounts of sediment stored behind these dams; Bombino et al. (2008) have done similar analyses in southern Italy. Conesa-Garcia and Garcia-Lorenzo (2009) have calculated the effects of the check dams in controlling scour and tested the results against field evidence from these same streams. Elsewhere in the Mediterranean region, Surian and Rinaldi (2003) have analysed the response of channels to channelisation and also gravel mining and have produced a conceptual model of the sequence of responses over time.

---

[1] http://www.port.ac.uk/research/recondes/

Table 4.1 Classes of events in 'ramblas' of SE Spain.

Class 1	Overbank flows which affect the whole fluvial system. In the case of the *ramblas*, they greatly modify the floodplain by producing at the same time both deposition and scour. Floodwater velocities over this surface are often spatially variable and may, in places, be high enough to produce scour rather than deposition. In Lower Segura, they contribute to the vertical accretion of the floodplain
Class 2	Dominant events controlling channel form. These comprise moderate discharges causing appreciable net changes in the bed and banks and flows which just fill the section of alluvial channels without overtopping the banks (bankfull conditions)
Class 3	Discharges sufficient to cover the whole of the main channel and produce bedload movement
Class 4	Very low energy flows, incapable of causing bedform adjustments

From Conesa-Garcia (1995).

The morphological changes are produced by sediment movement along the channels and are influenced by supply from hillslopes. Variation in sediment characteristics along one of the channels in SE Spain, the Nogalte, was the subject of ongoing research by Thornes. Direct measurements of either suspended or bedload sediment flux are still very few in these channels, and most evidence is from deposition after events. This is in spite of the major concerns with soil erosion on hillslopes and its downstream impacts (Poesen and Hooke 1997). Sediment flux has been measured at instrumented sites on other ephemeral channels and relations of flow dynamics analysed in detail (e.g. Hassan et al. 1999; Powell et al. 2007). One of the major developments and new focus of work in the last decade has been in the application of the concept of connectivity (e.g. Brierley and Fryirs 1998; Brierley and Stankoviansky 2002; Cammeraat 2004; Fryirs and Brierley 1999, 2001; Hooke 2003) both to runoff and sediment. Some of the development and application of the concept has been on these channels in SE Spain and in other semi-arid areas. Much of the Spanish work has been focused on hydrological connectivity and the role of soil characteristics and of vegetation patches on hillslopes (Bracken and Croke 2007; Kirkby et al. 2005; Mayor et al. 2008; Mueller et al. 2007; Okin et al. 2009; Reaney et al. 2007; Smith et al. 2010). The approach has the potential to provide valuable insights into the sources and storage of sediment along channels (Hooke 2003, 2006) and has been used in the development of spatial strategies to combat land degradation (Borselli et al. 2008; Hooke and Sandercock 2012; Sandercock and Hooke 2006). It focuses on the pathways of flow and influence of structures and features, topographic variations and vegetation.

Ecohydrology and vegetation–geomorphological integration have become a much more important component of research since 1996, as led and advocated by John (Thornes 1990). Much progress has been made and some useful reviews relating to Mediterranean vegetation in fluvial systems have been published including Sandercock et al. (2007) and Corenblit et al. (2010). Research in the American Southwest has provided some useful detailed data in environments similar to the dry Mediterranean (Bendix and Hupp 2000; Lite et al. 2005; Phillips and Hjalmarson 1994; Stromberg et al. 2005). A large literature exists on vegetation and processes on hillslopes in Mediterranean and similar environments, and initial work on vegetation interactions with runoff and sediment dynamics is now incorporated in developments and application of connectivity concepts (e.g. Boer and Puigdefábregas 2005; Ludwig et al. 2002). The EU RECONDES project, 2004–2007, aimed to examine and provide guidance on how vegetation could be used to combat desertification. The strategy is to reduce soil erosion and sediment flux by growth of vegetation in flow pathways such as to reduce connectivity within the system, both on hillslopes and in channels (Sandercock and Hooke 2006). Research was undertaken on the effects of specific plants on erodibility and sedimentation, particularly work on roots by De Baets (e.g. De Baets et al. 2007). Much of the specific work on vegetation properties and effects on flow and processes in the literature has been based on flume experiments, but some field evidence incorporating the complexity of conditions and dynamics was collected by Sandercock and Hooke (2010, 2011) in SE Spain. Feedback effects between vegetation and morphology have long been recognised (e.g. Graf 1978; Thornes and Brandt 1994) and are now very topical. Sandercock and Hooke (2010) measured the vegetation characteristics at cross sections in channels in the Murcia region and then used these in calculations of the variation in hydraulics across the channel and with different heights of flow. They also provided evidence of the effects of the vegetation on connectivity, particularly in producing sedimentation in vegetated zones (Sandercock and Hooke 2011). Much research has been undertaken in the last decade on the Tagliamento River in NE Italy showing in detail the role of vegetation in fluvial processes and changes (e.g. Bertoldi et al. 2009; Corenblit et al. 2010; Gurnell et al. 2001), but though called Mediterranean, this is a rather different environment with perennial flow and wide braided rivers. In more arid environments, Tooth and Nanson (1999, 2000) have demonstrated the significant role of vegetation in channels in central Australia, again of rather different character.

Remarkably little literature has been published in the last few years documenting morphological changes in ephemeral channels, and most literature relates to impacts of single flood events. Major gaps still exist in our understanding and quantification of morphological impacts of flow events, conditions affecting vegetation in channels and the feedback effects of vegetation in channels in moderating response. The relatively long-term data here represent a unique dataset of morphological and vegetation variations in ephemeral channels. It is not possible to present detailed analysis of all the data here, so changes are exemplified and outcomes synthesised.

Background and methods

The purpose of the channel modelling component in the MEDALUS III project was to produce a model that could simulate the effects of sequences of flows and impacts of other conditions on channels at reach scale. The idea was that predictions and scenarios of change in climate and/or land use resulting in changes in hydrological and runoff regime, produced elsewhere in the project, would provide the input to the model to allow predictions of the type, scale and spatial sensitivity of impacts in valleys. The cellular automata model simulated the effects of flow on morphology, sediment, vegetation and moisture status and allowed for feedback effects between all components. Details of the input conditions and processes producing changes in state are provided in Hooke et al. (2005). The monitoring programme was therefore designed to provide the data on input conditions and quantify changes and interactions between the major components.

In terms of the morphology, the model assumes an input of the channel and floodplain/valley floor topography. Erosion and deposition occur during flow events in accordance with relations of competence and capacity of flows; the Bagnold stream power relation was selected to model sediment flux (Hooke et al. 2005). These processes alter the morphology of the channel which has feedback effects on flow distribution and velocities in the next event. Distributions of particle sizes of sediment in zones within the reach are also altered. The conceptualisation of the vegetation component envisaged that vegetation of various types and states is present in different parts of the channel/valley floor and that these are affected by flows according to the characteristics of the event, the prior state of the plants and substrate, the properties of the plants in terms of hydraulic resistance and resistance to erosion and the ability to cause sedimentation. In between events, the plants grow according to temperature and moisture and have a life cycle from germination to mortality. They can also spread by processes such as suckering. Therefore, data on all these processes was needed with which to validate the model.

The lack of data available with which to calibrate or test the modelling led to the establishment of monitoring sites early in the project. The target area for the research was the Guadalentin basin in SE Spain (Mairota et al. 1998) (Figure 4.1). Nine sites for monitoring were selected to provide a range of lithological, hydrological and morphological settings and to ensure that some flow events occurred, given the high variability and infrequency of flows in these channels, typical of semi-arid areas. The sites comprise representative channel reaches of 100–200 m in length and are located in the upper, middle and lower parts of the channel systems of three catchments, Nogalte, Torrealvilla and Salada (Figure 4.1). Details of the sites are given in Table 4.2 and in Hooke and Mant (2000). The Nogalte catchment is in schist bedrock and the other two are in dominantly marls. The Torrealvilla Serrata site was replaced by Pintor in 2002. Salada 2 and Nog2 have been less intensively monitored than the other sites.

Methods of monitoring at each site were designed to measure the four basic components required for the modelling – flow occurrence and peak discharge, morphology, sediment characteristics and substrate and vegetation – and to detect change by repeat surveys. The details of monitoring methods have been described by Hooke (2007) and are summarised here. Methods were constrained by low funding for instrumentation and the need for methods to be accomplished in concentrated field periods and for the sites to be operable from a distant location. Practical difficulties of instrumentation also pertain to these highly flashy ephemeral flow regime channels.

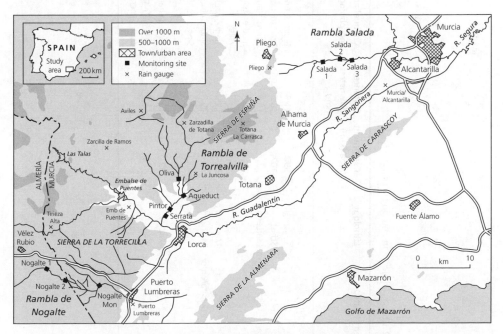

Figure 4.1 Locations of monitored channel reaches (sites) and rain gauges in SE Spain.

Table 4.2 Characteristics of monitoring sites in SE Spain.

Catchment	Site	Lithology	Catchment area (km²)	Distance downstream (km)	Channel and valley morphology	Gradient	Floodplain width (m)	Channel width (m)	Channel depth (m)	Vegetation Channel	Vegetation Floodplain
Nogalte	1	Schist	6.9	6.5	Shallow channel	0.0161	20	1.4–10.3	0.21–0.79	Sparse	Retama
	2	Schist	39.1	11.5	Braided, unconfined	0.0199	70	9.5–15.3	0.30–0.84	Shrubs	Retama
	Mon	Schist	102.7	18.5	Braided, unconfined	0.0188	120	12.7–24.6	0.34–1.15	Retama	Agriculture
Torrealvilla	Oliva	Marl and gravel	73.2	16.0	Single channel, mod-erately wide valley	0.0099	25	2.2–5.2	0.24–0.74	Bare	Tamarisk/agriculture
	Aqueduct	Marl and gravel	54.4	7.0	Single, confined	0.0072	25	2.6–8.1	0.40–1.55	Bare	Shrubs, tamarisk
	Serrata/Pintor	Marl and gravel	253.8	25.0	Single, embanked	0.0049	35	8.7–18.8	0.31–0.73	Bare	Shrubs
Salada	1	Marl	41.1	10.0	Single, confined	0.0072	20	1.0–5.0	0.32–0.66	Sparse	Tamarisk, shrubs, agriculture
	2	Marl	59.4	13.0	Incised channel	0.0063	45	2.5–5.4	1.40–1.61	Tamarisk	Dense, mixed
	3	Marl	82.0	15.0	Single, moderate confinement	0.0087	50	0.8–5.5	0.38–1.23	Herbs	Mixed, moderately dense

The key parameter in terms of flow events was to know whether flow had occurred in the period since last observation and to know at least peak stage/discharge. Ideally, it would have been desirable to know the timing of events, duration and characteristics of hydrographs, but installation of a continuous stream gauge at all nine sites was not possible. One pressure gauge was installed at the Aqueduct site on the Prado tributary of the Torrealvilla in cooperation with Leeds University, which provided data until 2002. The main method employed at other sites was the use of crest stage recorders using water-sensitive tape. These clearly indicate maximum height of flow in a period and, in such an environment where flow is only occasional, have worked well. The reading on the tape is always checked against flood marks in the channel, which are also surveyed. In a few cases, the pipe has become blocked but flood marks in the channel have been used. In one or two of the highest flows at some sites, the crest recorder was completely destroyed or bent over by the flood, but in these cases, the flood marks have been very clear. One discharge gauge is operated on the Salada by the regional hydrological authority, Confederación Hidrográfica del Segura, but data were not released by them in the measurement period. Flow events are discussed in terms of stage because this is of morphological importance at the sites and because considerable errors pertain to calculation of discharge in these channels (Sandercock and Hooke 2010).

Rainfall data were initially provided by an autographic rain gauge at the Prado site for the period 1998–2000, but neither these nor the flow gauge data are available for the latter part of the monitoring period. Daily rainfall data for the region from 10 available gauges located within or near the catchments (Figure 4.1) are obtained annually from the regional branch of the Spanish National Meteorological Office (INM/AEMET) and used to identify key events and to calculate hydrological conditions. Data from the Puentes rainfall gauge are used for overall analysis here since it is one of the most complete records and is central to a major group of the sites. Although individual stations depart from this, it is in the centre of the region covered by the three catchments, and it has a moderate relief, though rainfall daily maxima and annual totals are lower than most other stations in the region. It has a complete daily record since 1961. The flow records are analysed in relation to the two nearest gauges in each catchment: for Nogalte, Lorca Tirieza Alta and Puerto Lumbreras; for Torrealvilla sites, La Juncosa and Embalse de Puentes; and for Rambla Salada, Pliego and Alcantarilla. Annual and monthly data provide the context of the variability of conditions during the period 1996–2010. All days with greater than 10 mm rainfall have been identified and analysed for frequency of rainfall in various daily totals and the maximum daily rainfall in the year. The daily records were analysed to identify the highest flow events since the last reading of the crest stage recorder. In most cases, the rainfall events of greater than 20 mm are so rare that the generating events and dates of runoff can be easily identified. In a few cases, multiple events between readings took place and the exact date of peak runoff cannot be established.

Topography of the reaches has been surveyed, initially using a geodimeter (to 1998) and subsequently by differential GPS. Positions and height are measured to 20 mm accuracy. Major morphological features such as bank lines are surveyed by string lines to feed into DEM construction and points distributed across channel and floodplain according to topographical variation. In addition, several cross sections are surveyed in each site because it is found that these are more sensitive in detecting very small amounts of change. Amounts of erosion and deposition and changes in width, depth and channel cross-sectional area are calculated. Velocities and details of hydraulics have been calculated using WINX-SPRO (Sandercock and Hooke 2010).

Distribution of sediment patches is mapped in the field using the dGPS to delimit zones.

Sediment calibre is monitored by repeat photography of sample sediment quadrats in different zones and sediment size analysed from the photographs. This method has the advantage that it is relatively rapid in the field, is non-destructive and does not entail transport of sediment samples to the lab. It does only provide data on the surface material which may give a bias in terms of the overall size of material in transport, but it was still the most relevant data for assessment of resistance and initial entrainment. The sediment data are not analysed in detail here, but the evidence of characteristics is used to indicate the substrate and changes to indicate mobility.

Vegetation cover is very important because it has effects on flow resistance and on erosion and deposition processes. Vegetation cover is also a key indicator of desertification, with feedback effects (Sandercock et al. 2007). Vegetation is monitored by mapping and measurement in sample quadrats, mostly 3 m². Quadrats were established in different zones within each reach, for example, channel, bar and floodplain. Numbers, position and dominant plants in quadrats at each site are given in Table 4.3. On each visit, coverage of plants, species of each, height and state are mapped and the quadrats photographed. Vegetation was monitored more intensively in the period 1997–2000, and extra measurements and experiments were made to assess the strength and resistance of plants (Mant 2002). The quadrat approach is not suitable for highly vegetated reaches or for monitoring large trees, so ancillary surveys of specific vegetated cross sections have been made periodically. Repeat surveys of quadrats are made as a minimum once a year, in January, to maintain consistent season, though this does exclude increased cover in spring due to seasonal plants. Surveyed heights of the corners of the quadrats have also been used to calculate changes in topography and amounts of erosion and deposition.

Sites have been checked and measurements made after every major event as far as possible, particularly during the major project periods of 1997–2000 and 2004–2007. In the rest of the period, a minimum of annual surveys, usually in January, has been made. In this paper, the magnitude and occurrence of rainfall and runoff events and their temporal distribution in the period 1996–2010 are analysed to provide the context for analysis of impacts of individual events on morphology and vegetation and the wider context of possible conditions relating to desertification. The scale and types of morphological change in events are exemplified, and the amount of change in cross-sectional profiles is analysed in relation to magnitude of peak flow for key sites. Evidence for thresholds of impact and of trends and phases are investigated in relation to both morphological and vegetation variations.

Results

Hydrological conditions and events

Rainfall variations over the longer term provide the context of trends and phases in hydrological conditions for analysis of the events and changes taking place during the monitoring period. The rainfall gauge at Puentes reservoir is taken here as representative for the region to show general trends. No trend is discernible in the annual total rainfall for the period 1961–2009 (Figure 4.2a). Wetter phases occurred in 1968–1977 and 1985–1994. 1979–1983 was notably dry and 1994–1997 and 2004–2006 were less extended dry periods. The annual totals are important in terms of vegetation growth and agricultural production, but it is the maximum daily rainfalls which influence the occurrence of runoff events (Figure 4.2b). A period of high variability took place in 1964–1977 with several annual daily maxima near 80 mm. Daily maxima were relatively low from 1977 to 1985 and then higher in the mid-1980s with the maximum for the period, 92 mm, recorded in 1989. The period from 1990 was rather less variable with notably low years of less than 20 mm maximum occurring in 1995 and 2005. The monitoring period of 1995–2009 does exhibit a trend

Table 4.3 Vegetation quadrats: numbers at sites, position and composition.

Site	Quadrat no.	Position	Cover	Dominant vegetation		Anthyllis cytisoides	Artemesia barrelieri	Euphorbia pithyusa	Fumana thymifolia	Gramineae (sp)	Helichrysum italicum	Herbs (general)	Inula crithmoides	Inula viscosa (Dittrichia)	Juncus bulbosus	Limonium auric-ursifol	Limonium echioides	Lygeum spartum	Nerium oleander	Retama sphaerocarpa	Salicornia europaea	Santolina	Spergularia (sp)	Tamari parviflora	Thistles (general)	Thymelaea hirsuta
Torr Oliva	VQ1	Floodplain	Mod/sparse	Herbs		▓			▓	▓		▓		▓								▓	▓	▓	▓	
	VQ2	Channel	Bare	Bare																						
	VQ3	Floodplain	Moderate	Tamarisk	Herbs																				▓	
	VQ4	Floodplain	Dense	Grass						▓							▓							▓		
Torr Aqued	VQ1	Floodplain	Dense			▓				▓						▓		▓						▓		
	VQ2	Channel/bar	Moderate	Lygeum												▓										▓
	VQ3	Bar	Moderate	Lygeum														▓						▓		
	VQ4	Channel	Sparse	Bare																				▓	▓	
	VQ5	Floodplain	Dense	Thymylaea	Tamarisk	▓																		▓		▓
Torr Pintor	VQ1	Channel/bar	Sparse	Herbs											▓											
	VQ2	Outer channel	Bare	Bare												▓										
	VQ3	Inner channel	Sparse	Grass																					▓	
	VQ4	Bar	Bare	Bare																			▓			
Salada 1	VQ1	Bar/wake	Moderate	Herbs	Shrubs	▓						▓		▓									▓			
	VQ2	Channel	Sparse	Limonium									▓	▓												
	VQ3	Channel edge	Moderate	Tamarisk	Limonium							▓	▓	▓										▓		
	VQ4	Bar	Sparse	Tamarisk																				▓		
Salada 2	VQ1	Channel	Bare																							
Salada 3	VQ1	Bar	Dense	Tamarisk						▓				▓							▓	▓		▓		
	VQ2	Bar	Moderate	Dittrichia										▓												
	VQ3	Bar	Moderate	Dittrichia										▓		▓		▓								

(Continued)

Table 4.3 (*Continued*)

Site	Quadrat no.	Position	Cover	Dominant vegetation	Anthyllis cytisoides	Artemesia barrelieri	Euphorbia pithyusa	Fumana thymifolia	Gramineae (sp)	Helichrysum italicum	Herbs (general)	Inula crithmoides	Inula viscosa (Ditrichia)	Juncus bulbosus	Limonium auric-ursitol	Limonium echioides	Lygeum spartum	Nerium oleander	Retama sphaerocarpa	Salicornia europaea	Santolina	Spergularia (sp)	Tamari parviflora	Thistles (general)	Thymelaea hirsuta
Nogalte 1	VQ1	Bar	Moderate	Retama	■	■	■				■		■	■					■					■	
	VQ2	Channel:bar	Moderate	Retama	■	■	■				■		■	■			■		■				■	■	
	VQ3	Floodplain	Moderate	Retama		■	■		■				■	■					■					■	
	VQ4	Bar	Dense	Retama							■		■	■					■				■	■	
	VQ5	Channel	Sparse/mod	Herbs	■	■	■		■		■		■	■					■				■	■	
Nog Mon	VQ1	Bar	Moderate	Retama									■	■					■						
	VQ2	Channel	Sparse/bar	Bare																					
	VQ3	Bar	Dense	Retama							■			■					■					■	
	VQ4	Channel/bar	Moderate	Retama					■	■	■			■					■					■	

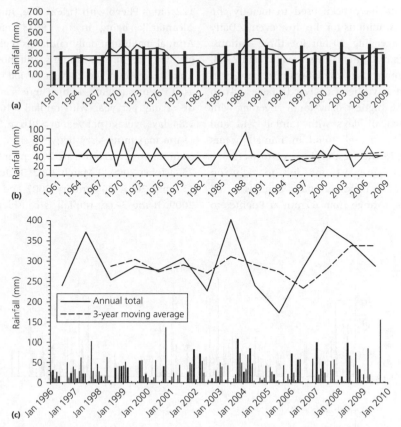

Figure 4.2 Puentes rain gauge record: (a) annual rainfall totals for 1961–2009 and trend, (b) annual maximum daily rainfall totals for 1961–2009 with trends for the whole period and for 1995–2009 only, and (c) annual and monthly rainfall totals for January 1996–January 2010.

(Figure 4.2b) of increasing maximum daily rainfall, but this is not statistically significant.

The annual totals for the calendar years 1996–2009 for Puentes (Figure 4.2c) give an average annual rainfall over that period of 292 mm. The highest total rainfall of 404 mm occurred in 2003, with 2007 and 1997 also having greater than 370 mm in the year. The lowest rainfall year was 2005 with 174 mm, but 1996, 1998, 2002 and 2004 were all below average for the period. The moving average indicates some decline in rainfall in the period 2004–2006 and then a rise. Other gauges do show some variation from this pattern, for example, the Juncosa record, in slightly higher relief, is similar in having a

relatively dry period for 1998–2002 but has a higher peak for 1997 and a lower one for 2002. Monthly totals for Puentes (Figure 4.2c) exhibit the seasonal regime expected of this Mediterranean climate to a large extent, and exceptional months (>80 mm) were September 1997, October 2000, December 2001, October 2003, April 2004, January 2007, May 2008, March 2009 and September 2009. This demonstrates the seasonal pattern of major rainfall in spring and autumn and also the extended drier period of 2005–2006 and the recent wet period. Average monthly rainfall over the period at Puentes was 25 mm, but for the winter months (September–April), it was 31 mm.

Daily data have been used to identify the dates and conditions for the flow events. Daily rainfalls exceeding 20 mm for three stations, one in each catchment, are shown in Figure 4.3, and maximum daily rainfalls for six gauges in the region have been analysed (Table 4.4) to identify the highest daily totals together with the number of days with rainfall >40 and >20 mm. An overall trend in rainfall occurs from drier in the west of the area to wetter in the east (Figure 4.1 and left to right on Table 4.4). The maximum daily rainfalls in the study period range from 65 mm at Puentes to 122 mm at Pliego with Tirieza Alta, Juncosa and Alcantarilla being in the 80–87 mm range. Annual maximum daily rainfall is highly variable over time at most stations, and most exhibit some increase in the 14-year period of record, as at Puentes (Figure 4.2b). There is an average of <1 day per year with rainfall >40 mm and 3.84 days average per year at >20 mm. The dates of the maximum vary between sites, but ones common to several sites are 29 September 1997, 23 October 2000, 3 March 2002, 30 June 2002, 15 April 2004, 26 January 2007 and 3 March 2009. If the 3-day rainfall is considered, then

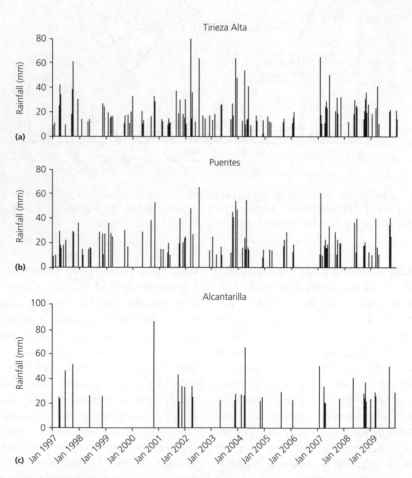

Figure 4.3 Daily data >20 mm at three rain gauges in the Guadalentin region, 1997–2010: (a) Tirieza Alta, (b) Puentes and (c) Alcantarilla.

Table 4.4 Daily rainfall statistics for six gauges in Guadalentin basin.

Rain gauges	190 Tirieza Alta	7211 Puerto Lumbreras	7205 Puentes	7207 Juncosa	170 Pliego	7228 Alcantarilla
Altitude (m)	790	465	450	580	381	85
Duration of record (years)	13.00	11.90	13.00	12.50	13.00	13.00
Maximum daily rainfall (mm)	79.6	67.4	65.0	84.6	121.5	86.6
Date	3 March 2002	26 January 2007	30 June 2002	30 June 2002	03 March 2009	23 October 2000
No. of days >40 mm	12	4	13	10	17	8
Mean no./year >40 mm	0.92	0.34	1.00	0.80	1.31	0.62
No. of days >20 mm	56	37	46	47	66	42
Mean no./year >20 mm	4.31	3.11	3.54	3.76	5.08	3.23

periods around 29 September 1997, 23 October 2000 and 26 January 2007 emerge as particularly high and widespread.

Stage

The crest stage readings over the monitoring period for all sites are given in Figure 4.4 for the date of reading, with the bankfull level of the main channel indicated. At the Torrealvilla sites, there is a single main channel below floodplain level; at the Salada sites, an inner channel is present below a bar area (outer channel) and a floodplain above that. On the Nogalte, the channel is very wide and braided, and the whole active channel below the more stable floodplain is differentiated. In this 14-year period, 23 flows over 1 m stage took place at the six marl sites and four flows of >0.25 m stage at the Nogalte (schist) sites; the highest flows reached 3.6 m above the channel bed and >1.5 m deep over floodplain on the Salada, 3.5 times inner bankfull elevation (Figure 4.4). At Oliva, the upstream site on the Torrealvilla, flows of >0.5 m stage have been recorded eight times since 1996 with the highest event being the September 1997 and other high flows in 2000, 2001, 2002, 2003 and then 2007, 2008, 2009. Flows on the Prado tributary (Aqueduct site) are rather more frequent, though few high flows have occurred since 1997, the highest events since once again being in 2000 and 2002.

The Pintor site replaced the nearby Serrata site which was affected by gravel extraction from the adjacent terrace. Serrata had high flows in 1997 and 2000; at Pintor, high flows occurred in 2002, 2003 and 2008. On the Rambla Salada, the highest flow was in 2003 when an exceptionally high stage occurred at all three sites. Other high flows occurred in 1997, 2000 and 2009. In the early part of the period, Salada 3 was one of the most responsive sites, but in 1999, a water purification plant was built nearby, and a minor amount of waste water has since flowed nearly continuously into the channel. Flows are much less frequent on the Nogalte with the major events being in 1997, 2002 and 2009. At most sites, only 2–4 flows have occurred above the bankfull level.

Daily rainfall records are the basis for identification of the occurrence and timing of specific flow events. Stage at the time of reading the crest recorder has been related to rainfall since the last reading by taking the highest daily rainfall in the intervening period. In many years, there is little doubt as to the date and event, but it is possible that multiple events have occurred within an observation period, and therefore, flow events missed as follows: June 1997 as well as September 1997, October 2003 as well as December 2003, 28 March 2004 as well as 15 April 2004 and early November 2006 as well as 31 May 2006 and in 2008, two other events of

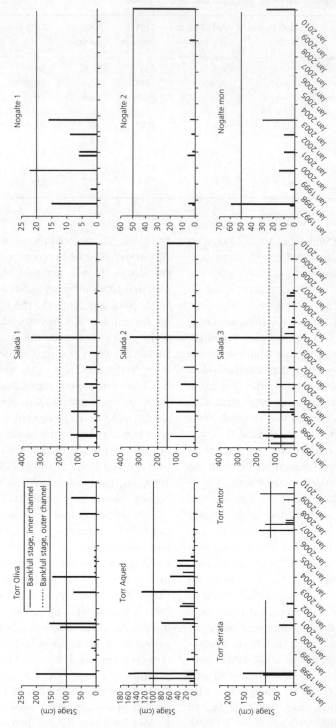

Figure 4.4 Recorded peak stage flow levels at monitored sites for 1997–2010. Stage is at the date of reading, not the date of flow.

comparable small magnitude, and in 2009, three large magnitude events, detected as one. In these cases, the morphological and vegetation changes measured may be the result of one or other of the events or cumulative of those within the period. Comparison of the stage records for sites within the same catchment indicates broadly similar patterns but spatial variability in the magnitude of events and absence of some. The largest events, for example, September 1997 and October 2003, tend to have occurred at all three locations in catchments. However, observations of channels between sites in a catchment indicate that flow was not always persistent down the channel and much flow is generated locally.

Peak stage recorded has been analysed in relation to the presumed generating daily rainfall to identify the thresholds and conditions for generation of flows in the channels (Figure 4.5). At most sites, a threshold is apparent; this at about 17 mm on the Nogalte, 28 mm on the main channel of the Torrealvilla, 14.5 mm on the Prado (Aqueduct) and about 16 mm on the Salada. The graphs also indicate the considerable scatter in the relationships and that some high daily rainfalls in the area do not produce any flow or only a small flow in the channel.

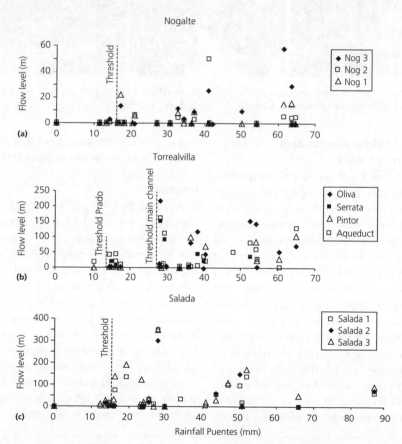

Figure 4.5 Peak flow stage measured at sites in (a) Nogalte, (b) Torrealvilla and (c) Salada catchments in relation to 24 h rainfall in the event; thresholds of rainfall for runoff generation of flow in main channels are indicated.

Figure 4.6 Photographs of monitored sites: (a) Nogalte 1, (b) Nogalte 2, (c) Nog Mon, (d) Torrealvilla Oliva, (e) Aqueduct, (f) Torrealvilla Pintor, (g) Salada 1, (h) Salada 2 and (i) Salada 3.

Impacts of flow events and conditions at sites

The morphological and vegetation changes taking place in each of the sites are presented in some detail to provide evidence of the impact of events and of the within- and between-site variability in the impact of individual events and sequences of conditions. Photographs of each of the sites are presented in Figure 4.6. Examples of cross-sectional change are shown in Figure 4.7. Major morphological changes took place at most sites in association with the September 1997 event (Hooke and Mant 2000, 2002). No changes on such a scale have taken place since then[2] (until September 2012). The high magnitude flow event at the Salada sites in 2003 did not produce much change. At most sites, there have been 2–4 occasions on which erosion or deposi-

tion of several centimetres has taken place. The following is a summary of changes at the sites.

Nogalte

This catchment is very different from the other two, being composed of schist bedrock and having an almost entirely gravel channel bed. It is also different in morphology in being dominantly braided, but the degree of valley confinement varies markedly through the catchment.

Nog1 (Figure 4.6a) is a site in the very upper headwaters of the Nogalte and is downstream of a check dam. It experienced a few minor flows, but these have not been enough to alter the morphology or to produce more than minor amounts of erosion and deposition of a few centimetre at most. It is rather different floristically from other sites since it has the highest elevation.

---

[2] Since this paper was written and before going to press, a very large flood has occurred in the region, on 28 September 2012, and this has produced very major morphological changes and has removed much vegetation, especially in the Nogalte.

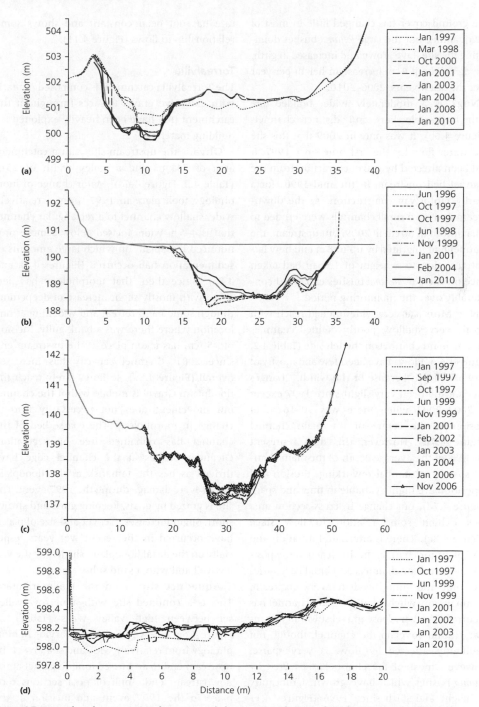

Figure 4.7 Examples of cross-sectional surveys and changes: (a) Oliva X3, (b) Sal1 X6, (c) Sal3 XCR and (d) Nog Mon X2 (part).

The ground cover has changed little in most of the quadrats, but the large *Retama* bushes dominating the site have grown and increased in girth, averaging about 8 cm increase in height per year over the last 4 years, 2006–2010.

Nog2 is a moderately wide, braided site with large-scale bars and distinct channels (Figure 4.6b). It was only in 2009 that this site had water flow for the first time since 1997. It had been affected by gravel extraction from the channel bed upstream in the mid-1990s for a nearby motorway construction. At the downstream end, two small channels were eroded to a maximum of 30 cm in 2009, but upstream, the channel had migrated by up to 8 m and new lateral accretion to a height of 1.65 m had taken place on the bar. *Retama* bushes expanded considerably over the monitoring period.

Nog Mon is a very wide braided reach with mostly very shallow, poorly defined channels but one main channel on the left side (Table 4.2, Figure 4.6c). Flows have been few and mostly of very low depth because of the width. Changes have mostly been very slight since 1997 except in 2002–2003 when there was 10–16 cm of erosion in the centre of the main channel (Figure 4.7d). However, this does represent 20–32% of the average depth of the main channel so is a substantial reworking. Erosion and deposition are highly variable in time and space (Figure 4.7d), but change in cross-section area does exhibit some relation to flow stage (Figure 4.8c). There is alternate formation and infill of minor channels in this reach as the position of the minor channels and braid bars varies over time and minor headcuts work upstream, but no major switching of the main channel has occurred. The very loose gravel substrate means that ground cover in the channel, though not frequently affected by flow, is very sparse. However, most of the channel is covered by *Retama* bushes which have grown significantly in height and girth since 1996 (Figures 4.11 and 4.133) with an average rate of growth of 19 cm per year. *Retama* is a phreatophyte tapping groundwater in the gravel at depth; growth

rate has not been constant and shows some relationship to flows (Figure 4.11).

Torrealvilla

The Torrealvilla catchment is composed of marls with extensive gravel terraces throughout the catchment that have been heavily exploited for building materials.

Oliva is the upstream site with a catchment area of 74 km^2 and a valley width of 25 m (Table 4.2, Figure 4.6d). Major change of morphology took place in 1997, from a relatively wide, shallow channel to a rectangular channel that is 4–5 m wide, incised below a more pronounced floodplain on which large amounts of sedimentation had occurred (Figures 4.7a and 4.8a). Since then, that morphology has persisted with mostly slight increase of deepening, though some bank retreat and widening at one location where there was a bank gully. Incision of >50 cm has taken place at the upstream end since 2001. Channel capacity has increased overall (Figure 4.8a), so flows rarely reach the floodplain. Gravel is mobile within the channel but movement does not necessarily lead to change in morphology. The coarse bed of the channel has remained free of vegetation. Oleander bushes at the channel edge have thrived as has the tamarisk in the floodplain since severe damage during the 1997 event. The site is grazed by goats, keeping grass and shrubs short. Slight increases in cover and size of plants have occurred in the recent wet years, especially on the distal floodplain, shaded by the valley wall and with a sand substrate.

Aqueduct site is on the Prado tributary. This is a confined site with a narrow valley set between high valley walls (Table 4.2, Figure 4.6e) and with a channel varying in morphology from relatively wide and shallow at the upstream end to an incised inner channel at the downstream end. Infill of pool sections took place in the 1997 event and incision at the downstream end, with significant deposition on the floodplain through the site. Further infilling of pools in the upstream part took place between

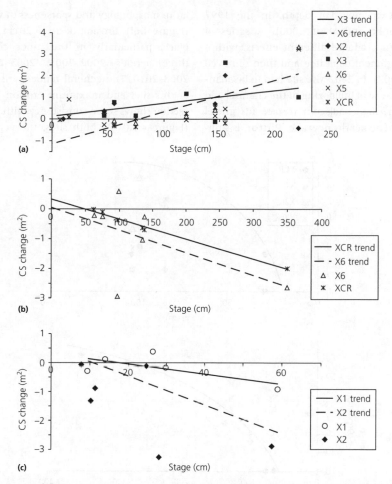

Figure 4.8 Amount of change in cross-section area in relation to flow stage: (a) Oliva, (b) Sal1 and (c) Nog Mon.

June 1999 and February 2002, but since then, changes have been relatively minor, only slightly modifying the existing morphology. Comparisons of monitored sediment quadrats indicate that sediment is mobile through this reach in events but that the same morphology re-forms. The site has also been affected by three major rockfalls of marl bedrock from the very steep valley walls. These partially blocked the downstream channel, but the large blocks in the channel and on the floodplain disintegrated within a few years. As with the Oliva site, the gravel channel has remained bare. On the bars, shrubs and grasses grow and have been affected by some sedimentation but have expanded in cover and height since 2001. Vegetation on the floodplain has remained virtually the same since 1997 though flow has lapped into the quadrats on two occasions. Vegetation looked less healthy and did not increase in height in the drought period, 2004–2006. Three of the quadrat sites were destroyed by excavations of exploratory pits at the site in 2009.

Serrata was a wide, shallow channel, sparsely vegetated, averaging 15 m wide and set in a wide valley but restricted by levées and irrigation channels (Table 4.2). It was scoured to bedrock by an average of 0.76 m, with potholes exceeding 1 m depth, in the 1997 event (Hooke and Mant 2000). A series of events then produced different effects, with a layer of sediment deposited and then removed (Hooke 2007). In 2002, the site had to be abandoned because of bulldozing of the channel and excavation of the adjacent terrace for gravel resources. The nearby new site, **Pintor**, is simi-lar in morphology and sparseness of vegetation (Figure 4.6f). Erosion of up to 20 cm has taken place, particularly in the inner channel in three periods 2002–2003, 2003–2004 and 2008–2010. The general tendency is erosional at this site, and maximum erosion in a cross section shows some consistency with flow stage (Figure 4.9a). At both sites, the vegetation is

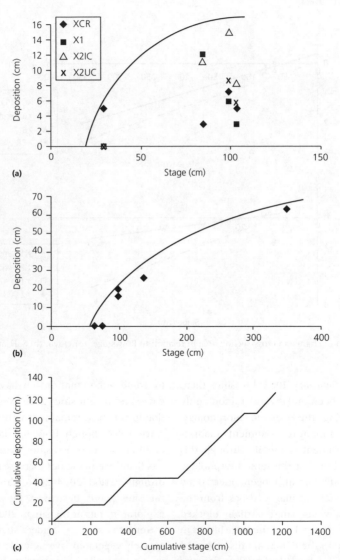

Figure 4.9 Relationship of processes to flow stage: (a) erosion at Pintor, with envelope curve (IC, inner channel; UC, upper channel); (b) deposition at Sal1; and (c) cumulative deposition in relation to cumulative stage at Sal1.

much affected by goat grazing. Tamarisk are present on the banks, with associated sandy wakes downstream of each plant.

Salada

The Salada is a dominantly marl catchment with extensive almond orchards but also much degraded land.

Sal1 is the upstream site and is set in a narrow, confined valley with high bedrock walls (Figure 4.6g). The channel bed morphology is rather variable. At the upstream end, a large pool was excavated in September 1997 and then infilled between September 1998 and June 1999, followed by very deep excavation again by January 2004 which has remained since but began to infill in 2009–2010. Excavation of this pool led to deposition downstream within the site. The deposition has mostly been at the edges of the channel (Figure 4.7b), but very large cobbles and boulders have been transported and deposited in the centre of the channel. Deposition is influenced by the presence of large tamarisk in places, for example, up to 114 cm has occurred on the upper corners of a quadrat (Figure 4.12), whereas little change occurred on the lower corners in the base of the channel, 3 m away. The October 2003 event, although larger than the September 1997 event at this site and reaching 3–4 m stage, had less effect morphologically overall, though there was localised deposition at the sides of the channel of up to 60 cm (Figure 4.7b). The tamarisk were battered by the 2003 high flow event but not destroyed and soon revived. One tamarisk bush in the channel remained upright and healthy in spite of water >3 m deep flowing over it. In the period 2007 to January 2010, cover has increased and a young tamarisk has grown rapidly in one of the wakes. The cobble channel only supports sparse *Limonium* occasionally, and these are obliterated by deposition in high flow events. Almond trees on the floodplain died after the October 2003 event. On average, deposition of 5–20 cm has taken place in the channel in several events since 1998 and shows a cumulative relation to flow stage (Figure 4.9c).

Sal2 site is very different in morphology, being a narrow incised inner channel set below a high floodplain (Figure 4.6h). Flows have relatively high stage through this reach; however, it is less sensitive and is less intensively monitored than other sites. The narrowest part of the channel has widened slightly, but slight deposition has taken place elsewhere in this long straight section. Dense tamarisk at the sides of the channel were damaged but not removed in the 1997 and 2003 events, and fine sedimentation has taken place in these events on the floodplain, elevated 2 m above the channel, forming marked vegetation wakes and dune features associated with herbs and grasses. Two large pine trees at the distal edge of the floodplain were destroyed in the 2003 event.

Sal3 is a relatively wide open site (Figure 4.6i) that experienced some scour and fill in the 1997 event. In November 1999, test flows or flushing out of a new waterworks, connected to an upstream tributary, began. The initial flows were very odorous and caused deposition of a black slime. Occasional flow and then more continuous low flows during working hours occurred from 2001 onwards. Very dense vegetation had developed on the low channel bars by 2003. The very high stage event of 2003 produced little change, though stage reached 3.5 m and discharge calculated using WINXSPRO and allowing for vegetation was 192 m^{3}s^{-1} (Sandercock and Hooke 2010). Maximum velocities of 3.32 m s^{-1} at an upstream section and 2.65 m s^{-1} at a downstream section have been calculated. Juvenile tamarisk bushes plants were bent and damaged in 1997 but quickly revived and since then have grown from 1.5 m in height to 4 m, mainly in the period 2001–2003 and 2008–2009. They were bent over in the 2003 event. The low bars have become much more densely vegetated since the nearly continuous low flow. Since 1999, there has been gradual aggradation throughout the site of the order of 20–30 cm at most locations (Figure 4.7c).

Morphological change and relations to flow

Analysis of amounts of change in relation to flow stage and interrelations and trajectories of morphology for all sites illustrates high variability of behaviour both within and between sites and at any one location over time with successive events. Statistical analysis is limited by the few data points of morphological change at many of the sites. Plots of amount of change in cross-sectional area, width and depth, and measures of maximum erosion and maximum deposition show high variation and complexity at many sites in relation to flow stage and complex trajectories. Further analysis is illustrated for one site from each catchment, selecting those sites least disturbed by direct human activity and which are most responsive and have the most data points, that is, Oliva, Sal1 and Nog Mon. Some consistency is apparent, but the only relationship which is significant with the few data points is that between cross-sectional area change and stage at Sal1 XCR cross section (Figure 4.8b), which exhibits net decrease in cross-sectional area in relation to stage of flow. The difference in trends of change for different sites is seen in Figure 4.8. Data for changes at Oliva in a series of cross sections through the reach indicate some general relation of magnitude of cross-sectional area change to flow stage (Figure 4.8a), with a net increase in cross-sectional area and channel capacity. The example of the cross-sectional profile changes (Figure 4.7a) showed the major changes resulting from the 1997 event and the gradual deepening since then, with also some broadening in this cross section. At Nog Mon, a braided section, the changes are highly variable, with position of channels and location of erosion and deposition varying (Figure 4.7d) but with a tendency for cross-sectional area to decrease in relation to flow stage, indicating increasing net aggradation with stage (Figure 4.8c).

At a few sites, a threshold for erosion and deposition is apparent and an envelope curve for activity may be suggested (Figure 4.9a and b). In some cases, it can be seen that, even if the process is not exactly related to flow stage, the cumulative effects are largely commensurate (Figure 4.9c). This relationship of cumulative deposition to flow stage at Sal1 is highly significant. However, the episodic nature of the processes is also apparent, with long periods of quiescence. Although magnitude of change is not great in most sites, in many individual events since 1997, the cumulative or net erosion or deposition on some cross sections amounts to about 60% of original bankfull depth even since 1997 (Figure 4.7).

Vegetation changes and variations

Table 4.5 provides the details of events when vegetation was removed or severely damaged. This indicates that herbs and ground cover are relatively easily removed, though Mant (2002), from the quadrat data and flow events, calculated a minimum depth of flow needed for decline of various species. Intensive monitoring in 1996–2000 illustrates the response to events. At a quadrat on the channel edge of the floodplain at Oliva site (Figure 4.10a), many herb and shrub plants were destroyed, and the floristic richness decreased by the occurrence of the September 1997 event but grass, thistles, *Spergularia* and *Helichrysum* increased, and they have remained dominant since then. The overall percentage cover decreased over that period, and only the channel bank is now vegetated; the floodplain surface in the quadrat has not become vegetated. Slight increases in cover and size of plants have occurred in the recent wet years, but cover and composition have remained very similar. Similarly, bar quadrats at Aqueduct site were severely affected by the September 1997 event, but the intensive monitoring (Figure 4.10b) shows how *Lygeum* appears to benefit in such events. Flow occurred over the bar quadrats in the year January 2001, 2003 and 2004, and some sedimentation occurred with some damage to plants, particularly in 2000–2001, but few plants were removed. Since then, the plants have expanded in cover and height, particularly the *Lygeum*.

Table 4.5 Details of the effects of flow events on vegetation at sites in SE Spain.

Site	Quadrat no.	Position	Event date	Stage (m)	Height flow over VQ (m)	Vegetation type[a]	Effect	Process	Subsequent changes
Oliva	Site		29/09/1997	2.20	1.00	*Tamarix*	Bent, battered and defoliated	Flow force, sedimentation	Regrowth
	VQ1	fp	29/09/1997	2.20	1.32	*Volutaria, Dittrichia, Juncus, Reichardia*	Removed/died	Sedimentation	Mortality
	VQ2	ch	05/05/2000	1.20	0.66	*Helichrysum*	Removed	Bank erosion?	
	VQ3	ch	22/08/2007	1.55	1.50	Small herbs	Removed	Bed erosion	Bare
	VQ3	ch	29/09/1997	2.20	2.00	Small herbs	Removed	Bed erosion	Rapid regrowth
	VQ4	fp	29/09/1997	2.20	0.39	Grass – Gramineae	Removed	Flow	Mortality
Aqueduct	VQ1	fp				*Oleander*	Gradual decline	Competition or moisture	
	VQ2	bar	29/09/1997	2.20	1.91	*Dittrichia, Lygeum*	Decrease	Flow and scour	
	VQ3	bar	29/09/1997	2.20	1.67	*Thymelaea*	Poor state	Flow and scour	Mortality
		bar	29/09/1997	2.20	2.02	*Artemisia, Dittrichia, Spergularia, Moricandia*	Removed	Scour	
	VQ4	ch	23/10/2000	?	0.58	*Limonium*	Removed	Scour	
		ch	19/11/2003	0.58	1.46	Small herbs	Removed		
		ch			0.91	Small herbs	Removed		Mortality
	VQ5	fp	29/09/2001	1.04	1.04	*Artemisia, Anthylis*	Stems broken	Flow	
Serrata	VQ3	ch	29/09/1997	0.99	0.99	*Juncus, Moricanda, Lygeum*	Removed	Scour	Mortality
Pintor	VQ1	bar	30/06/2002	1.04	1.04	Small herbs	Removed	Flow/scour	
	VQ4	bar	14/10/2008			Small herbs	Removed	Flow/scour	
		bar	30/06/2002			*Tamarix* (juv)	Bent		Mortality
Sal1	VQ1	bar	16/10/2003	3.5	2.8	*Salicornia, Thymeaea, Limonium*	Obliterated	Sedimentation	
	VQ2	ch	13/03/1999	1.35	1.31	*Limonium*	Obliterated	Sedimentation	
		ch	16/10/2003	3.5	3.4	*Limonium*	Removed		
	VQ3	ch	13/03/1999	1.35	1.42	*Limonium*	Obliterated	Sedimentation	
		ch	16/10/2003	3.5	3.45	*Limonium, Juncus. Lygeum*	Obliterated	Sedimentation, blocks	
Sal3	VQ1	bar	16/10/2003	3.5	3.3	*Salicornia*	Battered	Flow, siltation	Mortality
	VQ2	bar	16/10/2003	3.5	3.1	*Dittrichia*	Some removed	Flow, siltation	
Nog Mon	VQ2	ch	29/09/1997	0.59	0.25	*Tamarix* (juv), *Artemisia, Dittrichia,*	Removed		Bare
Torr channel	VXS	ch	29/09/1997	2	1.8	*Tamarix, Arundo, Phragmites*	Flattened	Flow	Regrowth

Position: ch, channel; fp, floodplain. 'Stage' = flow level above datum at crest stage recorder. 'Height' of flow over VQ = depth of water above vegetation quadrat.

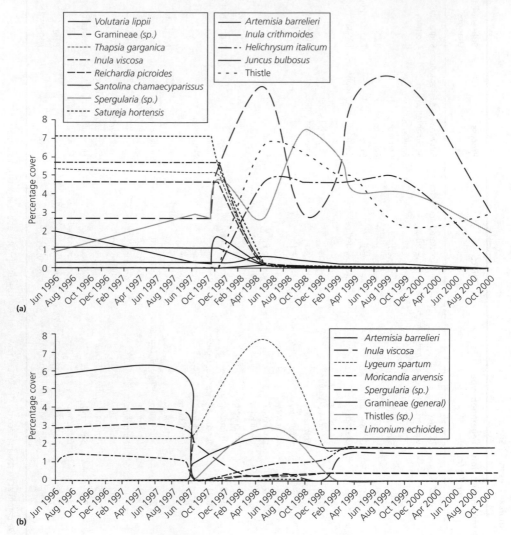

Figure 4.10 Changes in cover of individual species in quadrats, June 1996–October 2000: (a) Oliva VQ1, and (b) Aqueduct VQ3.

Further evidence from the quadrats suggests that it is erosion or sedimentation of plants that tends to obliterate or damage them as much or more than the fluid forces or inundation alone. *Lygeum* revives and even benefits from some inundation, but isolated small individuals may be destroyed, particularly if large particles are carried (e.g. Sal1, VQ3; October 2003). *Lygeum* withstood estimated velocities of 3.3 m s⁻¹ at Sal3 in October 2003 and 1.7 m s⁻¹ at Aqueduct in

September 1997 (Sandercock and Hooke 2010). These are velocities within the vegetation zone, not mean velocities for the section. Most shrubs survive even flows of this height, but *Dittrichia* are easily removed and *Thymelaea* was destroyed at Salada 3 at the upper margins of the 3.6 m high flow, with an estimated 2.6 m s⁻¹ velocity. The plants remained in place, still contributing to roughness. Tamarisk has been severely bent and defoliated in several events at several sites, but mature specimens have resprouted. Velocities withstood are calculated as up to $1.7\,\mathrm{m\,s^{-1}}$ (Sandercock and Hooke 2010) but highly influenced by the presence of the tamarisk affecting roughness. Only young specimens have been destroyed. Even exposure of over 20 cm of root has not killed plants. At additional, highly vegetated cross sections on Torrealvilla and Salada, the vegetation was affected but not destroyed in both the 1997 and 2003 events. *Tamarix* was severely bent and *Arundo* and *Phragmites* were nearly flattened, but all species revived quickly. At sections on Salada comprising dense *Lygeum*, this was thinned between plants but sedimentation increased within plants and they were not destroyed.

Cumulated evidence from the sites indicates that large, single-stemmed trees are, however, vulnerable to high flows. At Oliva, mature poplar trees were completely destroyed and removed by snapping of the trunks in the 1997 flood. At Oliva in the same event, almond and olive trees were mortally damaged, some by snapping of trunks, and at Sal1 in the October 2003 event, almond trees were inundated >1.5 m and covered in flood debris, and though still alive three months later, they subsequently died. Pine trees were destroyed at Sal2 in the October 2003 event. These trees were all located on the floodplain so did not experience maximum depth of flow but high velocities.

Evidence on growth rate of plants, particularly the large phreatophytes (*Tamarix* and *Retama*), indicates some relationship to rainfall and flows, particularly to rainfall. In the example in Figure 4.11, flows in 2000–2002

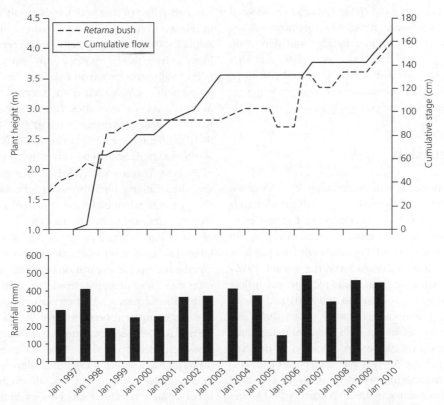

Figure 4.11 Growth of a *Retama* plant in relation to cumulative flow stage and to annual rainfall at Nog Mon.

Table 4.6 Numbers of quadrats in vegetation density classes in relation to morphological change.

Net change	Vegetation density	
	Bare and sparse	Moderate and dense
Stable	0	6
Erosion	10	2
Deposition	2	10

did not seem to have much effect, but the drought phase of 2004–2005 appears to have produced stasis.

To analyse the overall relationship between morphological change and vegetation cover, the quadrat states have been classified into bare/sparse and moderate/dense vegetation cover and into stable erosion and deposition in net morphological behaviour. The numbers of each have been counted (Table 4.6), and chi-squared analysis reveals statistically significant differences. It is seen that stability and deposition occur in moderate/dense vegetation and that sparsely vegetated or bare quadrats tend to be erosional. The association of deposition with vegetation cover and position is seen in Figure 4.12.

Discussion

This research was undertaken in association with projects focused on desertification trends and impacts and development of strategies to combat desertification. Major variations in occurrence of both high daily rainfalls and flow events are evidenced over the period 1996–2010, with wet phases and more frequent flow events in 1997–2000 and 2008–2010 but major drought periods, particularly 2004–2007. No net trend of decrease of total rainfall, which might contribute to desertification, is detected in either the 50-year rainfall record or the 14-year monitoring period. A trend of decreased flood frequency in the 20th century and longer dry periods and increased variability has been identified for the region (Machado et al. 2011). However, annual maximum daily rainfall does show an increase since 1996, but the variation in magnitude is within the range of variations in the last 50 years. Increased variability is predicted under climate change scenarios related to global warming (Hertig and Jacobeit 2008). Analysis of rainfall–runoff relations indicates thresholds of the order of 15–17 mm on the Nogalte, Prado and Salada and 28 mm on the Torrealvilla, but the graphs also indicate that some high daily rainfalls do not produce high flows in the channels. These results are very similar to those of Puigdefábregas et al. (1996) for the Rambla Honda in the nearby Almeria, a schist channel similar to the Nogalte. Antecedent conditions are known to be highly influential, and analyses of autographic records for the same areas have shown the complexity of relations (Bracken et al. 2008; Bull et al. 1999). The latter also showed that peak floods mostly occur in relation to storms of >18 h duration and total rainfall >50 mm, so use of daily rainfall record is likely to have picked up key events. Daily rainfalls of >50 mm occurred on 5 days in the monitoring period, giving recurrence intervals of the order of 3 years at most sites. This is comparable with a 5-year recurrence interval for 50 mm daily rainfall for the Rambla Honda site further south in Almeria and slightly drier, over a period of 30 years (Lazaro et al. 2001). It is suggested that the relatively high threshold for runoff in the channels exhibited by the Torrealvilla main channel sites may be due to the extensive areas of gravel but also at least partly related to the large-scale disruption of the drainage network and the widespread excavation of the gravel terraces that has occurred from the quarrying within the catchment. A difference in frequency of flows is apparent between sites (Figure 4.4), particularly between catchments, reflecting the east–west decrease in rainfall and the effects of lithology. Sites within a catchment have a generally similar pattern, but not all events do occur at all sites. Flows can be very localised, as is well known in semi-arid environments

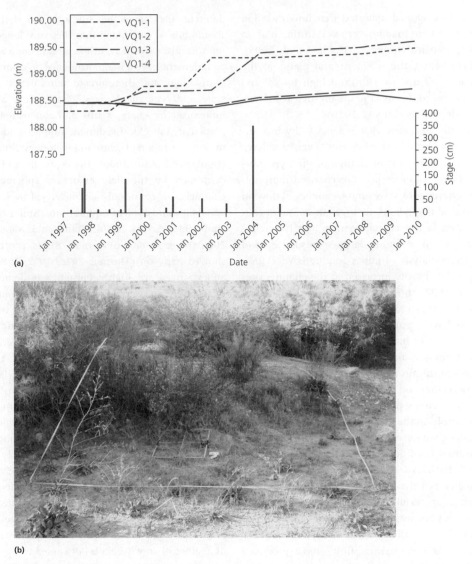

Figure 4.12 (a) Changes in height of the vertices of VQ1 quadrat over time at Sal1 and relationship to flow stage measured; (b) Photograph of quadrat Sal VQ1.

(Thornes 1994). Further analysis will examine the 3-day rainfalls and the relation to synoptic conditions.

The event of October 2003 is particularly interesting because it reached stages of 3.6 m in Salada 3 site but had little morphological or vegetation impact. The reason for this is not known without available continuous gauge records, but it is hypothesised that the flow

was of very short duration. This would also fit with the relatively low daily rainfall recorded and the highly localised nature and short duration of rainfall evident from radar images (G.G. Barbera, personal communication: radar images of October 2003). If this is so, it confirms the need for much more data and analysis of effects of flow duration as well as flow peaks. Preliminary runs of the CHANGISM simulation model developed indicated that flow duration and sediment loading were very influential on morphological response (Hooke et al. 2005). Nevertheless, the sediment and plants at the Salada still had to withstand high forces, and this helps to set upper limits on their resistance (Sandercock and Hooke 2010).

Some sites were changed markedly in morphology in the September 1997 event and have exhibited only minor change of the new morphology. At some sites, the changed morphology altered flow stages and frequency of flow on bars and floodplains. Changes in sediment calibre and in the vegetation may also have contributed to a more damped response, and further analysis of this is required. Vegetation cover has mostly increased at the channel sites since 1997; thus, thresholds for change may also have been raised as only smaller magnitude flows have occurred at some sites. At Sal3, it is likely that the increased vegetation cover resulting from the altered flow regime due to the water purification plant has altered the response. Observations on several channel headcuts and of connectivity patterns more widely, as well as the results of the 1997 flood event, indicate that major changes are often highly dependent on localised conditions, often influenced by structures such as tracks (Hooke and Mant 2000). Oscillatory behaviour is apparent in some cases, with scour and fill in different events. For example, a large pool was scoured out at the upstream end of Sal1, then infilled and then scoured again. At several sites, flows have produced sediment flux, as evidenced by change in the particles in the sediment quadrat photographs, but have produced the same morphology. It is suggested that this is particularly at confined sites such as Aqueduct where the hydraulics are tightly controlled and therefore forces and sediment flux are similar in events of similar magnitude. Scour and fill take place within the event as evidenced by particles being moved but final elevations remaining similar.

Differing trends and patterns of morphological trajectories are therefore apparent in the different sites over the study period, but it should not be concluded that these are longer-term trends. Obviously, lithology, substrate and confinement have a major influence on channel morphology and therefore to some extent on scale and type of changes. Reaches are also influenced by changes up- and downstream, particularly the local sediment supply, producing an association of scour and fill longitudinally (Hooke and Mant 2000). Conceptually and as confirmed by the field evidence, ephemeral channels are commonly characterised by high variability and lack of equilibrium (Bull 1997; Graf 1983; Schick 1974; Tooth and Nanson 2000; Wolman and Gerson 1978), as Thornes showed early on (Thornes 1976, 1977, 1980), and changes are highly episodic, as demonstrated here, so systematic quantitative relations are not necessarily expected. This is also why the model constructed by Hooke et al. (2005) involved many influencing factors and interactions, including positive feedback effects. The channels do not show marked changes in width over time as Thornes discussed, but this may be related to the scale of events, and this is much more applicable to the unconfined, braided, schist channels such as the Nogalte. A longer period of data is still needed to identify if equilibrium relations do apply at all given Richards' criteria (Richards 1982). The evidence of the variable behaviour here within the measurement period is that there is a range of responses which may represent a continuum of degrees of consistency as discussed by Tooth and Nanson (2000) and Nanson et al. (2002). Mostly, the responses are highly variable from one event to another, but the preceding morphology does constrain the response to moderate flows. Of course, the interpretations in relation to equilibrium depend very much on the timescale of analysis, and even a relatively lengthy period of monitoring such as here is insufficient to draw conclusions about the longer-term behaviour.

In relation to the events identified for these ramblas by Conesa-Garcia (1995) and discussed for the 1997 event by Hooke and Mant (2000) (Table 4.1), analysis of the flow stages and the morphological responses indicates that most of the flows are class 2 events, below or approaching bankfull and to which the channel form is adapted. Smaller events (class 3) do mobilise sediment. Several of the events were overbank and could therefore be classed as 1. They did produce some scour and deposition but could not be classed as catastrophic or producing major morphological change except at two sites for the September 1997 event.[2] Analysis of the events in relation to recurrence interval is not very meaningful in this environment given the tendency for persistence of drought over periods of a few years. However, the maximum daily rainfall records at Puentes and La Juncosa sites dating back to 1961 show the highest daily rainfalls occurred in 1989, though at Juncosa the 29 September 1997 rainfall was the second highest. Overall, the morphological changes tend to exhibit a general scaling relationship of maximum amount of change to flow stage, but the amount of change by erosion or deposition is highly variable both within cross sections and within sites in events of similar magnitude.

The limited and very episodic changes in morphology conform with major conceptual ideas applied to semi-arid and arid channels that it is large events producing the changes and there is a lack of restorative events in between (Wolman and Gerson 1978). Some of the changes appear to be ramped or can be interpreted as continued adjustment to the morphology created by the 1997 flood that was the highest flood in the period in the Torrealvilla and Nogalte, and thus, channels could be interpreted as created by that 1-in-15 year event. However, on the Salada, the 2003 event was higher but produced negligible change. As has been commonly noted, ephemeral channels tend to lack a consistency of form or equilibrium, so alterations cannot be interpreted as adjustment to a new equilibrium. In many cases, the changes at cross sections are much related to localised incision and small headcuts

progressing up channels. Therefore, care must be taken in upscaling and inferring widespread landscape changes from the localised data even with several sites. Lack of change in periods cannot be interpreted as signs of equilibrium condition on these timescales because it mostly represents simple lack of competent events. Overall, these sites, with their range of morphology and sedimentology, illustrate the range of sensitivity present within ephemeral channels even in one region in similar topography and even within single-channel systems. The degree of confinement and extent of vegetation cover are hypothesised to be key factors. These channels could be considered to be at various positions on a continuum from humid to arid channel morphology (Hooke and Mant 2002), but generalisations of typology even within such zones are seen to be dubious.

Few papers have been published on morphological change in Mediterranean channels or even other semi-arid channels since Hooke and Mant (2002) and Hooke (2006). The results here give an indication of the scale of variability to be expected under the range of conditions in this 14-year period. Such results may help us in scaling to longer-term changes and interactions, but the importance of thresholds and role of extreme events may limit this.

It has already been shown that a strong relationship exists in these channels between vegetation assemblages (composition and density) and position (Sandercock and Hooke 2010), with channels tending to remain bare, bars supporting herb and grass and shrub vegetation and riparian zones and floodplains tending to be the location of phreatophyte growth. There is a relationship to flow frequency and also to substrate, and so the upper locations, not in the bed of the channel, have experienced many fewer events and in some cases hardly any in the 14-year period. Changes in vegetation in the 30 monitored quadrats are not very great except for some obliteration of herbs and for damage, but not destruction, of tamarisk (Table 4.5). Evidence on thresholds for removal of various kinds and species of plant are scant in spite of the period of monitoring, mainly due to lack of events large enough to remove them, though it does indicate that thresholds are higher than the maximum values measured.[2] The lack of flows sufficient to destroy large plants has meant that shrubs, bushes and trees have increased in size over the period (Figures 4.11 and4.133). Experiments by Mant (2002) using board tests of bending characteristics against shear force indicated that for *Tamarix* and *Retama*, the force needed to bend individual juvenile plants is very closely related to the stem diameter and therefore thresholds will have increased. The flows that have occurred on the Nogalte have not been sufficient to destroy *Retama*[2], but ancillary evidence from measurements at the similar Ramble Honda, near Tabernas in Andalucia, indicate that they can be killed by high flow combined with large particles hitting the stems; some older *Retama* at the upstream ends of bars did not survive flows of >0.5 m high in September 1997, though most resprouted.

Longer-term evidence from air photographs indicates that events such as the 1973 event on the Nogalte (Thornes and Rowntree 2006) are needed to effectively zero the vegetation.[2] Without such events, the phreatophyte vegetation gradually increases in height and girth of plants over time. Figure 4.13 shows the increase in canopy cover due to growth of *Retama* at Nog Mon site. Shrub vegetation varies very little over periods of several years, and very little colonisation and spread of these plants have been detected. However, newspaper reports (e.g. Guardian, 7 June 2006) indicated that the 2004–2007 drought was being regarded as the worst for 10 years and possibly since 1947 in terms of water availability in this area, though the rainfall analysis alone, without temperature and examining sequences, indicates that 1991–1995 was more severe. Analysis of a drought index also indicated this for the western part of the region (Sanchez-Toribio et al. 2010). The change in hydrological

regime at Salada 3 did show that a persistent increase in moisture can have a marked effect on vegetation cover. Germination of large areas of new plants was rarely seen in the period, an exception being downstream of Pintor after an influx of fine sediment into a zone of ponding on which a bed of tamarisk germinated; however, the young plants were destroyed in a subsequent flow by influx of coarse material.

It has not been possible to identify that the drought conditions definitely caused mortality of plants other than the usual seasonal opportunistic variation in herbs and some grasses. Some herbs may have died or not grown, but they vary markedly with rainfall and season. Thresholds for effects of drought in spite of the prolonged, relatively dry period of 2004–2007 appear to be greater than in that period. This demonstrates the high level of adaptation and resilience of the natural vegetation in these channels. Many of the shrubs were brown and appeared in poor state, but all remained and revived in the subsequent wet period. Figure 4.11 indicates that the *Retama* showed a long period of stasis from 1998 to 2006. These results fit very well with the findings of Miranda et al. (2011) in which they manipulated rainfall in a nearby and similar area in SE Spain. They decreased rainfall and altered seasonality and found that neither community productivity nor growth of woody species significantly changed until 4 years after the rainfall decrease. They also interpret these results as demonstrating the very high resilience and adaptation of these plants to the inter-annual and intra-annual rainfall variability normal in this region and semi-arid ecosystems.

One of the major impacts on the channels that emerge is the effect of human activities, and these provide an index of the extent of various anthropogenic changes in southern Spain, especially in peri-urban areas in the period. Several sites have been disturbed or affected by human activity such that flows or morphology have been altered. These include gravel extraction

Figure 4.13 Comparative photos of Nog Mon site showing increase in vegetation cover and size of *Retama*: (a) 1998 and (b) 2010.

within the site (Serrata) or nearby (Nog2) or within the catchment channel network (Torrealvilla), digging of exploratory pits (Aqueduct), construction of water purification plant and alteration of flows (Sal3 and Serrata later), land-fill sites nearby (Aqueduct, Sal3) and destruction, rebuilding and new construction of check dams (Torrealvilla system, Nog1). Sal2 was affected some years ago by the construction of the water transfer canal from the Tagus. Nog1 site is also threatened with a golf course, and near the Aqueduct site in Torrealvilla, a solar power 'farm' has been constructed on former irrigated fields. The impacts reflect the scale and nature of activity in SE Spain over the 14-year period, which was one of great economic growth and much construction, with increasing urbanisation and also construction of infrastructure. Cumulatively, this may have more impact than floods as Hooke (1999) also found for the United States. This scale of human impact has implications for the implementation of the Water Framework Directive in this part of Spain as Thornes and Rowntree (2006) also pointed out. A full analysis of possible alterations of run-off response is the subject of further research.

The changes that have occurred at each of the nine monitored reaches are summarised in Table 4.7. The variability both within and between sites in the impact of events has been demonstrated. It has confirmed the need for combinations of state of channels, sequences of events and conditions, sediment supply from catchments and upstream and feedback effects to be incorporated in modelling since all can have an effect on outcome. The CHANGISM model was designed to do that (Hooke et al. 2005), and data derived from this measurement programme will enable validation of that modelling. In spite of a period of reported severe droughts, a clear signal is not detectable in the channels over the period but may be seen in longer-term comparison as a period of stasis. The few flows have mostly had limited morphological impact since the 1997 event, strengthening the idea that it is the large infrequent flows that are most effective[2] but also that there is a feedback effect of changed morphology and vegetation cover. However, the 2003 Salada flow event indicates that even high, winter season, flows may not necessarily produce much response. Likewise, the vegetation appears relatively static, except for increase in phreatophyte (*Tamarix* and *Retama*) size and cover, and is well adapted to the environment. It may be that impacts will be greater when thresholds are eventually crossed.[2] The impact of desertification depends much on the effects of climate and land use on runoff generation and on sediment mobilisation in the catchment and on hillslopes but also its connectivity down the channel. Even where there appear to be persistent changes over the period of monitoring, this does not necessarily indicate a long-term response, for example, of aggradation as a response to increased soil erosion and sediment loading from the catchment but may be more localised.

It is apparent from the data presented here that even with so many sites and potential observation dates, the number of data points it has been possible to collect is few. This is mainly due to the lack of significant flows in some years, though also the infrequency of surveys in some wet years. It illustrates the considerable challenges in accumulating much data on morphological responses in this environment and the need for long-term data. The study period has coincided with severe droughts, meaning a lack of flows and possibly some stasis in plant growth rates but little actual decline in cover. However, signs of effects of increased desertification impact such as decrease in vegetation cover and increase in sediment delivery to the channels are also not readily evident in the channels, except possibly in plant growth rates. Catchment signals may be rather different. The high variability within and between sites illustrates the importance of having a range of sites monitored and that, even with more sophisticated instrumentation, care would need to be taken in relating changes at one point in the channel to wider environmental changes. Intensive instrumentation is also seldom possible at a large number of sites, and an extensive approach is particularly suitable for this environment. The need for detailed data on rainfall events that allow analysis of duration and intensities is also essential, but changes in data policy by catchment management authorities in Spain are now easing that situation and making autographic rainfall data available.

Conclusions

In this type of environment with average annual rainfall of ca.300 mm p.a., rainfall events of >20 mm occur on average 4 days a year, and events of >40 mm one or less days per year. No trends in annual totals are evident in the 14-year measurement period, and a tendency to increase in annual maximum daily rainfall is not statistically significant. Wetter and drier phases of 2–4 years duration are apparent. The data on runoff responses suggest a threshold of the order of 20 mm per day for flow in main

Table 4.7 Summary of changes and impacts at channel monitoring sites in SE Spain.

Site	Major flow events	Morphological change	Vegetation change	Human impacts
Torrealvilla Oliva	29 January 1997, 5 May 2000, 23 October 2000, 30 June 2002, 18 October 2003, 22 August 2007, 8 May 2008, 27 September 2009	Major incision 9/97, further incision since	Growth on floodplain, none in channel	Goats
Torrealvilla Aqueduct	18 June 1997, 29 January 1997, 23 October 2000, 30 June 2002, 19 November 2003	Slight sedimentation in wider upper part, some bed and bank erosion in narrow downstream part	Channel influenced by flows; growth on bars, floodplain static	Dumping of rubbish, pits dug in channel
Torrealvilla Serrata	18 June 1997, 29 January 1997, 4 September 2000, 23 October 2000	Deep scour to bedrock 9/97; sedimentation and erosion in successive events	Removal of sparse channel vegetation; little regrowth	Check dam upstream; banks excavated for gravel and bed used for access; goats
Torrealvilla Pintor	3 March 2002, October/November 2003, 14 October 2008, 3 March 2009	Slight erosion of bed through most of reach	Sparse vegetation declined further	Check dam upstream destroyed 9/97, goats
Salada 1	18 June 1997, 29 January 1997, 27 February 1999, 4 October 1999, 23 October 2000, 30 June 2002, 18 October 2003, 3 March 2009	Erosion in pool at upstream end; deposition through rest of reach	Removal and regrowth of herbs in events; growth of tamarisk	Pylons built on slopes
Salada 2	18 June 1997, 29 January 1997, 23 October 2000, 22 September 2001, October/November 2003, 3 March 2009	Erosion and deposition; sedimentation on the left floodplain, gullying on the right floodplain	Abundant growth of dense riparian and floodplain vegetation	Influenced by pipeline and bridge upstream, constructed in the 1970s
Salada 3	8 April 1997, 18 June 1997, 29 January 1997, 27 February 1999, 4 October 1999, 23 October 2000, October/November 2003	Progressive fine sedimentation	Prolific growth since flow regime altered	Water purification plant effluent flows daily since 1999
Nogalte 1	No significant flows	Very little change	Slight growth in vegetation	Golf course planned
Nogalte 2	30 March 2009	Erosion and movement in braid channels in single event	Not monitored	Gravel excavation upstream in the 1990s
Nogalte Mon	29 September 1997, 30 June 2002, 30 March 2009	Variable erosion and deposition as braid channels cut and fill	Growth of *Retama*	Gravel road through middle of site, on least active part

channels. Flow events at all the sites average <1 per year and much less at some, with a major difference in the number of events on the channels in schist bedrock compared with marl. The channels in this region are zones of activity, and erosion and deposition have occurred several times in the 14-year measurement period in spite of the low number of events. Much of the morphological change was concentrated in the few, relatively high flow events, but even flows of similar peak magnitude produced differential changes within as well as between sites. Some events produce sediment flux but with little morphological effect. Different types of morphological behaviour are apparent between differing channel reaches, partly related to channel shape and confinement and partly to substrate but also to sequences of events and spatial relations of erosion, deposition and sediment flux. This illustrates that care should be taken over generalisation of response of semi-arid channels, which should be analysed in the context of typology of channel and connectivity in the systems. At a few locations, a more consistent relation of erosion or deposition to magnitude of flow is apparent, set within the context created by a large flow event. Overall, these long-term measurements at multiple sites demonstrate the high spatial and temporal variability in morphological change in channels in semi-arid areas, even within one region.

The vegetation is shown to be highly resilient, both to the effects of flows and to drought periods, demonstrating the adaptation to this semi-arid environment and the episodicity of flash floods and of moisture inputs. Vegetated zones are subject to much less erosion than bare areas and tend to be zones of sedimentation, with positive feedback effects from the presence of plants. The overall growth of perennial vegetation has taken place in this period of lack of extreme floods. Human activity related to urbanisation and increase in infrastructure, reflecting the major changes in this part of Spain in the period, has had a major impact on the catchments and some direct effects in channels.

References

Bendix, J., Hupp, C. R. (2000) Hydrological and geomorphological impacts on riparian plant communities. *Hydrological Processes* **14**, 2977–2990.

Bertoldi, W., Gurnell, A., Surian, N., Tockner, K., Zanoni, L., Ziliani, L., Zolezzi, G. (2009) Understanding reference processes: linkages between river flows, sediment dynamics and vegetated landforms along the Tagliamento River, Italy. *River Research and Applications* **25**, 501–516.

Boer, M., Puigdefábregas, J. (2005) Effects of spatially structured vegetation patterns on hillslope erosion in a semiarid Mediterranean environment: a simulation study. *Earth Surface Processes and Landforms* **30**, 149–167.

Boix-Fayos, C., de Vente, J., Martinez-Mena, M., Barbera, G. G., Castillo, V. (2008) The impact of land use change and check-dams on catchment sediment yield. *Hydrological Processes* **22**, 4922–4935.

Bombino, G., Gurnell, A. M., Tamburino, V., Zema, D. A., Zimbone, S. M. (2008) Sediment size variation in torrents with check dams: effects on riparian vegetation. *Ecological Engineering* **32**, 166–177.

Borselli, L., Cassi, P., Torri, D. (2008) Prolegomena to sediment and flow connectivity in the landscape: a GIS and field numerical assessment. *Catena* **75**, 268–277.

Bracken, L. J., Croke, J. (2007) The concept of hydrological connectivity and its contribution to understanding runoff-dominated geomorphic systems. *Hydrological Processes* **21**, 1749–1763.

Bracken, L. J., Kirkby, M. J. (2005) Differences in hillslope runoff and sediment transport rates within two semi-arid catchments in southeast Spain. *Geomorphology* **68**, 183–200.

Bracken, L. J., Cox, N. J., Shannon, J. (2008) The relationship between rainfall inputs and flood generation in south-east Spain. *Hydrological Processes* **22**, 683–696.

Brierley, G., Fryirs, K. (1998) A fluvial sediment budget for upper Wolumla creek, south coast, New South Wales, Australia. *Australian Geographer* **29**, 107–U124.

Brierley, G., Stankoviansky, M. (2002) Geomorphic responses to land use change: lessons from different landscape settings. *Earth Surface Processes and Landforms* **27**, 339–341.

Brookes, C. J., Hooke, J. M., Mant, J. (2000) Modelling vegetation interactions with channel flow in river valleys of the Mediterranean region. *Catena* **40**, 93–118.

Bull, W. B. (1997) Discontinuous ephemeral streams. *Geomorphology* **19**, 227–276.

Bull, L. J., Kirkby, M. J., Shannon, J., Hooke, J. M. (1999) The impact of rainstorms on floods in ephemeral channels in southeast Spain. *Catena* **38**, 191–209.

Cammeraat, E. L. H. (2004) Scale dependent thresholds in hydrological and erosion response of a semiarid catchment in southeast Spain. *Agriculture, Ecosystems & Environment* **104**, 317–332.

Castillo, V. M., Mosch, W. M., Garcia, C. C., Barbera, G. G., Cano, J. A. N., Lopez-Bermudez, F. (2007) Effectiveness and geomorphological impacts of check dams for soil erosion control in a semiarid Mediterranean catchment: El Carcavo (Murcia, Spain). *Catena* **70**, 416–427.

Conesa-Garcia, C. (1995) Torrential flow frequency and morphological adjustments of ephemeral channels in southeast Spain. In: Hickin, E. J. (ed.) *River Geomorphology*. John Wiley & Sons, Ltd, Chichester, pp. 169–192.

Conesa-Garcia, C., Garcia-Lorenzo, R. (2009) Effectiveness of check dams in the control of general transitory bed scouring in semiarid catchment areas (South-East Spain). *Water and Environment Journal* **23**, 1–14.

Corenblit, D., Steiger, J., Tabacchi, E. (2010) Biogeomorphologic succession dynamics in a Mediterranean river system. *Ecography* **33**, 1136–1148.

De Baets, S., Poesen, J., Knapen, A., Galindo, P. (2007) Impact of root architecture on the erosion-reducing potential of roots during concentrated flow. *Earth Surface Processes and Landforms* **32**, 1323–1345.

Fryirs, K., Brierley, G. J. (1999) Slope-channel decoupling in Wolumla catchment, New South Wales, Australia: the changing nature of sediment sources following European settlement. *Catena* **35**, 41–63.

Fryirs, K., Brierley, G. J. (2001) Variability in sediment delivery and storage along river courses in Bega catchment, NSW, Australia: implications for geomorphic river recovery. *Geomorphology* **38**, 237–265.

Graf, W. L. (1978) Fluvial adjustments to the spread of tamarisk in the Colorado Plateau region. *Geological Society of America Bulletin* **89**, 1491–1501.

Graf, W. L. (1983) Flood-related channel change in an arid-region river. *Earth Surface Processes and Landforms* **8**, 125–139.

Gurnell, A. M., Petts, G. E., Hannah, D. M., Smith, B. P. G., Edwards, P. J., Kollmann, J., Ward, J. V., Tockner, K. (2001) Riparian vegetation and island formation along the gravel-bed Fiume Tagliamento, Italy. *Earth Surface Processes and Landforms* **26**, 31–62.

Hassan, M. A., Schick, A. P., Shaw, P. A. (1999) The transport of gravel in an ephemeral sandbed river. *Earth Surface Processes and Landforms* **24**, 623–640.

Hertig, E., Jacobeit, J. (2008) Assessments of Mediterranean precipitation changes for the 21st century using statistical downscaling techniques. *International Journal of Climatology* **28**, 1025–1045.

Hooke, R. L. (1999) Spatial distribution of human geomorphic activity in the United States: comparison with rivers. *Earth Surface Processes and Landforms* **24**, 687–692.

Hooke, J. (2003) Coarse sediment connectivity in river channel systems: a conceptual framework and methodology. *Geomorphology* **56**, 79–94.

Hooke, J. M. (2006) Human impacts on fluvial systems in the Mediterranean region. *Geomorphology* **79**, 311–335.

Hooke, J. M. (2007) Monitoring morphological and vegetation changes and flow events in dryland river channels. *Environmental Monitoring and Assessment* **127**, 445–457.

Hooke, J. M., Mant, J. M. (2000) Geomorphological impacts of a flood event on ephemeral channels in SE Spain. *Geomorphology* **34**, 163–180.

Hooke, J., Mant, J. (2002) Morpho-dynamics of ephemeral streams. In: Bull, L. J., Kirkby, M. J. (eds.) *Dryland Rivers: Hydrology and Geomorphology of Semiarid Channels*. John Wiley & Sons, Ltd, Chichester, pp. 173–204.

Hooke, J., Sandercock, P. (2012) Use of vegetation to combat desertification and land degradation: recommendations and guidelines for spatial strategies in Mediterranean lands. *Landscape and Urban Planning* **107**, 389–400.

Hooke, J. M., Brookes, C. J., Duane, W., Mant, J. M. (2005) A simulation model of morphological, vegetation and sediment changes in ephemeral streams. *Earth Surface Processes and Landforms* **30**, 845–866.

Kirkby, M. J., Bracken, L. J., Shannon, J.(2005) The influence of rainfall distribution and morphological factors on runoff delivery from dryland catchments in SE Spain. *Catena* **62**, 136–156.

Lazaro, R., Rodrigo, F. S., Gutierrez, L., Domingo, F., Puigdefábregas, J. (2001) Analysis of a 30 year rainfall record (1967–1997) in semi-arid SE Spain for implications on vegetation. *Journal of Arid Environments* **48**, 373–395.

Lite, S. J., Bagstad, K. J., Stromberg, J. C. (2005) Riparian plant species richness along lateral and longitudinal gradients of water stress and flood disturbance, San Pedro River, Arizona, USA. *Journal of Arid Environments* **63**, 785–813.

Ludwig, J. A., Eager, R. W., Bastin, G. N., Chewings, V. H., Liedloff, A. C. (2002) A leakiness index for assessing landscape function using remote sensing. *Landscape Ecology* **17**, 157–171.

Machado, M. J., Benito, G., Barriendos, M., Rodrigo, F. S. (2011) 500 years of rainfall variability and extreme hydrological events in southeastern Spain drylands. *Journal of Arid Environments* **75**, 1244–1253.

Mairota, P., Thornes, J. B., Geeson, N. (1998) *Atlas of Mediterranean Environments in Europe, the Desertification Context*. John Wiley & Sons, Ltd, Chichester.

Mant, J. (2002) Vegetation in the ephemeral channels of southeast Spain: its impact on and response to morphological change. PhD thesis. University of Portsmouth, Portsmouth.

Mayor, A. G., Bautista, S., Llovet, J., Bellot, J. (2008) Measurement of the connectivity of runoff source areas as determined by vegetation pattern and topography: a tool for assessing potential water and soil losses in drylands. *Water Resources Research* **44**, W10423.

Miranda, J. D., Armas, C., Padilla, F. M., Pugnaire, F. I. (2011) Climatic change and rainfall patterns: effects on semi-arid plant communities of the Iberian Southeast. *Journal of Arid Environments* **75**, 1302–1309.

Mueller, E. N., Wainwright, J., Parsons, A. J. (2007) Impact of connectivity on the modeling of overland flow within semiarid shrubland environments. *Water Resources Research* **43**, W09412.

Nanson, G., Tooth, S., Knighton, A. D. (2002) A global perspective on dryland rivers: perceptions, misconceptions and distinctions. In: Bull, L. J., Kirkby, M. J. (eds.) *Dryland Rivers: Hydrology and Geomorphology of Semi-arid Channels*. John Wiley & Sons, Ltd, Chichester, pp. 17–54.

Okin, G. S., Parsons, A. J., Wainwright, J., Herrick, J. E., Bestelmeyer, B. T., Peters, D. C., Fredrickson, E. L. (2009) Do changes in connectivity explain desertification? *Bioscience* **59**, 237–244.

Phillips, J. V., Hjalmarson, H. W. (1994) Flood flow effects on riparian vegetation in Arizona. In: Proceedings of the 1994 National Conference on Hydraulic Engineering (Vol. 1). Buffalo, NY, 1–5 August 1994. American Society of Civil Engineers (Hydraulics Division), New York, pp. 707–711.

Poesen, J. W. A., Hooke, J. M. (1997) Erosion, flooding and channel management in Mediterranean environments of southern Europe. *Progress in Physical Geography* **21**, 157–199.

Powell, D. M., Brazier, R., Parsons, A., Wainwright, J., Nichols, M. (2007) Sediment transfer and storage in dryland headwater streams. *Geomorphology* **88**, 152–166.

Puigdefábregas, J., Alonso, J. M., Delgado, D., Domingo, F., Cueto, M., Gutierrez, L., Lazaro, R., Nicolau, J. M., Sanchez, G., Sole, A., Vidal, S. (1996) The Rambla Honda field site: interactions of soil and vegetation along a catena in semiarid Spain. In: Thornes, J., Brandt, J. (eds.) *Mediterranean Desertification and Land Use*. John Wiley & Sons, Ltd, Chichester, pp. 137–168.

Reaney, S. M., Bracken, L. J., Kirkby, M. J. (2007) Use of the Connectivity of Runoff Model (CRUM) to investigate the influence of storm characteristics on runoff generation and connectivity in semi-arid areas. *Hydrological Processes* **21**, 894–906.

Richards, K. S. (1982) *Rivers: Form and Process in Alluvial Channels*. Methuen, London.

Sanchez-Toribio, M. I., Garcia-Marin, R., Conesa-Garcia, C., Lopez-Bermudez, F. (2010) Evaporative demand and water requirements of the principal crops of the Guadalentin valley (SE Spain) in drought periods. *Spanish Journal of Agricultural Research* **8**, S66–S75.

Sandercock, P., Hooke, J. (2006) Strategies for reducing sediment connectivity and land degradation in desertified areas using vegetation: the RECONDES project. *Sediment Dynamics and the Hydromorphology of Fluvial Systems* **306**, 287–294.

Sandercock, P. J., Hooke, J. M. (2010) Assessment of vegetation effects on hydraulics and of feedbacks on plant survival and zonation in ephemeral channels. *Hydrological Processes* **24**, 695–713.

Sandercock, P. J., Hooke, J. M. (2011) Vegetation effects on sediment connectivity and processes in an ephemeral channel in SE Spain. *Journal of Arid Environments* **75**, 239–254.

Sandercock, P. J., Hooke, J. M., Mant, J. (2007) Vegetation tin dryland river channels and its interaction with fluvial processes. *Progress in Physical Geography* **31**, 107–129.

Schick, A. P. (1974) Formation and obliteration of desert stream terraces – a conceptual analysis. *Zeitschrift für Geomorphologie Supplementband* **21**, 88–105.

Smith, M. W., Bracken, L. J., Cox, N. J. (2010) Toward a dynamic representation of hydrological connectivity at the hillslope scale in semiarid areas. *Water Resources Research* **46**, W12540.

Stromberg, J. C., Bagstad, K. J., Leenhouts, J. M., Lite, S. J., Makings, E. (2005) Effects of stream flow intermittency on riparian vegetation of a semiarid region river (San Pedro River, Arizona). *River Research and Applications* **21**, 925–938.

Surian, N., Rinaldi, M. (2003) Morphological response to river engineering and management in alluvial channels in Italy. *Geomorphology* **50**, 307–326.

Thornes, J. B. (1976) *Semi-arid Erosional Systems.* Department of Geography, Occasional Papers No. 7. London School of Economics, London. 79pp.

Thornes, J. B. (1977) Channel changes in ephemeral streams: observations, problems and models. In: Gregory, K. J. (ed.) *River Channel Changes.* John Wiley & Sons, Ltd, Chichester, pp. 317–355.

Thornes, J. B. (1980) Structural instability and ephemeral channel behaviour. *Zeitschrift für Geomorphologie Supplementband* **36**, 233–244.

Thornes, J. B. (1990) The interaction of erosional and vegetational dynamics in land degradation: spatial outcomes. In: Thornes, J. B. (ed.) *Vegetation and Erosion.* John Wiley & Sons, Ltd, Chichester, pp. 41–53.

Thornes, J. B. (1994) Channel processes, evolution and history. In: Abrahams, A., Parsons, A. J. (eds.) *Geomorphology of Desert Environments.* Chapman and Hall, London, pp. 288–317.

Thornes, J. B., Brandt, J. (1994) Erosion-vegetation competition in a stochastic environment undergoing climatic-change. In: Millington, A. C., Pye, K. (eds.) *Environmental Change in Drylands: Biogeographical and Geomorphological Perspectives.* John Wiley & Sons, Ltd, Chichester, pp. 305–320.

Thornes, J. B., Rowntree, K. M. (2006) Integrated catchment management in semiarid environments in the context of the European Water Framework Directive. *Land Degradation and Development* **17**, 355–364.

Tooth, S., Nanson, G. C. (1999) Anabranching rivers on the Northern Plains of arid central Australia. *Geomorphology* **29**, 211–233.

Tooth, S., Nanson, G. C. (2000) Equilibrium and non-equilibrium conditions in dryland rivers. *Physical Geography* **21**, 183–211.

Wolman, M. G., Gerson, R. (1978) Relative scales of time and effectiveness of climate in watershed geomorphology. *Earth Surface Processes and Landforms* **3**, 189–208.

CHAPTER 5

Stability and instability in Mediterranean landscapes: A geoarchaeological perspective

John Wainwright

Department of Geography, Durham University, South Road, Durham, UK

Introduction

> We still inhabit a conceptual world which is mainly linear in response.
>
> Thornes 1980, p. 243

Landscape evolution, like war, has been described as being formed of 'long periods of boredom interrupted by brief moments of terror' (Brunsden 1990, p. 17). Unlike war (or perhaps exactly like war *reporting*), some of the boredom in landscape evolution may be more apparent than real, because of issues of recording events in the sedimentary record, and in some cases their subsequent loss due to erosion in more recent events (Ager 1973; Briant et al. 2008). While over geological timescales, this pattern has remained relatively consistent (Ager 1973; Willenbring and von Blanckenburg 2010), the effects of human activity over the Holocene have meant that erosion rates have accelerated significantly. Indeed, some authors argue that people are now the most significant geomorphic agent based on the amount of sediment moved (Hooke 2000; Wilkinson and McIlroy 2007). While these studies are useful in defining the larger-scale

problem, they suffer from very simple assumptions, not least that the acceleration is simply due to 'more of the same' but more of it because of the increase in global population. This sort of assumption is exactly the 'linear conceptual world' noted in the quotation at the head of this section. It also ignores other feedbacks due to anthropic climate change. Furthermore, given the debate instigated by Earth scientists that we are no longer living in the Holocene but within the Anthropocene (Zalasiewicz et al. 2011, but see commentary by Brown et al. 2013), being able to understand the processes that underpin the evidence used in this debate is fundamental, with major disciplinary, political, economic and social implications.

The purpose of this chapter is not to suggest that frequency and magnitude of events are not central to the episodicity of landscape evolution (see Thornes and Brunsden 1977; Brunsden and Thornes 1979); far from it. It is, however, to suggest that what we see in the geomorphic record is *also* a consequence of more complicated and largely non-linear dynamics. The interpretation of these dynamics takes its inspiration from John Thornes'

Monitoring and Modelling Dynamic Environments, First Edition. Edited by Alan P. Dykes, Mark Mulligan and John Wainwright.
© 2015 John Wiley & Sons, Ltd. Published 2015 by John Wiley & Sons, Ltd.

own work, evaluating the non-linear feedbacks between vegetation and erosion (Thornes 1985), as well as from his work on catastrophe theory as applied to dryland channels (Thornes 1980). To these aspects of stability and instability, more recent developments of the understanding of connectivity of geomorphic systems (Turnbull et al. 2008; Bracken et al. 2013, 2015) will be added to build a stability–instability–connectivity (SIC) conceptual model for landscape evolution. The SIC model aims to demonstrate why we get periods – and, as will be seen, locations – of stability (boredom) and instability (terror) as well as periods of apparent stability (in a disconnected landscape) and apparent instability (in a stable landscape that becomes connected). The context in which this conceptual model will be evaluated is the geoarchaeological record of the Mediterranean region. This choice reflects John Thornes' own contributions to the understanding of the dynamics of the landscapes around archaeological sites in Spain (Wise et al. 1982; Thornes and Gilman 1983; Gilman and Thornes 1985), as well as my own early work with him on the erosion of archaeological sites (Wainwright and Thornes 1991) and subsequent interpretations of Mediterranean environments in general (Wainwright and Thornes 2004).

SIC in landscape evolution

... our understanding of the role of vegetation in soil erosion has been rather badly neglected, despite its importance in historical as well as contemporary studies. Historical geomorphologists have tended to incorporate vegetation relationships only implicitly through the supposed effects of ill-specified changes of climate. Process geomorphologists ..., have dealt with soil erosion processes on bare areas or in terms of the results from field experiments where the vegetation element has been poorly specified.

Thornes 1985, p. 222

From a dynamical systems perspective, the interactions between erosion and vegetation growth show a complex interplay of stable and unstable equilibria (Thornes 1983, 1985). Depending on the nature of the feedback (linear or exponential, vigorously responding vegetation), unstable or stable equilibria may form at different points in the vegetation–erosion phase space (Figure 5.1). In all cases, it is possible to obtain stability when the maximal vegetation biomass is present, reducing erosion rates to zero, but in the cases of both linear and exponential vegetation feedbacks, there is also a metastable equilibrium at an intermediate stage. The important aspect of this equilibrium is that perturbations may lead to conditions of runaway erosion. Thornes' only suggestion for stability with less than full vegetation cover was in where erosion is strongly self-regulating and vegetation regrows rapidly, as might be possible gullied landscapes with a high drainage density (Figure 5.1c). In terms of human-induced landscape evolution, the first two cases provide the potential for interventions to produce significant amounts of erosion that may show up in the sedimentary archive or management to mitigate erosion rates. The third case produces self-regulation even in environments that are typically considered highly erodible (cf. Wise et al. 1982). In all cases, there are shown to be multiple stable states in the system, demonstrating that simple interpretations of the interactions are not possible, a point that has been increasingly made in evaluating grazed dryland systems (Illius and O'Connor 1999; Sullivan and Rohde 2002). A recent study by Moreno-de las Heras et al. (2011) has demonstrated that it is possible to develop and parameterize this type of interaction model for disturbed sites in Teruel, Spain. They used a simple power-law approximation to rill erosion as a function of discharge and related discharge to slope length, with a simple threshold-based vegetation feedback; vegetation growth is logistic with an exponential erosion feedback producing the phase space shown in Figure 5.2. Although their model

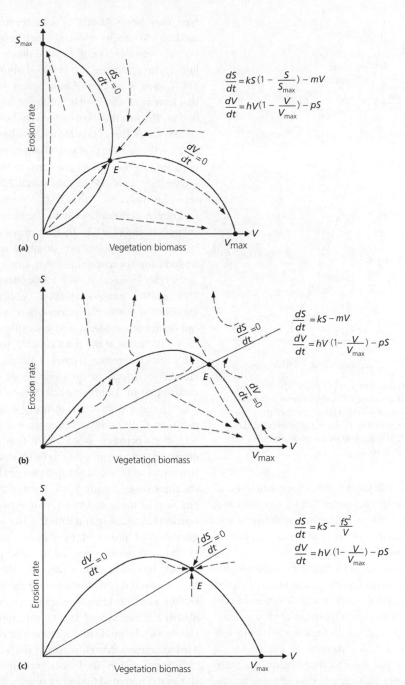

Figure 5.1 Vegetation–erosion phase spaces as defined by Thornes (1985): (a) linear feedback between erosion and vegetation, (b) exponential feedback between erosion and vegetation and (c) vigorous vegetation growth in a self-stabilizing landscape. See text for further discussion. V is the vegetation biomass, S the erosion rate, t is time, V_{max} is the maximal vegetation biomass in an area, S_{max} is the maximum soil erosion possible in an area (soil thickness), k is an erodibility parameter, m and f reflect the effect of vegetation in reducing the erosion, h is the net biomass growth, and p reflects the effect of erosion in reducing vegetation growth.

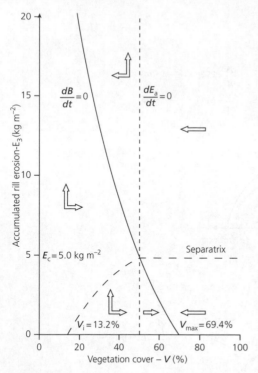

Figure 5.2 Vegetation–erosion phase space for rill erosion on recovering mine spoil heaps in Teruel, Spain, showing a practical application of the type of model illustrated in Figure 5.1. (Moreno-de las Heras et al. 2011. Reproduced with permission of John Wiley & Sons.)

tended to overpredict the exact amounts of erosion over a 16-year period of measurement from 1987/1988 to 2002/2003, it did represent the trajectories of erosion and vegetation responses on the slopes well.

Paths to stability in landscape evolution in this type of interaction model, then, are contingent and depend on the extent and direction of perturbations. The superimposition of climate variability and change and land-use modifications will produce a complex pattern of evolution for which it is not easy to find indicators. More recent studies also suggest that changes in the system may be sudden rather than gradual as a response to perturbations. Scheffer et al. (2001) reviewed the now-extensive ecological literature that suggests that sudden shifts in vegetation

type may be explained by catastrophe-theory models. In such models, the relationship between variables is not a simple (linear or non-linear) curve, but one that is folded in some way. This folding allows two fundamental properties that have been observed in studies of land degradation. The first is that of sudden changes, as the system jumps from one state to the other (Figure 5.3a), reflecting the dramatic changes in vegetation, for example, from grassland to shrubland (Schlesinger et al. 1990; van Auken 2000). The second is hysteresis – that is, that transitions in one direction do not occur in the same way as changes in the other, so that returning a shrubland to a grassland is not simply a matter of reproducing the conditions that existed previously (Schlesinger et al. 1990; Brandt and Thornes 1996; Rango et al. 2006). Adding a third variable can produce a system where transitions can occur both suddenly and smoothly, depending on the value of the third variable, producing the cusp catastrophe (Figure 5.3b: see further discussions later). In an example of the fold catastrophe by Holmgren and Scheffer (2001) relating to the Mediterranean-type environments in Chile, it has been suggested that the transitions between woodland (or shrubland), grassland and bare ground can be represented by two sets of folds, allowing rapid changes between the three states (Figure 5.3c). They relate different parts of these states to resource (water and nutrient) availability and thus the 'islands of fertility' model proposed by Charley and West (1975) and widely applied to desertification studies following Schlesinger et al. (1990). Notwithstanding the limitations of the islands of fertility model at landscape scale (see Barbier et al. 2014; Stewart et al. 2014), this approach is clearly a useful guide to understanding change in Mediterranean environments in general.

Putting erosion feedbacks into this model shows the potential for very complex dynamics depending on local conditions and vegetation types (Figure 5.3d). On the figure, curves a–c are erosion isoclines of either linear or exponential type. Curve a allows the possibility of

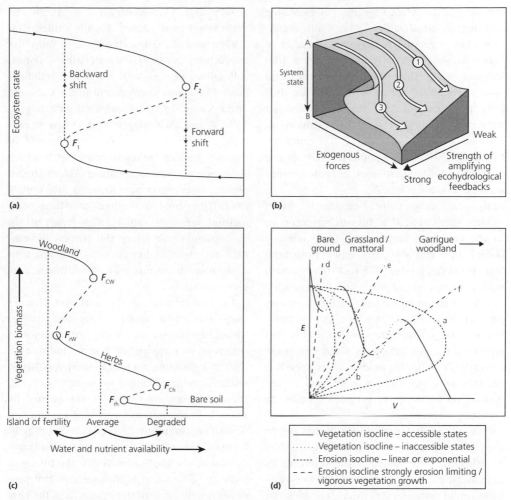

Figure 5.3 The use of catastrophe theory to demonstrate interaction between environmental variables: (a) a simple fold and (b) cusp catastrophes, (c) complex fold-catastrophe model of Holmgren and Scheffer (2001) to account for land degradation in Mediterranean-type environments in Chile and (d) modification of the Holmgren and Scheffer model to account for vegetation interactions.

metastable states under all three surface conditions (bare ground, grass/matorral, garrigue/woodland), although an exponential feedback would cause runaway erosion conditions to be prevalent, at least under the bare or grass/matorral conditions. As the erosion parameters cause the curve to shift first from a to b and then to c, the equilibria under woodland/garrigue and then grassland/matorral disappear, making these curves intuitively less likely. The strongly

limited erosion case is shown by the erosion isoclines d–f on the figure. In d, a single equilibrium is attainable under 'bare ground' conditions, but paradoxically the form of the fold suggests it may only occur when vegetation is increasing. A single equilibrium is possible in curve e, but this time where the vegetation is grassland/matorral and attainable from either trajectory. Multiple equilibria occur with curve f, although again the case for grassland/matorral can only occur when vegetation is increasing and the woodland/garrigue case is only attainable when vegetation is *de*creasing. If the curve continued to have a shallower slope (not shown), only the woodland/garrigue case would remain as having an equilibrium, first only in the case of decreasing vegetation, but ultimately in both cases. The existence of densely wooded badlands in Barcelonnette, France, at least at some points in the Holocene (Ballais 1996), suggests that in some circumstances, this case is feasible.

The implications of these catastrophe-theory models when applied to human interventions in the landscape are that there are multiple potentially metastable states, and perturbations from them can produce different trajectories, in which some parts of the 'normal' vegetation development are missed out. Perturbations may produce pulses of sediment as a result of runaway erosion or lead to the development further equilibrium states. Put simply, the definitions of the causes of sediment production as a result of perturbations are not straightforward and depend on knowing the previous state of the system.

One of the first applications of catastrophe theory in geomorphology was by Thornes (1980), where he looked at the behaviour of ephemeral channels in drylands. In this cusp-catastrophe model, the three variables are stream power, ratio of sorting to particle size and sediment load (Figure 5.4a). When particles are well sorted in relation to their size, changes in stream power produce gradual changes in entrainment and conversely deposition. However, for poorly sorted material, entrainment stays on the lower limb of the cusp until reaching a specific threshold (cf. the broader debates on selective entrainment), while on the waning flow, material stays in motion until the stream power reaches a lower threshold. The effect of transmission losses along the channel will produce pulses of deposition followed by narrower channels, in a discontinuous pattern as described in numerous examples by Bull

(1997) and equivalent to the 'beads' of Wainwright et al. (2002). Slightly earlier, Graf (1979) used the cusp catastrophe to show the interactions between channel stream power, valley-floor biomass and amount of sediment moved in arroyo development in the Colorado Front Range, based on empirical data (Figure 5.4b). These data suggest that at low stream power, as biomass decreases, the amount of channel incision remains relatively low but changes in a continuous manner. At high stream power, however, erosion remains low until a critical threshold is reached, at which point incision increases rapidly. The hysteresis in the system ensures that the erosion remains high even with higher amounts of biomass, in a situation analogous with the hillslope case given earlier.

The combination of the Graf and Thornes cusp-catastrophe models suggests that at timescales beyond the event, perturbations to the vegetation (whether human induced or related to climate) can lead to sudden pulses of sediment being produced, while during events, these sediment pulses will be represented by spatially discontinuous patterns of erosion and deposition. In terms of preservation of sediments that we can use to evaluate landscape evolution, the implication is that the timing of sediment production is both variable and path dependent because of the hysteresis in the Graf model and that its eventual location is variable and path dependent because of the hysteresis in the Thornes model.

The third element of SIC develops from the rapidly expanding literature on connectivity, describing the ways that catchment elements and processes are interlinked (Bracken and Croke 2007; Bracken et al. 2013). Although multiple definitions of connectivity have been used, a major aspect is the distinction between structural and functional (or process-based; see Bracken et al. 2013) connectivity. Turnbull et al. (2008) used these terms to describe the dynamics of dryland systems in terms of the structure (e.g. spatial patterns of vegetation) that controls

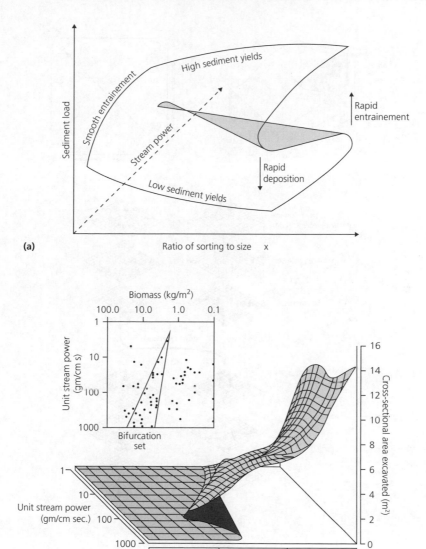

Figure 5.4 Catastrophe-theory models in fluvial geomorphology: (a) Thornes' (1980) cusp catastrophe for the entrainment and deposition of sediments in ephemeral river channels and (b) Graf's (1979) cusp-catastrophe model for the interaction between vegetation and stream power on arroyo formation in the Colorado Front Range, United States.

surface properties such as infiltration and surface roughness, which in turn control the behaviour of the system, and thus feedbacks on the structure (see also Wainwright et al. 2011). In particular, Turnbull et al. (2008) described a cusp-catastrophe model for structural and functional connectivity, which suggests that the passage from one vegetation type to another can be gradual or sudden and demonstrate hysteresis (Figure 5.5). Thus, the erosion produced as a result of vegetation change on hillslopes may connect directly to the channel or it may

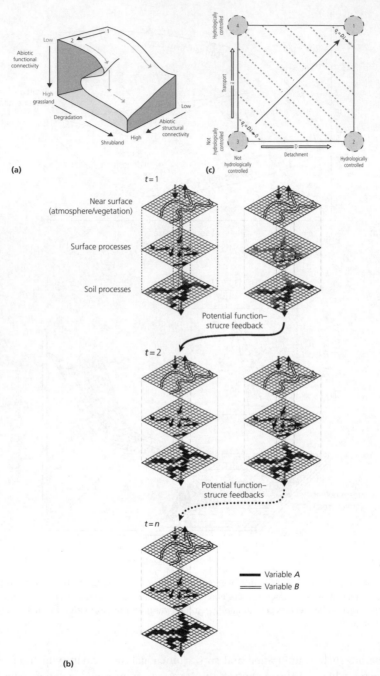

(a)

(b)

(c)

Figure 5.5 Definitions of connectivity: (a) cusp catastrophe showing the links between structural and functional connectivity on land degradation (Turnbull et al. 2008. Reproduced with permission of John Wiley & Sons.) (b) conceptual model of structural and functional connectivity feedbacks (Wainright et al. 2011. Reproduced with permission of Elsevier.) (c) link between hydrological and sediment connectivity and transport distance as a way of linking conceptual approaches in geomorphology (Bracken et al. 2015) (qs is sediment flux, D is detachment rate, and λ is mean particle travel distance).

not, depending on patterns of vegetation structure and the trajectory of the system. In a further development, Bracken et al. (2015) have linked the understanding of hydrological connectivity with that of sediment connectivity. They argue that a unified concept of sediment connectivity allows us to evaluate both hillslope–channel coupling in the traditionally used sense and the effects of dynamics within landscape units in a more holistic way. Furthermore, they demonstrate that the approach is closely linked with the transport-distance models of Parsons et al. (2004, 2006) and Wainwright et al. (2008) in a way that allows significant problems of scale in geomorphic process models to be overcome (Figure 5.5c). Thus, investigations of sediment connectivity that focus on travel distance of the sediment involved by the observed transport processes allow us to evaluate not just when sediment will be produced by human interventions in the landscape but also where it is likely to end up. Of course, the Thornes (1980) catastrophe-theory model of event entrainment and deposition described earlier means that this is a non-trivial process once sediment makes it into the channel system. Similarly, the extent of remobilisation of floodplain material implied by the Graf (1979) model means that interpretation of the record (in particular the absences of deposits) requires us to understand the trajectory of the system over the entire time period from initial deposition to observation.

The dynamical systems approach as applied through the SIC framework suggests that understanding patterns of past erosion in dryland environments is a non-trivial exercise. At each step of sediment production, transfer and deposition, simple feedbacks produce complex systems dynamics. Sediment may be produced in both boring and terrifying ways following the catastrophe-theory models of slope behaviour and may be preserved in both boring and terrifying ways following the catastrophe-theory models of channel behaviour. Furthermore, understanding the patterns of what is happening relies on our knowing the past history of the system. This path dependency implies that studies based on individual snapshots of specific time periods are likely to be misinterpreted, and case studies should involve as holistic and longitudinal an approach as possible. While these arguments also hold for perturbations of a climatic nature (see Holmgren and Scheffer 2001; Wainwright 2007), the emphasis here will be to evaluate the extent to which they can be seen within the existing geoarchaeological record of the Mediterranean basin.

Mediterranean geoarchaeology

> Because of the varying sensitivity of the landscape it is evident that this [episodicity of landscape evolution] is a spatial as well as a temporal dictum.
>
> Brunsden and Thornes 1979, p. 481

In Wainwright and Thornes (2004), we assembled the evidence for human impacts on the Mediterranean landscape in the Holocene in an attempt to elucidate what McNeill (1992, p. 147) called 'the endless pas de deux' between population history and landscape evolution. We also developed an argument as to why climate variability (but *not* climatic events) was relatively unimportant in comparison to the effects of land-use change over this timescale. Table 5.1 summarizes the data in terms of whether or not land degradation could be observed at different sites at different times in the Holocene, as a function of whether archaeological and/or historical data suggested whether local population was increasing or decreasing. While further case studies could be employed from the past decade of research, they would not fundamentally change the overall pattern of the analysis. Examples from each of the different types of behaviour will be used to evaluate the SIC approach.

In the case of increasing human pressure/landscape degradation, the case of Laval de la Bretonne, Aude, France, is informative. Wainwright (1994)

Table 5.1 Evidence of apparent stability and instability in the Mediterranean landscape from geoarchaeological studies.

Human impact	Landscape status	
	Degradation	**No/little degradation**
Increasing pressure	Israel: Kebaran/Geom. Kebaran, Early Bronze Age, Later Mediæval?	Argolid: Classical and Late Mediaeval
	Negev: Later Roman	
	Argolid: Early Bronze Age, Middle Byzantine	Epirus: Late Bronze Age–Iron Age
	Argive Plain: Middle–Late Neolithic, Early Bronze Age, Late Bronze Age	
	Melos: Late Bronze Age	Vera Basin: Visigothic/Byzantine, Moorish
	Nemea: Early Neolithic	
	Euboea: Early Neolithic	Rio Aguas: Roman–C15
	Thessaly: Middle–Late Neolithic	
	Macedonia: Early Neolithic, Late Neolithic, Classical	Rio Aguas: Roman–C15
	Epirus: C19–Modern	
	Santorini: Early-mid Bronze Age?	Valdaine: Early Neolithic, Late Bronze Age, Early Roman
	Crete: Roman	
	Turkey: Late Bronze Age	Haut Comtat: Early Neolithic–Chalcolithic
	Alpujarras: C18–C19	
	Vera Basin: Chalcolithic, Argaric, Phoenician/Punic, Roman, C18→	Etang de Berre: Bronze Age–Roman
	Rio Aguas: Chalcolithic, Argaric, C18→	
	Valdaine: Mid/Late Neolithic, Chalcolithic, Iron Age, C14→	Etruria: Iron Age
	Alpilles: Early Neolithic–Chalcolithic, Iron Age	
	Haut Comtat: Iron Age, Early Roman, Late Middle Ages	Biferno Valley: Middle Bronze Age
	South France: generally relating to C17–C18→ deforestation	
	Croatia: Bronze Age	
	Etruria: Late Bronze Age, Classical	
	Biferno Valley: Late Neolithic/Early Bronze Age?, Samnite/Early Roman, Modern	
	Laval de la Bretonne: Early Bronze Age	
	Teruel: Early Bronze Age	
	Languedoc/Provence/Pyrenees: Iron Age	
	Etang de Berre: Late Neolithic	
	Morocco: Roman, Early Mediæval	
Decreasing pressure	Argolid: Hellenistic/Early Roman, Modern	Vera Basin: post-Argaric
	Epirus: post-Roman	Valdaine: Early Mediæval
	Alpujarras: post-*Reconquista*	Haut Comtat: Bronze Age
	Vera Basin: post-*Reconquista*	Crete: Modern (other periods probably fall within this category, but are not well enough documented for specific inclusion)
	Rio Aguas: Late (post-Argaric) Bronze Age, post-*Reconquista*	
	Valdaine: Early Bronze Age, Late Roman	
	Alpilles: Bronze Age, Late Roman, Early Mediæval	
	Haut Comtat: post-Roman	
	Etang de Berre: post-Roman	
	South France: generally relating to post-Roman depopulation	
	Etruria: post-Roman depopulation	
	Biferno Valley: Early Mediæval, C13, C19	
	Algeria, Tunisia, Libya: Late Roman	

Wainwright and Thornes 2004. Reproduced with permission from Taylor and Francis.

and Gascó et al. (1996) argued that the likely reason for the deep burial of the site on the lower hillslopes above a tributary of the Aude river (Figure 5.6) was most likely related to the occurrence of a single, extreme storm event when the area above the site had been cleared for agriculture. In terms of the SIC approach, this interpretation can be understood in terms of an instability caused by the modification of the vegetation for agriculture producing a large sediment source. Connectivity *on* the slope meant that this sediment, travelling a reasonable distance because of the size of the inferred event, could increase in magnitude downslope, but the disruption of that connectivity (by the build-up of floodplain sediments on the Bretonne valley floor) allowed the accumulation of sediment above the site. In the Early Bronze Age of the Argolid, van Andel ct al. (1986; also van Andel and Runnels 1987) noted the presence of extensive debris flows that appear to coincide with the first major population increase in the area. They argued that a reduction of the fallow period as a response to trying to feed the expanding population meant that surfaces where left exposed for longer in the late summer and early autumn when major storms occur. Again, in SIC terms, this pattern is

one of destabilization of the upper slopes leading to the major mobilisation of sediment in large events and the combination of the transport mechanism (debris flows) and the lack of connectivity into the channel systems. Similar examples elsewhere in Early Bronze Age Greece can be found in the rapid accumulation and valley floor colluvium build-up around Tiryns (Zangger 1994) (Figure 5.7), and at Markiani on the island of Amorgos in the Cyclades (Whitelaw 2000), as well as in the Early Bronze Age of the Vera Basin in southern Spain (Courty et al. 1994; Castro et al. 1996).

Conversely, a number of examples are found where human pressure is increasing, but there is little or no evidence of contemporary land degradation. In the case of the Argolid, van Andel et al. (1986) and van Andel and Runnels (1987) suggest that this situation arises in the Classical and Late Mediaeval periods. The very fact that the steepest slopes are being cultivated at these times requires the construction of terraces. The terraces essentially act to break up the hydrological and sediment connectivity, but also reduce the local slope so that although major perturbations of the vegetation occur, the overall erosion rate goes down (e.g. see measurements in Kosmas et al. 1997). Similar

Figure 5.6 View of Laval de la Bretonne, Aude, France, with Holocene terraces visible in the middle ground and typically eroded garrigue slopes in the foreground.

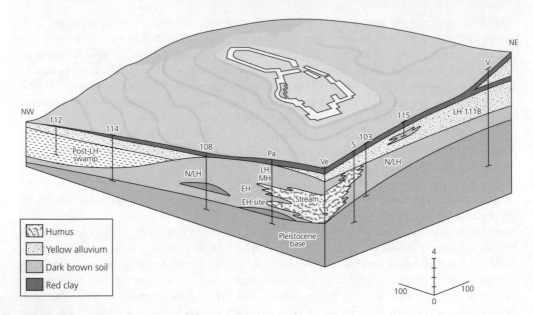

Figure 5.7 Block diagram showing the Holocene stratigraphy west and south of Tiryns, Greece (Zangger 1994. Reproduced with permission of Archaeological Institute of America).

examples can be found in the Middle Bronze Age in the Biferno Valley of central Italy (Hunt 1995) and in numerous parts of Spain in the post-Roman/Moorish period (McNeill 1992; Courty et al. 1994; Castro et al. 1996).

When human impacts decrease, it is also possible to find a number of examples where land degradation – perhaps paradoxically – seems to increase. Examples of the post-*Reconquista* in Spain are common. McNeill (1992) has argued the case for the Alpujarras, while Courty et al. (1994) have suggested it for the Vera and Castro et al. (1996) for the Aguas. The suggestion is that the collapse of terraces in these cases, perhaps coupled with increased fires due to the higher fuel availability as agriculture is replaced by matorral, produces pulses of erosion that are highly connected due to gully formation. Similar processes seem to occur in the Late Roman period in southern France (Provansal et al. 1994) and northern Africa (Vita-Finzi 1969).

The coincidence of decreasing pressure and little or no degradation is probably frequent but

under-reported as a negative result (and thereby apparently uninteresting). Specific examples can be found in Crete in the Modern period (Grove and Rackham 2001) and have been suggested for the Middle Bronze Age in the Vera Basin (Courty et al. 1994) and in parts of southeastern France (Berger et al. 1994). In these cases, the likely explanation is that terracing put in place to stabilize slopes during phases of expansion does not collapse following abandonment and that vegetation regrowth simply allows the source of erosion to be switched off, with the remaining terrace structures breaking sediment connectivity. Van Andel and Runnels (1987) made a similar argument for phases in the Argolid. The key aspect in this scenario is likely the extent to which large storm events do not occur in the period immediately after abandonment, emphasising the importance of stochastic events, or simply that the soil and vegetation conditions in some areas of the Mediterranean are simply more resilient, for example, because of higher infiltration rates, or rapidly recolonizing vegetation.

In summary, the spatial variability in the episodicity of the Mediterranean landscape can be explained well by the SIC approach. It emphasizes the need to understand the sources, pathways and sinks of material in linking process understanding of landscape evolution and in so doing recognises the need to relate to the importance of understanding place in geomorphological studies. In doing so, we can improve the understanding of where and when land degradation does and – equally as importantly in terms of identifying where to employ scarce resources in mitigation measures – does not take place.

Modelling the emergence of SIC

[we] need to go beyond the dominantly physical study of erosion to take into account forage ecology. The dynamics of these systems should be modelled to provide information on their stability to socio-economic and biophysical perturbations.

Thornes 2007, p. 25

One of the problems with the wider applicability of the dynamical systems models seen earlier is that while they are useful for understanding the general behaviour of a system, they are paradoxically rather static when applied to any specific context. In particular, they require the evaluation of 'perturbations' to the system that can be rather general and ad hoc. To overcome these problems, the development of agent-based models (ABMs) to geomorphological questions shows great promise (Wainwright 2007; Wainwright and Millington 2010) and indeed allows us to integrate approaches across the different aspects of human and physical geography (O'Sullivan et al. 2012; Millington and Wainwright, submitted). ABMs can contain direct representations of people (and other animals) through the landscape and thus allow the dynamic interactions between people, vegetation and soils to be represented. In this section, a brief overview of an ABM that

represents these dynamics is presented in order to evaluate the extent to which SIC can also emerge from a bottom-up representation of landscape processes, rather than from the top-down representation of the dynamical systems approach.

CybErosion[eag] is a simplified version of the CybErosion model of Wainwright (2007), designed to evaluate the impact of agricultural populations entering and moving across a landscape previously inhabited by hunter-gatherers, for example, as occurred during the Neolithic of the Mediterranean (e.g. Guilaine 2003). The model represents the human populations as two types of agent: hunter-gatherer and agriculturalist. Both are represented as aggregate agents at the level of the settlement. Each settlement contains a number of individuals of specific ages, but individual decision-making is not explicitly represented in this version of the model. Settlements may have a minimal viable size and a maximal size, both reflecting ethnographic examples (Johnson and Earle 1987); settlements smaller than the minimum threshold will attempt to amalgamate with a nearby settlement of the same type, while settlements bigger than the threshold will split into two and one move away from the other. For simplicity, agricultural settlements cultivate a prescribed area and have a herd of generic domesticated animals, while hunter-gatherer settlements have prescribed areas in which they hunt and gather. An agricultural settlement will decide to move if the soil fertility (a function of length of time under agriculture) in the immediately surrounding area declines below a threshold, and a hunter-gatherer settlement will decide to move based on the availability of wild animals in their vicinity as well as the type of vegetation (providing resources to be gathered based on the examples of Alling Gregg [1988]. Different rates of appearance of new agricultural settlements and their impact on the environment can also be modified. Vegetation growth uses a state-and-transition approach representing bare, grass, shrub and forest vegetation types (similar to that of Millington et al. 2009) with fires represented using the model of Perry and Enright (2002) and Millington et al. (2009), so that fire frequency is a function of climate. Feedback to erosion uses the exponential feedback based on vegetation cover (Thornes 1990) to a simplified version of the MAHLERAN model of Wainwright et al. (2008; see also Wainwright 2006, 2007). A typical mosaic of vegetation types simulated around an agricultural settlement and related erosion rates are shown in Figure 5.8.

Three scenarios are discussed here to demonstrate some general characteristics of the model (Table 5.2): the first has a low invasion rate, low pressure on the environment and an extensive mode of cultivation. In the second scenario, a moderate invasion rate and population pressure and extensive production are simulated. The third scenario evaluates the

effects of a rapid invasion rate, high population pressure and intensive production. For each scenario, the cell size used in the model is 100×100 m, and the extent of the example landscape is 15×15 km. Summary results are shown in Figure 5.9. Because each simulation contains a number of stochastic elements (choice of direction of movement or of areas to cultivate or graze, transitions in the vegetation model), these results are intended to be illustrative rather than absolute representations of model behaviour. In scenario 1, the forest cover remains high throughout the 500-year simulation, and the rest of the landscape is made up of patches of grass and shrub vegetation recovering from the localised, infrequent agriculture and from fire scars. Erosion is limited to the areas where agriculture is or has recently been practised, and the net erosion at the landscape scale shows oscillations through time which relate to the crossing of

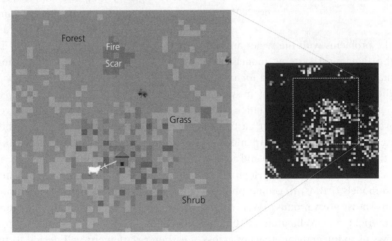

Figure 5.8 A typical mosaic of vegetation types and fire scars around an agricultural settlement and related erosion rates in the CybErosion[eag] model. The different shades of green are annotated to show which vegetation type they represent; brown areas are bare areas that have previously been cultivated around the settlement ('house' symbol), while blue areas are currently under cultivation; and areas in red are fire scars that have not yet started to recover. It is assumed each settlement has a (generic) flock of domesticated animals, the location of which is shown by the 'sheep' symbol; the location of wild animals is shown by the 'rabbit' symbol. Each pixel is 100×100 m, allowing the relative scales of the two images to be compared. The highest erosion rates are shown in white, and zero erosion is in black, with intermediate rates in brown. The slightly larger area of the erosion map shows how, as the settlement has moved northwards, there remains a legacy of higher erosion rates in the formerly settled area, which has yet to fully recover (For colour detail, please see colour plate section).

Table 5.2 Parameters used in the three scenarios to show different characteristics of the CybErosion[eag] model and their relationships to the SIC approach.

Parameter	Scenario 1	Scenario 2	Scenario 3
Maximum hunting radius (m)	1000	1000	1000
Maximum gathering radius (m)	1000	1000	1000
Minimum population of hunter-gatherer settlements	10	10	10
Maximum population of hunter-gatherer settlements	100	100	100
Maximum hunting efficiency*	0.5	0.5	0.5
Maximum gathering efficiency*	0.5	0.5	0.5
Initial number of hunter-gatherer settlements	5	5	5
Birth-rate sensitivity of hunter-gatherers[†]	0	0.05	0.05
Maximum flock radius (m)	1000	1000	1000
Maximum radius of cultivation (m)	200	200	200
Maximum mobility threshold (m)	0.5	0.5	0.5
Birth-rate sensitivity of agricultural settlements[†]	0	0.05	0.25
Minimum population of agricultural settlements	20	20	20
Maximum population of agriculturalist settlements	200	200	200
Initial number of agricultural settlements	2	2	2
Dominant direction of movement of agriculturalists	Northern quadrant	Northern quadrant	Northern quadrant
Initial location of agriculturalist settlements	South side	South side	South side
Soil-fertility threshold[‡]	0.5	0.5	0.5
Runoff coefficient of cultivated areas (—)	0.74	0.74	0.74
Runoff coefficient of bare areas (—)	0.8	0.8	0.8
Runoff coefficient of burned areas (—)	0.85	0.85	0.85
Runoff coefficient of grass areas (—)	0.2	0.2	0.2
Runoff coefficient of shrub areas (—)	0.41	0.41	0.41
Runoff coefficient of forest areas (—)	0.01	0.01	0.01
Erodibility of cultivated areas	0.85	0.85	0.85
Erodibility of bare areas	0.5	0.5	0.5
Erodibility of burned areas	0.6	0.6	0.6
Erodibility of grass areas	0.05	0.05	0.05
Erodibility of shrub areas	0.25	0.25	0.25
Erodibility of forest areas	0.01	0.01	0.01
Transport-distance parameter	0.5	0.5	0.5
Average rainfall (mm)	600	600	600
Average temperature (°C)	21	21	21
Initial vegetation type	Forest	Forest	Forest
Initial number of wild animals	184	184	184

*Efficiencies are calculated as $0.1 + (U[0, \text{efficiency parameter}] - 0.1)$ giving the likelihood of success of the particular task (hunting, gathering) each time that they are attempted, where $U[\text{min, max}]$ is a uniformly distributed random number on the range from min to max.

[†]Sensitivities are used to increase or decrease the birth rates according to whether the food supply for the particular community is above a specific surplus threshold or deficit threshold, respectively.

[‡]If the average soil fertility in the cells surrounding the agriculturalist settlement, a decision will be taken to move the settlement in the direction specified by the dominant direction of movement parameter.

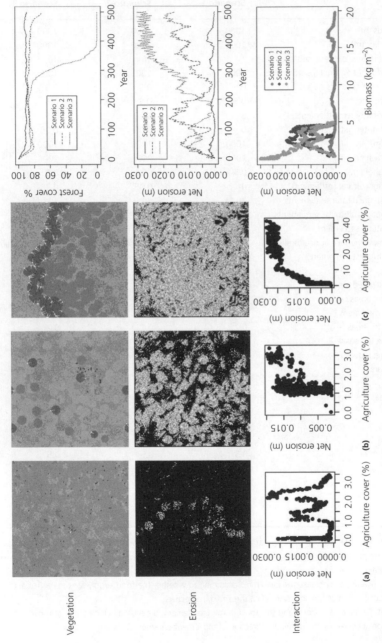

Figure 5.9 Results of CybErosion[eag] for the three scenarios (see Table 5.2 for details of parameters used): (a) low invasion rate, low pressure on the environment and an extensive mode of cultivation; (b) moderate invasion rate and population pressure and extensive production; (c) rapid invasion rate, high population pressure and intensive production (colours and symbols used in the top two panels of a–c are described in Figure 5.8); and (d) shows time series of the evolution of forest cover through time (top panel) and of net erosion through time (middle panel), as well as the relationship between average biomass density across the landscape and average net erosion (For colour detail, please see colour plate section).

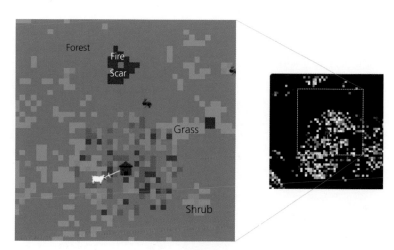

Figure 5.8 A typical mosaic of vegetation types and fire scars around an agricultural settlement and related erosion rates in the CybErosion^{cag} model. The different shades of green are annotated to show which vegetation type they represent; brown areas are bare areas that have previously been cultivated around the settlement ('house' symbol), while blue areas are currently under cultivation; and areas in red are fire scars that have not yet started to recover. It is assumed each settlement has a (generic) flock of domesticated animals, the location of which is shown by the 'sheep' symbol; the location of wild animals is shown by the 'rabbit' symbol. Each pixel is 100×100 m, allowing the relative scales of the two images to be compared. The highest erosion rates are shown in white, and zero erosion is in black, with intermediate rates in brown. The slightly larger area of the erosion map shows how, as the settlement has moved northwards, there remains a legacy of higher erosion rates in the formerly settled area, which has yet to fully recover.

Monitoring and Modelling Dynamic Environments, First Edition. Edited by Alan P. Dykes, Mark Mulligan and John Wainwright.
© 2015 John Wiley & Sons, Ltd. Published 2015 by John Wiley & Sons, Ltd.

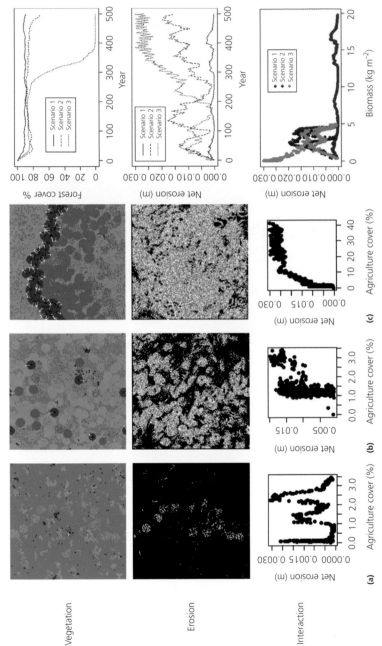

Figure 5.9 Results of CybErosion[eaf] for the three scenarios (see Table 5.2 for details of parameters used): (a) low invasion rate, low pressure on the environment and an extensive mode of cultivation; (b) moderate invasion rate and population pressure and extensive production; (c) rapid invasion rate, high population pressure and intensive production (colours and symbols used in the top two panels of a–c are described in Figure 5.8); and (d) shows time series of the evolution of forest cover through time (top panel) and of net erosion through time (middle panel), as well as the relationship between average biomass density across the landscape and average net erosion.

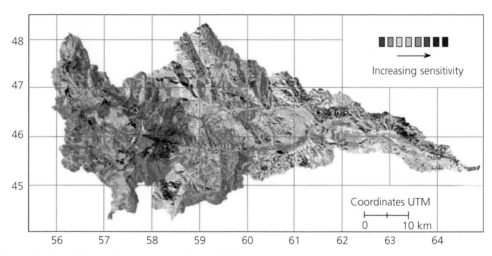

Figure 6.6 Map of the Agri basin showing desertification sensitivity (map courtesy of A. Ferrara).

the locations of the settlements and the larger channels in the landscape. In other words, the amount of sediment lost is a function of both the sediment generated by human disturbance and the connectivity of those sources with pathways of flow across the landscape. This pattern is further demonstrated by the oscillations in the relationship between total biomass and net erosion and by the percentage of the landscape biomass contributed by agriculture and the net erosion rates. Although there is a slight increase in net erosion under this scenario, it is unlikely to be detectable in an actual landscape, apart from under very specific conditions of preservation (e.g. long slopes with no connectivity to the channel system). In scenario 2, the forest cover remains high for most of the 500 years but decreases in the last 50 years or so to an overall cover of 71%. There are still relatively continuous stands of forest, but the larger number of agricultural settlements has produced a more open landscape, which has led to the patches of higher erosion rates being more connected. The net erosion at landscape scale still shows distinctive oscillations relating to the movement of settlements relative to the dominant flow pathways. The link between catchment biomass and net erosion shows multiple values of erosion for any value of average biomass and a highly non-linear relationship. There remains a threshold at about 1% landscape agriculture and net erosion, but the values of erosion are higher and continue to be about an order of magnitude higher than with scenario 1. In scenario 3, there are major differences. After about 200 years, scenarios 2 and 3 diverge, with the far more intensive activity in the latter, reinforced by the more intensive occupation of the landscape, causing the forest cover to be totally lost over a period of about 180 years. Under these conditions, the erosion rate is high across the landscape, and although there are still oscillations in the net erosion rate, they are of much smaller amplitude, suggesting that the connectedness of sources across the landscape is far

more important than the connectivity to different pathways. The relationship between biomass and net erosion shows some similarities with scenario 2, but once the behaviour diverges, there is a further, different, non-linear relationship between low biomass and net erosion rates. At low percentages of agricultural activity (at the start of the simulation), a similar threshold of net erosion occurs, but as agricultural expansion continues to much higher covers, the net erosion increases exponentially, eventually converging to oscillate around a mean rate by between 25 and 30% of the landscape being under agriculture.

Thus, these simulations suggest that it is possible to produce the emergent characteristics of the SIC model without directly incorporating them in the model. Instabilities within the landscape are a function of local conditions, and there are a number of thresholds with respect to this landscape, for example, the pulse of sediment produced once agricultural activity is more than 1% of the landscape biomass. The links between biomass and erosion at landscape scale are masked by the effects of connectivity, which becomes a more dominant factor in affecting the landscape response. Thus, it is important to capture the patterns of change in the landscape as much as the relative changes of specific variables. This pattern-orientated approach to modelling is receiving increasing attention in the ecological literature (e.g. Grimm et al. 2005). The effects of path dependency (e.g. because of the removal of nutrients in agricultural areas leading to changes in infiltration) mean that just because vegetation recovers at a particular point in time, it does not follow that the erosion will return to its original rate. Indeed, in scenario 3, there is a suggestion that the final removal of forest cover leads to a much bigger pulse of sediment than in previous periods of low vegetation in this simulation, suggesting that at high settlement densities, a threshold may be crossed where the landscape is no longer resilient to the human activity.

Conclusions

> The recognition that under a constant or near constant climate very small changes in environment may produce dramatic changes in behaviour implies that, conversely, large changes in behaviour can have many causes other than climate. However, with the adoption of a dynamical approach we may begin to identify the relationships between equilibrium and transient behaviour and to mobilise these in building fresh models of geomorphological evolution.
>
> Thornes 1983, p. 234

The Mediterranean landscape through the Holocene is dominated by human activity. It has shown a series of responses that are difficult to interpret using simple relationships between population and erosion or indeed vegetation and erosion. This chapter has attempted to demonstrate how a more sophisticated conceptual model can be used to interpret the changes that have occurred. The SIC approach takes the dynamical systems theory first expounded by John Thornes for interactions between vegetation and erosion and links them into other aspects of non-linear systems approaches introduced by him and others to the discipline and links these ideas to more recent developments in connectivity theory. The key aspect of SIC is that it suggests that patterns of landscape evolution will be picked out once it moves away from periods of boredom, that is, from stability to instability, and that these shifts occur both on hillslopes and in channels at multiple spatial and temporal scales. The effect of connectivity is to make the moments of terror more or less visible in the subsequent landforms and sediments. Thus, it should be possible to link the narratives of archaeological and historical settlement of the landscape with the narrative of landscape evolution, but recognising the processes that produce the latter is not just those that produce the former. We can thus link the process and historical/evolutionary approaches to geomorphology. The fact that it is possible to simulate the emergent properties of landscape evolution using this approach, in this chapter illustrated using the CybErosion^{eag} model, suggests that there is a strong potential beyond producing 'just-so' stories of landscape evolution.

As with the original vegetation–erosion work of Thornes (1985), there needs to be a recognition that these conceptual models require geomorphologists to develop different modes of investigation, something which they have largely attempted to avoid (Wainwright and Parsons 2010). One of the major difficulties of this sort of approach is in the very definition of what equilibria are or are not. As Scheffer et al. (2001, p. 591) noted in their review of the ecological literature, it has previously been 'easier to demonstrate shifts between alternative stable states in models than in the real world'. Bracken and Wainwright (2006, 2008) have discussed the issue at length in a geomorphological context and suggest that not least because of the complexity of geomorphic systems, it is difficult to make definitions of equilibria on *a priori* grounds. If we do not recognise this underlying complexity and develop complex systems approaches to the discipline, we are likely to be stuck within the (still-dominant) linear perspectives highlighted by the quotation at the head of this paper. We are also likely to be stuck in simplistic, climatically controlled interpretations (see also Wainwright 2006) and unable to contribute to the debate undergoing in relation to the definition of the Anthropocene.

In a broader context, the major landscape changes that may result from relatively small fluctuations around stable conditions due to human activity have major implications for understanding landscape resilience and hence management. The simulations carried out here demonstrate that care is needed in assuming that models can provide deterministic predictions that can be used to underpin management strategies in settings where the underlying processes and their connexions are inherently stochastic. There is a need to look at patterns of change and how those patterns reflect conditions where key thresholds seem to be crossed.

They do emphasise the need for an understanding of place and that thus adaptive management strategies are required to respond to environmental change in a coherent way (e.g. Osbahr et al. 2010). A greater emphasis on the study of the instabilities and connectivity that underpin the breakdown of resilience and allow the definition of better management strategies especially for the extensive and largely impoverished drylands of the world would be a fitting legacy for the advances made by the geomorphological studies of John Thornes.

Acknowledgements

The figures have been redrawn by Chris Orton. I would like to thank (normal caveats notwithstanding) two anonymous reviewers for their constructive comments in contributing to an improved version of this chapter.

References

Ager, D. U. (1973) *The Nature of the Stratigraphic Record.* John Wiley & Sons, Ltd, Chichester.

Alling Gregg, S. (1988) *Foragers and Farmers. Population Interaction and Agricultural Expansion in Prehistoric Europe.* University of Chicago Press, Chicago, IL.

Ballais, J-L. (1996) L'âge du modelé de roubines dans les Préalpes du Sud: l'exemple de la region de Digne. *Géomorphologie: Relief, Processus, Environnement* **4**, 61–68.

Barbier, N. J., Bellot, J., Couteron, P., Parsons, A. J., Mueller, E. N. (2014) Short-range ecogeomorphic processes in dryland systems. In: Mueller, E. N., Wainwright, J., Parsons, A. J., Turnbull, L. (eds.) *Patterns of Land Degradation in Drylands: Understanding Self-Organized Ecogeomorphic Systems.* Springer, Dordrecht, pp. 85–101.

Berger, J-F., Beeching, A., Brochier, J-L., Vital, J. (1994) Le basin de la Valdaine. In: van der Leeuw, S. E. (ed.) *Understanding the Natural and Anthropogenic Causes of Soil Degradation and Desertification in the Mediterranean Basin. Volume 3: Dégradation et Impact Humain dans la Moyenne et Basse Vallée du Rhône dans l'Antiquité (Part I).* Final Report on Contract EV5V-CT91-0039. EU, Brussels, pp. 159–256.

Bracken, L. J., Croke, J. (2007) The concept of hydrological connectivity and its contribution to understanding runoff dominated geomorphic systems. *Hydrological Processes* **21**, 1749–1763.

Bracken, L. J., Wainwright, J. (2006) Geomorphological equilibrium: myth and metaphor? *Transactions of the Institute of British Geographers NS* **31**, 167–178.

Bracken, L. J., Wainwright, J. (2008) Equilibrium in the balance? Implications for landscape evolution from dryland environments. In: Gallagher, K., Jones, S., Wainwright, J. (eds.) *Landscape Evolution: Temporal and Spatial Scales of Denudation, Climate and Tectonics.* Geological Society Special Publication, London, pp. 29–46.

Bracken, L. J., Wainwright, J., Ali, G. A., Tetzlaff, D., Smith, M. W., Reaney, S. M., Roy, A. G. (2013) Concepts of hydrological connectivity: research approaches, pathways and future agendas. *Earth-Science Reviews* **119**, 17–34.

Bracken, L. J., Turnbull, L., Wainwright, J., Bogaart, P. (2015) Sediment connectivity: a framework for understanding sediment transfer at multiple scales. *Earth Surface Processes and Landforms* **40**, 177–188.

Brandt, C. J., Thornes, J. B. (eds.) (1996) *Mediterranean Desertification and Land Use.* John Wiley & Sons, Ltd, Chichester.

Briant, R. M., Gibbard, P. L., Boreham, S., Coope, G. R., Preece, R. C. (2008) Limits to resolving catastrophic events in the Quaternary fluvial record: a case study from the Nene valley, Northamptonshire, UK. In: Gallagher, K., Jones, S. J., Wainwright, J. (eds.) *Landscape Evolution: Denudation, Climate and Tectonics Over Different Time and Space Scales.* Special Publications 296. Geological Society, London, pp. 79–104.

Brown, A. G., Tooth, S., Chiverrell, R. C., Rose, J., Thomas, D. S. G., Wainwright, J., Bullard, J. E., Thorndycraft, V., Downs, P. (2013) ESEX commentary. The Anthropocene: is there a geomorphological case? *Earth Surface Processes and Landforms* **38**, 431–434.

Brunsden, D. (1990) Tablets of stone: towards the ten commandments of geomorphology. *Zeitschrift für Geomorphologie Supplementband* **79**, 1–37.

Brunsden, D., Thornes, J. B. (1979) Landscape sensitivity and change. *Transactions of the Institute of British Geographers NS* **4**, 463–484.

Bull, W. B. (1997) Discontinuous ephemeral streams. *Geomorphology* **19**, 227–276.

Castro, P. V., Chapman, R. W., Gili, S., Lull, V., Micó, R., Rihuete, C., Risch, R., Sanahuja Yll, M. E. (eds.) (1996) *Aguas Project. Palaeoclimatic Reconstruction and the Dynamics of Human Settlement and Land-Use in the*

Area of the Middle Aguas (Almería), in the South-East of the Iberian Peninsula. European Commission, Directorate-General, Science, Research and Development, Luxembourg.

Charley, J. L., West, N. E. (1975) Plant-induced soil chemical patterns in some shrub-dominated semi-desert ecosystems of Utah. *Journal of Ecology* **63**, 945–964.

Courty, M. A., Federoff, N., Jones, M. K., McGlade, J. (1994) Environmental dynamics. In: Castro, P., Colomer, E., Courty, M. A., Federoff, N., Gili, S., González Marcén, P., Jones, M. K., Lull, V., McGlade, J., Micó, R., Montón, S., Rihuete, C., Risch, R., Ruiz Parra, M., Sanahuja Yll, M. E., Tenas, M. (eds.) *Temporalities and Desertification in the Vera Basin, South East Spain. Archaeomedes Projects*, Vol. **2**, pp. 19–84 EU, Brussels.

Gascó, J., Carozza, L., Wainwright, J. (1996) Un petit habitat agricole de l'age du bronze ancien en Languedoc occidental: Laval de la Bretonne (Monze, Aude). Hypothèses et conséquences d'un enfouissement sur la «courte durée» de l'occupation humaine. In: Mordant, C., Gaiffe, O. (eds.) *Fondements Culturels, Techniques, Économiques et Sociaux des Débuts de l'Âge du Bronze, 117ᵉ Congrès national des Sociétés savantes*. Comité des Travaux Historiques et Scientifiques, Paris, pp. 373–385.

Gilman, A., Thornes, J. B. (1985) *Land Use and Prehistory in South East Spain*. George Allen & Unwin, London.

Graf, W. L. (1979) Catastrophe theory as a model for change in fluvial systems. In: Rhodes, D. D., Williams, G. P. (eds.) *Adjustments of the Fluvial System*. Dubuque Kendall/Hunt Publishers, Dubuque, IA, pp. 13–32.

Grimm, V., Revilla, E., Berger, U., Jeltsch, F., Mooij, W. M., Railsback, S. F., Thulke, H. H., Weiner, J., Wiegand, T., Deangelis, D. L.(2005) Pattern-oriented modeling of agent-based complex systems: lessons from ecology. *Science* **310**, 987–991.

Grove, A. T., Rackham, O. (2001) *The Nature of Mediterranean Europe: An Ecological History*. Yale University Press, New Haven, CT.

Guilaine, J. (2003) *De la Vague à la Tombe. La Conquête Néolithique de la Méditerranée*. Seuil, Paris.

Holmgren, M., Scheffer, M. (2001) El Niño as a window of opportunity for the restoration of degraded arid ecosystems. *Ecosystems* **4**, 151–159.

Hooke, R. L. B. (2000) On the history of humans as geomorphic agents. *Geology* **28**, 843–846.

Hunt, C. O. (1995) The natural landscape and its evolution. In: Barker, G. (ed.) *A Mediterranean Valley: Landscape Archaeology and Annales History in the Biferno Valley*. Leicester University Press, Leicester, pp. 62–83.

Illius, A. W., O'Connor, T. G. (1999) On the relevance of nonequilibrium concepts to arid and semiarid grazing systems. *Ecological Applications* **9**, 798–813.

Johnson, A. W., Earle, T. (1987) *The Evolution of Human Societies: From Foraging Group to Agrarian State*. Stanford University Press, Stanford, CA.

Kosmas, C., Danalatos, N., Cammeraat, L. H., Chabart, M., Diamantopoulis, J., Farand, R., Gutierrez, L., Jacob, A., Marques, H., Martinez-Fernandez, J., Mizara, A., Moustakas, N., Nicolau, J. M., Oliveros, C., Pinna, G., Puddu, R., Puigdefabregas, J., Roxo, M., Simao, A., Stamou, G., Tomasi, N., Usai, D., Vacca, A. (1997) The effect of land use on runoff and soil erosion rates under Mediterranean conditions. *Catena* **29**, 45–59.

McNeill, J. R. (1992) *The Mountains of the Mediterranean World: An Environmental History*. Cambridge University Press, Cambridge, UK.

Millington, J. D. A., Wainwright, J. (submitted) 'Through thick and thin: the additive process of agent-based modelling for understanding geography', *Progress in Human Geography*.

Millington, J. D. A., Wainwright, J., Perry, G. L. W., Romero Calcerrada, R., Malamud, B. D. (2009) Modelling Mediterranean landscape succession-disturbance dynamics: a landscape fire-succession model. *Environmental Modelling and Software* **24**, 1196–1208.

Moreno-de las Heras, M., Díaz-Sierra, R., Nicolau, J. M., Zavala, M. A. (2011) Evaluating restoration of man-made slopes: a threshold approach balancing vegetation and rill erosion. *Earth Surface Processes and Landforms* **36**, 1367–1377.

O'Sullivan, D., Millington, J. D. A., Perry, G. L. W., Wainwright, J. (2012) Agent-based models – because they're worth it? In: Heppenstall, A. J., Crooks, A. T., See, L. M., Batty, M. (eds.) *Spatial Agent-based Models: Principles, Concepts and Applications*. Springer, Berlin, pp. 109–123.

Osbahr, H., Twyman, C., Adger, W. N., Thomas, D. S. G. (2010) Evaluating successful livelihood adaptation to climate variability and change in southern Africa. *Ecology and Society* **15**, 27.

Parsons, A. J., Wainwright, J., Powell, D. M., Kaduk, J., Brazier, R. E. (2004) A conceptual model for understanding and predicting erosion by water. *Earth Surface Processes and Landforms* **29**, 1293–1302.

Parsons, A. J., Brazier, R. E., Wainwright, J., Powell, D. M. (2006) Scale relationships in hillslope runoff and erosion. *Earth Surface Processes and Landforms* **31**, 1384–1393.

Perry, G. L. W., Enright, N. J. (2002) Humans, fire and landscape pattern: understanding a maquis-forest complex, Mont Do, New Caledonia, using a spatial 'state-and-transition' model. *Journal of Biogeography* **29**, 1143–1158.

Provansal, M., Bertucchi, L., Pelissier, M. (1994) The swampy grounds in western Provence, indicators of the Holocene morphogenesis. *Zeitschrift für Geomorphologie NF* **38**, 185–205.

Rango, A., Tartowski, S., Laliberte, A., Wainwright, J., Parsons, A. J. (2006) Islands of hydrologically enhanced biotic productivity in natural and managed arid ecosystems. *Journal of Arid Environments* **65**, 235–252

Scheffer, M., Carpenter, S., Foley, J. A., Folke, C., Walker, B. (2001) Catastrophic shifts in ecosystems. *Nature* **413**, 591–596.

Schlesinger, W. H., Reynolds, J. F., Cunningham, G. L., Huenneke, L. F., Jarrell, W. M., Virginia, R. A., Whitford, W. G. (1990) Biological feedbacks in global desertification. *Science* **247**, 1043–1048.

Stewart, J., Parsons, A. J., Wainwright, J., Okin, G. S., Bestelmeyer, B., Fredrickson, E., Schlesinger, W. H. (2014) Modelling emergent patterns of dynamic desert ecosystems as a function of changing landscape connectivity. *Ecological Monographs* **84**, 373–410.

Sullivan, S., Rohde, R. (2002) On non-equilibrium in arid and semi-arid grazing systems. *Journal of Biogeography* **29**, 1595–1618.

Thornes, J. B. (1980) Structural instability and ephemeral channel behaviour. *Zeitschrift für Geomorphologie NF Supplementband* **36**, 233–244.

Thornes, J. B. (1983) Evolutionary geomorphology. *Geography* **68**, 225–235.

Thornes, J. B. (1985) The ecology of erosion. *Geography* **70**, 222–235.

Thornes, J. B. (1990) The interaction of erosional and vegetational dynamics in land degradation: spatial outcomes. In: Thornes, J. B. (ed.) *Vegetation and Erosion: Processes and Environments*. John Wiley & Sons, Ltd, Chichester, pp. 41–55.

Thornes, J. B. (2007) Modelling soil erosion by grazing: recent developments and new approaches. *Geographical Research* **45**, 13–26.

Thornes, J. B., Brunsden, D. (1977) *Geomorphology and Time*. Methuen, London.

Thornes, J. B., Gilman, A. (1983) Potential and actual erosion around archaeological sites in south-east Spain. In: de Ploey, J. (ed.) *Rainfall Simulation, Runoff and Soil Erosion*. Catena Supplement 4. Catena, Cremlingen, pp. 91–113.

Turnbull, L., Wainwright, J., Brazier, R. E. (2008) A conceptual framework for understanding semi-arid land degradation: ecohydrological interactions across multiple-space and time scales. *Ecohydrology* **1**, 23–34.

van Andel, T. H., Runnels, C. (1987) *Beyond the Acropolis: A Rural Greek Past*. Stanford University Press, Stanford, CA.

van Andel, T. H., Runnels, C., Pope, K. O. (1986) Five thousand years of land use and abuse in the southern Argolid, Greece. *Hesperia* **55**, 103–128.

van Auken, O. W. (2000) Shrub invasions of North American semiarid grasslands. *Annual Review of Ecology and Systematics* **31**, 197–215.

Vita-Finzi, C. (1969) *The Mediterranean Valleys: Geological Changes in Historical Times*. Cambridge University Press, Cambridge, UK.

Wainwright, J. (1994) Anthropogenic factors in the degradation of semi-arid regions: a prehistoric case study in Southern France. In: Millington, A. C., Pye, K. (eds.) *Effects of Environmental Change on Drylands*. John Wiley & Sons, Ltd, Chichester, pp. 285–304.

Wainwright, J. (2006) Degrees of separation: hillslope-channel coupling and the limits of palaeohydrological reconstruction. *Catena* **66**, 93–106.

Wainwright, J. (2007) Can modelling enable us to understand the rôle of humans in landscape evolution? *Geoforum* **39**, 659–674.

Wainwright, J., Millington, J. D. A. (2010) Mind, the gap in landscape-evolution modelling. *Earth Surface Processes and Landforms* **35**, 842–855.

Wainwright, J., Parsons, A. J. (2010) Classics in physical geography revisited. Thornes, J.B. 1985: The ecology of erosion. Geography 70, 222–235. *Progress in Physical Geography* **34**, 399–408.

Wainwright, J., Thornes, J. B. (1991) Computer and hardware modelling of archaeological sediment transport on hillslopes. In: Lockyear, K., Rahtz, S. (eds.) *Computer Applications and Quantitative Methods in Archaeology 1990*. BAR International Series 565, Oxford, pp. 183–194.

Wainwright, J., Thornes, J. B. (2004) *Environmental Issues in the Mediterranean: Processes and Perspectives from the Past and Present*. Routledge, London.

Wainwright, J., Parsons, A. J., Schlesinger, W. H., Abrahams, A. D. (2002) Hydrology–vegetation interactions in areas of discontinuous flow on a semi-arid bajada, southern New Mexico. *Journal of Arid Environments* **51**, 319–330.

Wainwright, J., Parsons, A. J., Müller, E. N., Brazier, R. E., Powell, D. M., Fenti, B. (2008) A transport-distance approach to scaling erosion rates: 1. background and model development. *Earth Surface Processes and Landforms* **33**, 813–826.

Wainwright, J., Ibrahim, T. G., Lexartza-Artza, I., Turnbull, L., Thornton, S. F., Brazier, R. E. (2011) Linking environmental régimes, space and time: interpretations of structural and functional connectivity. *Geomorphology* **126**, 387–404.

Whitelaw, T. (2000) Settlement instability and landscape degradation in the southern Aegean in the third millennium BC. In: Halstead, P., Frederick, C. (eds.) *Landscape and Land Use in Postglacial Greece*. Sheffield Studies in Aegean Archaeology, 3. Sheffield Academic Press, Sheffield, pp. 135–161.

Wilkinson, B. H., McIlroy, B. J. (2007) The impact of humans on continental erosion and sedimentation. *Geological Society of America Bulletin* **119**, 140–156.

Willenbring, J., von Blanckenburg, F. (2010) Long-term stability of global erosion rates and weathering during late-Cenozoic cooling. *Nature* **465**, 211–214.

Wise, S., Thornes, J. B., Gilman, A. (1982) How old are the badlands? A case study from south-east Spain. In: Yair, A., Bryan, R. (eds.) *Badland Geomorphology and Piping*. Geo Books, Norwich, pp. 29–56.

Zalasiewicz, J., Williams, M., Haywood, A., Ellis, M. (2011) Stratigraphy of the Anthropocene. *Philosophical Transactions of the Royal Society A* **369**, 1036–1055.

Zangger, E. (1994) Landscape changes around Tiryns during the Bronze Age. *American Journal of Archaeology* **98**, 189–212.

CHAPTER 6

Desertification indicator system for Mediterranean Europe: Science, stakeholders and public dissemination of research results

Jane Brandt and Nichola Geeson

Fondazione MEDES, Sicignano degli Alburni (SA), Italy

Introduction

DESERTLINKS (Combating Desertification in Mediterranean Europe: Linking Science with Stakeholders) was the last in the series of collaborative EU-funded projects that John Thornes coordinated, building on the considerable wealth of data, scientific expertise and detailed local knowledge of study sites throughout Mediterranean Europe that had been accumulated over the previous 10 years or so.

As Kirkby et al. (2015) describe, by the time of the EU's Fifth Framework Programme in 2000 (known as 'FP5'), focus was on supporting options for sustainable development of desertification-affected areas and, importantly, the research programme required projects to include interaction with stakeholders (European Commission 2000). The DESERTLINKS project, started in 2001, worked with a range of stakeholders who had roles in land management of desertification-affected areas including those drawn from local communities, from different levels of political and governmental decision-making and the United Nations Convention to Combat Desertification (UNCCD) national committees. The project developed a dialogue between the scientists and the stakeholders with the aim of providing research results that matched their need for information and that were available in an accessible style and format to a much wider readership than hitherto. The importance of using stakeholder-friendly methods for disseminating research results has increased since FP6, with specific dissemination strategies being a requirement for current projects. At least in the field of European desertification, DESERTLINKS, with its desertification indicator system for Mediterranean Europe (DIS4ME) website, was an important forerunner of the public dissemination undertaken by more recent projects such as DESIRE, LEDDRA and CASCADE, each of which also has websites dedicated to the public dissemination of research issues and results (Table 6.1).

One of the central themes of DESERTLINKS was the identification and use of desertification indicators to promote more general understanding and awareness of desertification issues. The results were published on the website DIS4ME (DESERTLINKS 2004) which was designed to convey complex information about the causes, processes and consequences of desertification in a more easily understood manner, presenting the

Monitoring and Modelling Dynamic Environments, First Edition. Edited by Alan P. Dykes, Mark Mulligan and John Wainwright.

Table 6.1 Websites of publicly available research projects and related resources.

Project	Theme	Website
DESERTLINKS	Identification and use of desertification indicators to promote more general understanding and awareness of desertification issues	www.kcl.ac.uk/projects/desertlinks/ indicator_system/introduction.htm (username: desertlinks; password: dis4me)
DESIRE	A web-based information system of 'best management practices' for responsible land use	www.desire-his.eu
LEDDRA	Provide support to sustainable land management and to responsive policymaking at national, EU and international level	leddris.aegean.gr
CASCADE	'CASCADE will develop ways to predict the proximity of the CASCADE's dryland ecosystems to *thresholds* in such a way that these predictions can be used by policymakers and land users for more sustainable management of drylands worldwide' (website)	www.cascade-project.eu
OLIVERO	The future of olive plantations systems on sloping and mountainous land; scenarios for production and natural resource conservation	http://mp.mountaintrip.eu/uploads/media/ deliverable/olivero_d9.pdf (final project report only; title in previous column)
DESERTNET	'Scientific network for international research on desertification' (website)	www.desertnet-international.org

research in a clear narrative at varying levels of detail and complexity to suit different audiences.

DIS4ME was designed to be a contribution to the work of the UNCCD, particularly in the Annex IV subregion (Portugal, Spain, Italy and Greece). DIS4ME contains around 150 descriptions of indicators derived from (i) the National Action Programmes (NAPs) for Portugal, Spain, Italy and Greece; (ii) over a decade of European research; and (iii) the people who live in desertification-affected areas. The uses of each indicator are fully described using a common format. A database allows indicator selection according to various logical frameworks and temporal and spatial scales.

Within Europe, there are many causes and consequences of desertification, which have wide-ranging local variability. A different set of indicators may be required to understand the bigger picture in each landscape. DESERTLINKS drew on reports and documents from the NAPs and organised a workshop with local stakeholders to identify the main desertification issues for Europe as land abandonment; intensive irrigated farming; overgrazing; deforestation; littoralisation (increasing concentration economic

activities on coastal areas); inappropriate agricultural practices; changes in the economic activity; degradation of the physical environment; availability of water resources; changes in the social structure; and capacity of institutional organisations to combat desertification. Each of these issues is explored separately in DIS4ME in a synthesis of knowledge about desertification processes and how they are known to affect different parts of Mediterranean Europe. On the website, the text is available in Portuguese, Spanish, Italian and Greek, as well as in English.

DIS4ME also contains several examples of indices calculated by combining a number of different indicators. The expert system for evaluating the Environmental Sensitivity Index (ESI) of a local area is calculated from 13 different indicators. It allows the land user to analyse the properties of their land and identify which of the factors under their control could be altered to improve the situation. The expert system and the accompanying method for mapping environmental sensitivity to desertification have already been applied at municipality, national and Mediterranean-wide scales as part of the National and Regional Action Programmes.

This chapter has two aspects. First, it reviews the identification and use of socioecological characteristics and variables that can be considered to be desertification indicators, as presented on the DIS4ME website. Secondly, it reflects on the exercise of developing the DIS4ME indicator system as a website and as an important tool for research dissemination.

DESERTLINKS and DIS4ME

When the DESERTLINKS project was originally conceived in the late 1990s, plans were made to disseminate the results as widely as possible through methods used at that time: a variety of academic publications, an edited book 'Manual on Desertification Indicators', themed journal special issues plus, maybe, a well-illustrated pamphlet aimed at a more general readership. However, as the project progressed, it became clear that there was emerging potential for at least some of these dissemination goals to be met through the use of a website. In particular, information could be written at different levels of detail or complexity to suit the interests of different readers and could also be available in Greek, Italian, Portuguese and Spanish, the languages of our stakeholders.

Therefore, the DESERTLINKS project developed the web-based DIS4ME as part of the contribution by the European Commission to the work of the UNCCD and in particular the Annex IV subregion countries of Portugal, Spain, Italy and Greece. A vital element of the project was the link with stakeholders at the local level to discover their perception of desertification and their needs for indicators. Each country had begun work on their NAPs, so collaboration with the national committees and local stakeholders in four desertification-affected study areas was the first step to establish the specifications for a useful indicator system. Because past research activities had built up a considerable body of information about them, DESERTLINKS focused on study areas in the

Alentejo region of Portugal, the Guadalentín basin of Spain, the Agri basin in Italy and the Greek island of Lesvos.

The completed indicator system, DIS4ME, is available online at http://www.kcl.ac.uk/projects/desertlinks/indicator_system/introduction.htm (username, desertlinks; password, dis4me). Figure 6.1 shows a screenshot of the DIS4ME introduction page. The DIS4ME site is divided into four main sections:

1 *Desertification and DIS4ME*: Providing general information about desertification and an invitation to provide feedback on the contents of the site and the site map
2 *Indicators*: Providing background information about indicators, access to the database of indicator descriptions and descriptions of how subsets of indicators relate to specific desertification issues
3 *Using and combining indicators*: Showing how indicators can be aggregated to map desertification or degradation risk over large areas and providing an assessment of the risk of erosion or salinisation
4 *Linking science with stakeholders*: Describing the work done with the stakeholders and how DIS4ME has been used and evaluated

Indicator database

At the heart of DIS4ME is the database of indicator descriptions. A list of some 220 candidate indicators was compiled, gathering information from the following sources:

• Indicators already used to map desertification at a national scale by the Annex IV Focal Points (MEDRAP 2002). These indicators tend to be those for which data is available at the national scale (such as wildfire incidence, drought index, land-use index).
• All indicators described by Enne and Zucca (2000).
• All indicators described in Kosmas et al. (1999a).
• All indicators listed in Greeuw et al. (2001).

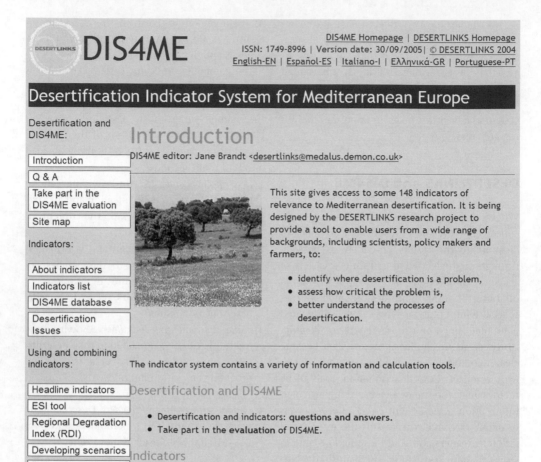

Figure 6.1 Screenshot of DIS4ME website introduction page.

- Indicators suggested by the expert knowledge of members of the DESERTLINKS project. These have come from over a decade of European research into the causes and consequences of desertification in Mediterranean Europe. These indicators range from the Mediterranean-wide scale (vegetation cover from remote sensing; regional degradation index) to the sub-national (employment index; deforested area; effective precipitation) to the plot scale (soil depth; tillage operations).
- Indicators suggested by the expert knowledge of local stakeholders who themselves live in desertification-affected areas (land abandoned from agriculture; fragmentation of land parcels; groundwater exploitation; net farm income).

In addition, a review of other environmental indicator systems was undertaken, including those of:

- Commission on Sustainable Development (CSD)
- Organisation for Economic Cooperation and Development (OECD)
- European Environment Agency (EEA)
- Indicator Report on the Integrated Environmental Concerns into Agricultural policy (IRENA)

- Towards European Pressure Indicators (TEPI)
- Land Degradation Assessment of Drylands (FAO–LADA)
- Agri-environmental indicators for sustainable development in Europe (ELISA)
- Proposal on agri-environmental indicators (PAIS)
- International Institute for Sustainable Development (IISD)

According to the OECD (2001), in order to be useful, an indicator should be specific, measurable, achievable, relevant and time-bound. The EEA suggests that indicators should demonstrate policy relevance and utility for users and analytical soundness as well as measurability (Gentile 1998). It should be cost effective to compile the necessary data and should not include specialist measurements with limited availability. Indicators should also be comprehensible, easy to interpret and suitable for indicating changes over time. With these criteria in mind, an iterative review of the candidate indicator list was conducted. Indicators that could not be described, measured or used in a practical way were abandoned. The exercise also revealed gaps, particularly in the social and institutional areas. The indicator list was optimised, retaining around 150 indicators that were fully described (Table 6.2).

Each indicator is described using the same format as that used by Enne and Zucca (2000), and the description includes:
- Definition
- Target and political pertinence
- Methodological description and basic definitions (including the levels of any relevant benchmarks)
- Evaluation of data needs and availability
- Author
- Bibliography

More powerfully, the indicators are also available in an online database (NRD 2009), allowing the user to select indicators according to various logical frameworks and temporal and spatial scales. Figure 6.2 shows a screenshot of the DIS4ME database website.

Different indicators for different issues

Within Europe, the causes and consequences of desertification and land degradation are manifold, with wide-ranging local variability. For example, a particular issue associated with desertification in the Alentejo is rural depopulation driven by the search of young people for better incomes and standards of living in the cities, which is resulting in an ageing rural population. In the Guadalentín, the issues relate to changing land uses in response to available EU land-use subsidies, resulting in large-scale, easily erodible contouring of hillslopes and planting of almond trees. The land degradation in these two situations has different causes and consequences. A different set of indicators is required to understand the bigger picture in each.

Examination of reports and documents from the NAPs (CIPE 1999; Portugal National Focal Point for Combating Desertification 1999; Greek National Committee for Combating Desertification 2001; MEDRAP 2002) and a survey and workshop in which local stakeholders were asked what their perception of desertification was and how it affected them highlighted the main desertification issues for which indicators were required. Figure 6.3 shows a screenshot of the desertification issues section of DIS4ME.

For each main issue, a number of sub-themes were also identified from the same data and information sources (Table 6.3). They broadly relate to processes or state dynamics affecting the particular issue, reasons for its occurrence (or factors influencing the issue or driving forces and pressures) and its consequences (or impacts and responses). Each of these issues, with their sub-themes, is explored separately in DIS4ME. For example, in the land abandonment section:
- Abandoned land is described as an 'area previously cultivated but where farming or keeping livestock has ceased and the natural vegetation has been allowed to grow under various intensities of grazing'.

Table 6.2 List of desertification indicators described and used in DIS4ME.

Physical and ecological indicators

Climate:	Air temperature; aridity index (1); aridity index (2); climate quality index; drought; drought index; effective precipitation; potential evapotranspiration; rainfall; rainfall erosivity; rainfall seasonality; wind speed
Water:	Groundwater depth (change in); water quality
Runoff:	Area of impervious surface; dam sedimentation; drainage density; erosivity (RDI); flooding frequency; floodplain and channel morphology; rainfall–runoff relationship; runoff threshold (RDI); soil permeability
Soils:	Acidified area; drainage; erosion risk (RDI); infiltration capacity; organic matter in surface soil rs; organic matter in surface soil; organic matter mixing with depth; parent material; rock fragments; salinisation potential; slope aspect; slope gradient; soil crusting; soil depth; soil erosion (USLE); soil erosion (measured); soil loss index; soil quality index; soil structure; soil surface stability; soil texture; soil type; water storage capacity
Vegetation:	Area of matorral; biodiversity conservation; deforested area; drought resistance; ecosystem resilience; erosion protection; forest fragmentation; vegetation cover; vegetation cover rs; vegetation cover type; vegetation quality index
Fire:	Burned area; fire frequency; fire risk; forest and wildfires; fuel models; wildfire incidence

Economic indicators

Agriculture:	Expenditure on water; family size; farmer's age; farm ownership; farm size; forest productivity; fragmentation of land parcels; gross margin index; traditional agricultural products; net farm income; parallel employment
Land management:	Agri-environmental management; fire protection; forest management quality; management quality index; organic farming; reclamation of affected soils; reclamation of mining areas; soil erosion control measures; soil water conservation measures; sustainable farming; terraces (presence of)
Land use:	Area of cultivated and semi-natural vegetation (rs); area of marginal soil used; land abandoned from agriculture; land use evolution; land use intensity; land use type; natural vegetation; period of existing land-use type; Shannon's diversity index; urban sprawl
Cultivation:	Area of hillslope cultivated; fertiliser application; mechanisation index; tillage direction; tillage depth; tillage operations
Husbandry:	Grazing; grazing control; grazing impact; grazing intensity; husbandry intensity
Water use:	Aquifer over exploitation; external water resources; groundwater exploitation; hydrological regulation (artificial); irrigated area; irrigation intensity and seawater intrusion; irrigation percentage of arable land; irrigation potential realised; runoff water storage; water consumption by sector; water leakage; wastewater recycling; water scarcity; water availability
Tourism:	Penetration of tourist eco-labels; tourism contribution to local GDP; tourism change; tourism intensity
Macroeconomics:	Employment index; GDP per capita; accessibility; unemployment rate; value added by sector

Social indicators

Adult education level; depopulation caused by degradation; Gini index; human poverty index; number of technicians with a knowledge of desertification; old age index; population density; population growth rate; public perception of desertification

Institutional indicators

EU production subsidies; hydrological and forestry plans; internal resources mobilisation; Local Agenda 21; NGO contribution; policy enforcement; protected areas; recycled waste; R&D expenditure; river basin management plan; water use policy/law

Composite indicators

Environmental Sensitivity Index (ESI)

For full details of each, see http://www.kcl.ac.uk/projects/desertlinks/indicator_system/indicators_list.htm.

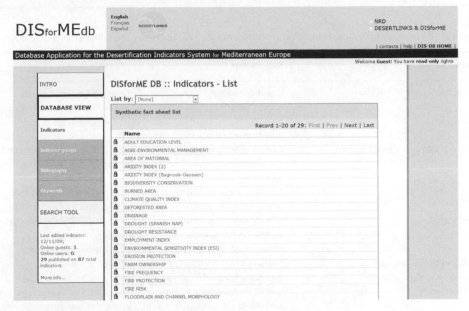

Figure 6.2 Screenshot of online DIS4ME database.

Figure 6.3 Screenshot of the desertification issues section of DIS4ME.

Table 6.3 The most important issues in desertification in Mediterranean Europe identified by DESERTLINKS.

Issues	Sub-themes
Land abandonment	Climatic conditions, soil conditions, water availability, employment opportunities in agriculture and elsewhere, income from land, changing rural population, availability of alternative choices, change in land management, change in vegetation cover
Increase in intensive irrigated farming	Climatic conditions, soil conditions, water availability, income from land, soil salinisation, deterioration of water availability, change in cultivation techniques
Overgrazing	Climatic conditions, number of animals, soil and vegetation conditions, land tenure, EU subsidy policies pre-Agenda 2000
Deforestation	Climatic conditions, drought tolerance of forest, forest destruction by fire, forest productivity, impact of grazing on deforestation, role of forest management, impact of human population, change in erosion risk
Littoralisation (concentration of economic and social activity in coastal areas)	Economy by sectors (coast), GDP inland/coast and rate of change, tourism development (coast), population and rate of change (inland/coast), agriculture development (coast), expansion of artificial areas and tourism settlements in coastal zones, role of planning and land-use policy (coast), employment opportunities (inland/coast), existence of subsidies for economic activities (coast), tourism demand (housing, services, etc.) (coast), tourism and irrigation demand for water (coast), land abandonment (inland), water consumption (coast), soil consumption due to urban expansion (coast), biodiversity loss (coast), water pollution (coast), land/land-use change (inland), sustainable policies
Inappropriate dry farming agricultural practices on marginally productive land	Farm income, including subsidies, income from off-farm employment, land management changes, climate conditions, soil conditions, economic and socio-economic conditions, farming practices, land use, land abandonment, soil degradation, policy enforcement, agri-environmental management
Changes in the economic activity in desertification-affected areas	Climatic conditions, ecosystem conditions, benefits and subsidies, change in farm income, development of tourism, changing rural population, expansion of use of irrigation, exploitation of resources, progressive decline in traditional agriculture, development of new activities (apart from irrigated agriculture)
Degradation of the physical environment	Off-site impacts, changing land use, vegetation cover, soil conditions, control of erosion, fire, water availability, population, biodiversity change, climate conditions, production methods, productivity change, salinisation
Changes in the availability of water resources	Climatic conditions, soil conditions, changes in land use, increase in urban water use, increase in irrigated agriculture, intensification of agriculture, reduction in water reserves and quality, change in flooding frequency, farm income
Changes in the social structure	Changing agricultural system, changing opportunities outside agriculture, changes in the rural population, littoralisation and urbanisation, changing land use
Capacity of institutional organisations to combat desertification	National assessment of desertification risk, national policy and strategy framework, local capacity for combating desertification, local use of best practices

- Information is given on the reasons for land abandonment. They tend to be because the land no longer provides a sufficient income for the farmer, sometimes because the land has become degraded, sometimes because there is insufficient labour to maintain the cultivation structures such as terraces but more often because the traditional cultivation is no longer profitable.
- Examples of reasons for land abandonment in each of the study areas are given. For example, in Portugal, intensive cereal production begun in 1900 caused such extensive soil degradation that by 1950 agricultural productivity

started to decline and there were large-scale population migrations to the cities and abroad. Similar examples are given for the other study areas.
- Finally, an overview of which of the indicators in the database are of particular relevance to land abandonment is given, together with a discussion about how they interrelate. The indicators fall into four groups, relating to soil, climate, socio-economic and management conditions (Figure 6.4).

All the other issues have similar pages, bringing together a synthesis of knowledge about desertification processes and how they are

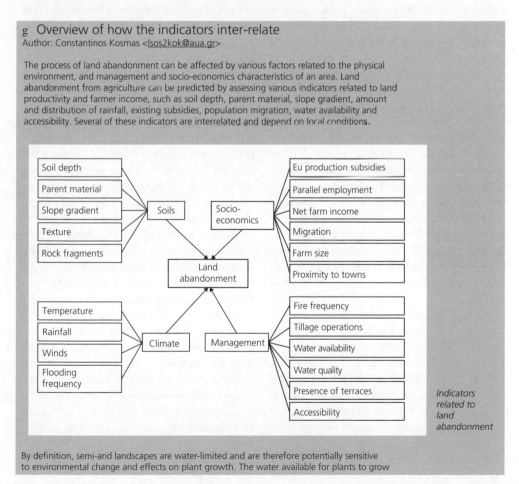

g Overview of how the indicators inter-relate
Author: Constantinos Kosmas <lsos2kok@aua.gr>

The process of land abandonment can be affected by various factors related to the physical environment, and management and socio-economics characteristics of an area. Land abandonment from agriculture can be predicted by assessing various indicators related to land productivity and farmer income, such as soil depth, parent material, slope gradient, amount and distribution of rainfall, existing subsidies, population migration, water availability and accessibility. Several of these indicators are interrelated and depend on local conditions.

Soil depth
Parent material
Slope gradient
Texture
Rock fragments

Soils

Socio-economics

Eu production subsidies
Parallel employment
Net farm income
Migration
Farm size
Proximity to towns

Land abandonment

Temperature
Rainfall
Winds
Flooding frequency

Climate

Management

Fire frequency
Tillage operations
Water availability
Water quality
Presence of terraces
Accessibility

Indicators related to land abandonment

By definition, semi-arid landscapes are water-limited and are therefore potentially sensitive to environmental change and effects on plant growth. The water available for plants to grow

Figure 6.4 Screenshot of the section of DIS4ME discussing the indicators of relevance to land abandonment.

known to affect different parts of Mediterranean Europe. In this way, an integrated approach is brought to each of the desertification issues.

Combining indicators into an index of desertification

Although every area may be affected by different factors, some are clearly more badly affected by desertification than others. Ways of comparing the perceptions of the intensity of desertification issues, desertification risk or the spatial scale of desertification issues would be useful to highlight priorities for action. With some caution, desertification indicators can be used in combination to suggest these priorities. It is easy to become overenthusiastic in the search for a single, desertification indicator or index that is universally applicable. The problem is that as indicators are aggregated, great care has to be taken with interpreting the resulting single number. Clearly, there are indices such as the aridity index (the ratio of annual rainfall to potential evapotranspiration) which are in widespread use and easily understood, but as the number of indicators combined increases, so interpretation becomes less clear. However, within DIS4ME, there are several examples of indices calculated by combining a number of different indicators. These indices have been developed using field data collected from field sites in each of the study areas over 15 years of research.

One of these indices is the expert system for evaluating the ESI of a local area. The ESI has been developed from the method for mapping environmentally sensitive areas that are vulnerable to desertification developed during the MEDALUS III project (Kosmas et al. 1999b). It is calculated from 13 different indicators associated with vegetation, climate, soil and management. They are as follows:

- Fire risk, soil erosion protection, drought resistance, vegetation type and plant cover together determine the *vegetation* quality.

- Rainfall, aridity index and aspect determine the *climate* quality.
- Drainage, soil depth, rock fragments, soil texture, parent material and slope gradient determine the *soil* quality.
- Policy enforcement and land-use intensity determine the *management* quality.

Together, these four qualities determine the ESI. Full details of the methodology are given in DIS4ME.

For any local area, DIS4ME allows the user to calculate the ESI (Figure 6.5). The user selects values for each of the parameters from the drop-down menus and then clicks the 'Evaluate ES Index' button. The results table shows the values of each of the vegetation, soil, climate and management qualities separately as well as the index itself. However, it also provides additional feedback on those indicators that particularly influence the result. For the combination of indicator values in the example in Figure 6.5, the expert system reports that 'The type of vegetation characterised by a very low erosion protection and resistance to drought. Very shallow soil. The arid climate with a very low annual rainfall'. With caution, values of the ESI at different locations can be compared, as a tool not only to determine current desertification status but also to suggest the likely success of mitigation measures. For example, for a degraded area, it might be possible to compare ESI calculations for different land uses or different management options. This expert system is potentially of great interest to land users, enabling them to analyse the properties of their own land and to understand what its sensitivity to desertification is and why and which of the factors under his control could be altered to improve the situation.

The method has already been used to map desertification sensitivity over areas the size of municipalities or river basins. Two brief examples demonstrate this. (i) The Agri basin in Southern Italy is about $1700\,\mathrm{km^2}$. Figure 6.6 shows that areas to the north and east of the basin are particularly susceptible to desertification. The ability of the ESI to analyse, investigate

Complete the table and the System will analyse the Environmental Sensitivity to desertification of your local area

Vegetation			Climate		
Vegetation type	Mediterranean macchia	▾	Mean annual rainfall	280 - 650 mm	▾
Plant cover	low (10 to 40%)	▾	Slope aspect	S, SW, SE	▾
Soil			Aridity index	100-125	▾
Soil Depth	shallow (15-30 cm)	▾	**Management**		
Slope Gradient	gentle (6 to 18%)	▾	Land use intensity	medium (~ sustainable)	▾
Texture	sandy (SC, SiL, SiCL)	▾	Policy enforcement	complete (>75%)	▾
Parent Material	shale, schist	▾			
Drainage	well drained	▾			
Rock Fragments	very stony (>60%)	▾		Evaluate ES Index	

	Quality class		Critical factors, %	Quality score
Vegetation quality	■	Low	49	1.39
Soil quality	□	Medium	22	1.28
Climate quality	□	Medium	51	1.78
Management quality	■	Good	10	1.1

	Sensitivity class	Sensitivity index	Sensitivity score
ES Index to desertification	▨ Area with medium environmental sensitivity (Fragile)	32	1.36

Main risk factors of the area are:
The type of vegetation characterised by high risk of fire and a low plant cover. Shallow soil. The climate with a low annual rainfall, in a south-facing slope.

Figure 6.5 Appearance of the ESI tool page in DIS4ME.

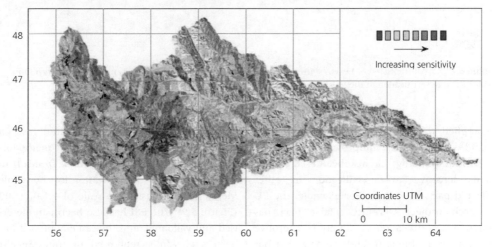

Figure 6.6 Map of the Agri basin showing desertification sensitivity (map courtesy of A. Ferrara). (For colour detail, please see colour plate section).

and identify the causes or sources that contribute to each land unit's score is important. It is possible to compare and characterise areas with the same levels of sensitivity but caused by very different combinations of critical factors. (ii) The Greek island of Lesvos has been mapped under its present conditions (Figure 6.7), showing that the western part of the island is in a critical situation. However, in DIS4ME, the map alternates with another map showing an alternative scenario in which (i) animal grazing of pastures is reduced to a sustainable number according to the land productivity, (ii) olive grove terraces are protected from collapse and (iii) adequate fire protection is given to pine and oak forests. This is an example of how the ESI methodology can also be used to demonstrate that, were such sustainable land uses to

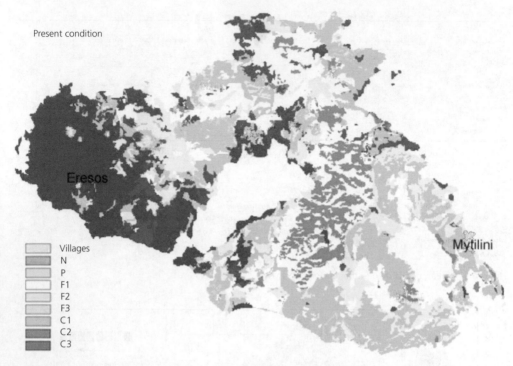

Present condition

Eresos

Mytilini

Villages
N
P
F1
F2
F3
C1
C2
C3

Figure 6.7 Map of the island of Lesvos showing desertification sensitivity under current conditions with most widespread desertification in the west, around the area of Eresos (map courtesy of C. Kosmas).

be applied, the desertification sensitivity on the island would be greatly improved.

This methodology has also been used to map areas sensitive to desertification in other affected parts of Europe. For example, the EU research project OLIVERO (Table 6.1) has mapped desertification sensitivity in the olive grove belt of Chania (Crete), an area that has been selected as pilot area by the Local Focal Point of the Greek National Committee for Combating Desertification. Figure 6.8 shows that most of the area is in a fragile state, and therefore, future land use should be chosen and monitored carefully so as not to exacerbate the particular desertification issues found there.

As a result of the NAP in Italy, administrative regions have been requested to identify their sensitive areas, and many of them have also applied this same methodology. Environmentally sensitive area maps were produced of Basilicata (DESERTNET 2004) and Toscana (DESERTNET 2005) (Table 6.1). The methodology has been used and adapted by DISMED to map sensitivity to desertification for the whole Mediterranean basin at a scale of 1:1,000,000 (Figure 6.9). The ESI has also been parameterised and validated, and the same approach has been successfully replicated in Iran (Sepehr et al. 2007) and in SW Spain (Lavado Contador et al. 2008), and the cost of the necessary data collection and analysis has been assessed (Ferrara et al. 2012).

Stakeholder participation in the development of DIS4ME

The final part of DIS4ME highlighted in this chapter is the section describing how a participatory, as well as a scientific, approach has

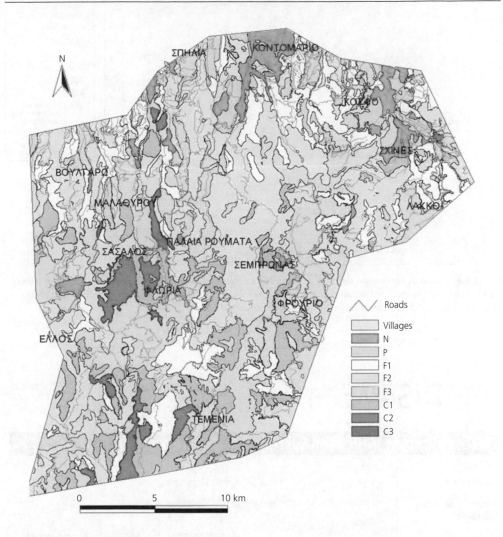

Figure 6.8 Map of part of Crete showing desertification sensitivity (map courtesy of C. Kosmas, N. Moustakas, J. Metzidakis, G. Papathanasiou, D. Kosma, M. Tsara and X. Sergedani).

been taken to developing the indicator system. Stakeholder groups were formed in each of the study areas (Alentejo, Guadalentín, Agri and Lesvos). These stakeholder groups included:

- Representatives from different socio-professional groups and organisations drawn from local communities
- Representatives from different levels of political and governmental decision-making, including representatives of the National Committees to Combat Desertification

- Members of the scientific community in the different fields related to desertification, from the natural and the social sciences

The stakeholders were invited to take part in a series of four workshops which took place over a period of 3 years in each of the study areas. Figure 6.10 shows a screenshot of the section of DIS4ME where the stakeholder consultation workshops are presented:

- The first workshop examined the impact of desertification as it is perceived by the local

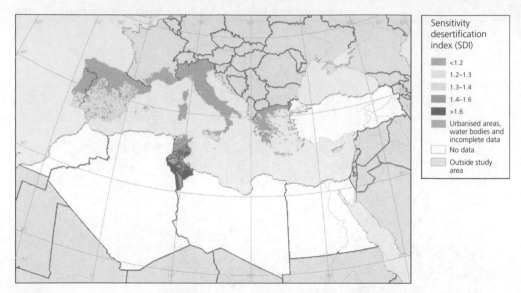

Figure 6.9 Map of desertification sensitivity for the Mediterranean basin showing low values in most of Greece and Italy and in northern Spain and Portugal; mid-range values in Sicily , central and southern Spain and southern Portugal; and high values in Tunisia (map courtesy of EEA).

Figure 6.10 Results of the stakeholder consultation workshops presented in DIS4ME.

stakeholders, and the results were used in the identification of the list of desertification issues.
- The second workshop analysed the effects of different types of land management on land degradation. The results were used in the development of a tool for Agricultural Management Practices Assessment which is also part of DIS4ME (but which is not presented in this chapter).
- The third workshop analysed the factors affecting land-use decision-making in order to identify indicators related to driving forces and pressures imposed on the natural resources. The results were used to revise and adjust indicators included in the database.
- In the final workshop, the stakeholders were given an opportunity to spend some time looking in detail at DIS4ME and to provide feedback on the contents. In order to facilitate this process, the system was translated into Spanish, Italian, Greek and Portuguese.

The feedback from most stakeholders was very positive. Through DIS4ME, some success has been had in promoting awareness of the issues of desertification. The potential has been highlighted for the use of indicators by different groups such as NGOs, community-based organisations and technicians, as well as by individual land users by themselves.

Conclusions and applications of the DIS4ME approach

DESERTLINKS, the last of the series of EU-funded research projects in which John Thornes played a major role, was a pioneer project in several ways. First, it recognised the need to involve a wide range of stakeholders in the science. Traditionally, scientists planned research investigations in relative isolation, and the results were confined mainly to papers in academic journals and reports to funding bodies such as the EU. However, by involving stakeholders, a two-way dialogue was set up, so that researchers could learn from traditional

knowledge and also discuss the practical implications of their research. Therefore, in DESERTLINKS, a sequence of four stakeholder workshops was held with stakeholders, so that at the end everyone would fully understand how the research had developed and how they could use it.

The second advance was to depart from reporting in purely scientific language, in journals and in books, and develop ways of presenting information in less-scientific language. In this way, the results could be understood and used by a much greater range of people. Special effort was required to do this, and the scientists, experienced in writing closely reasoned arguments throughout their careers, suddenly needed to find and use different creative writing skills to deliver their core messages.

The final innovation was to exploit newly emerging media and technologies. By choosing to publish DIS4ME as a website, new options to present material to different users and in different languages became possible. Books and reports have to adhere to a linear sequential structure, but on a website, it is possible to signpost sections more clearly, and readers can be given options to read in more detail, or not. The main advantage of a website over paper is the inclusion of instant and automatic interactive tools. The expert system for evaluating the ESI of a local area is there to be used. It can be played like a simulation game, changing values for different indicators, or it can be used to evaluate and compare real situations.

Although DIS4ME was developed for use in Mediterranean Europe, there are certainly many physical and ecological indicators described in DIS4ME, which would be useful in any desertification-affected environment. The individual indicator descriptions are a step towards consensus in the definition of the indicators, which is an important advance. All users of desertification indicators, especially those using comparison tools, need to have the same idea about what an indicator is actually indicating. For example, farmers know that degraded lands with 'shallow soils' overlying consolidated bedrock cannot be used for productive agriculture, but without a definition, they might not know precisely what is meant by 'water availability'. In addition, the approach that has been taken to breaking down the overall problem of desertification into separate issues or facets, such as land abandonment or water resources, may well be one which is transportable to other areas of environmental science.

Acknowledgements

While Brandt and Geeson are the authors of this chapter, the work reported results from the efforts of the entire DESERTLINKS consortium. The DESERTLINKS project (Combating Desertification in Mediterranean Europe: Linking Science with Stakeholders, 2001–2005) was funded by the European Commission (Framework Programme 5, contract number EVK2-CT-2001-00109), whose support is gratefully acknowledged.

References

CIPE (1999) National Action Programme to Combat Drought and Desertification. Inter-Ministerial Committee for Economic Planning, Italy. Available at: www.unccd.int/ActionProgrammes/italy-eng2000.pdf (accessed 15 April 2013).

DESERTLINKS (2004) DIS4ME: A Desertification Indicator System for Mediterranean Europe. Available at: www.kcl.ac.uk/projects/desertlinks/indicator_system/introduction.htm (accessed 15 April 2013).

DESERTNET (2004) Carta delle Aree Sensibili alla desertificazione (ESAs). Available at: http://oldwww.unibas.it/desertnet/ (accessed 15 April 2013).

DESERTNET (2005) Monitoraggio e azione di lotta alla desertificazione nella regiona mediterranea europea. Programma INTERED 111 B MEDOCC, Asse 4 – Misura A4. Available at: www.ibimet.cnr.it/Case/desertnet/ (accessed 15 April 2013).

Enne, G., Zucca, C. (2000) *Desertification Indicators for the European Mediterranean Region: State of the Art and Possible Methodological Approaches*. ANPA, Roma and NRD, Sassari, 261pp.

European Commission (2000) Work Programme for Energy, Environment and Sustainable Development. Work programme for research, technological development and demonstration under the fifth framework programme. C(2000)3118. Office for Official Publications of the European Communities, Luxembourg. Available at: http://cordis.europa.eu/eesd/calls/a_adv_200001.htm (accessed 19 July 2013).

Ferrara, A. F., Salvati, L., Sateriano, A., Nolè, A. (2012) Performance evaluation and cost assessment of a key indicator system to monitor desertification vulnerability. *Ecological Indicators* **23**, 123–129.

Gentile, A. R. (1998) From national monitoring to European reporting: the EEA framework for policy relevant environmental indicators. In: Enne, G., d'Angelo, M., Zanolla, C. (eds.) *Proceedings of the International Seminar on Indicators for Assessing Desertification in the Mediterranean*, 18–20 September 1998. ANPA (the Italian Agency for the Protection of the Environment), Rome, pp. 16–26.

Greek National Committee for Combating Desertification (2001) Greek National Action Plan for Combating Desertification (Extended Summary). Available at: www.unccd.int/ActionProgrammes/greece-eng2001.pdf (accessed 15 April 2013).

Greeuw, S., Kok, K., Rothman, D. (2001) MedAction Deliverable 1. Factors, Actors, Sectors and Indicators: The Concepts and Application in MedAction. Internal report of the MedAction Project I01-E004. International Centre for Integrated Assessment and Sustainable Development, Maastricht University, Maastricht.

Kirkby, M. J., Bracken, L., Brandt, C. J. (2015) John Thornes and desertification research in Europe. In: Dykes, A. P., Mulligan, M., Wainwright, J. (eds.) *Monitoring and Modelling Geomorphological Environments*. John Wiley & Sons, Ltd, Chichester, pp. 317–326, this volume.

Kosmas, C., Kirkby, M. J., Geeson, N. (1999a) The MEDALUS Project: Mediterranean desertification and land use. Manual on key indicators of desertification and mapping environmentally sensitive areas to desertification. European Commission Project Report, EUR 18882. Office for Official Publications of the European Communities, Luxembourg, 87pp.

Kosmas, C., Ferrara, A., Briassoulis, H., Imeson, A. (1999b) Methodology for mapping Environmentally Sensitive Areas (ESAs) to desertification. In: Kosmas, C., Kirkby, M., Geeson, N. (eds.) *The MEDALUS*

Project: Mediterranean desertification and land use. Manual on key indicators of desertification and mapping environmentally sensitive areas to desertification. European Union Project Report 18882. Office for Official Publications of the European Communities, Luxembourg, pp. 31–47.

Lavado Contador, J. F., Schnabel, S., Gómez Gutiérrez, A., Pulido Fernández, M. (2008) Mapping sensitivity to land degradation in Extremadura, SW Spain. *Land Degradation and Development* **20**, 129–144.

MEDRAP (2002) Concerted action to support the northern Mediterranean Regional Action Programme to combat desertification. Internal Report on the Second MEDRAP Workshop, 6–8 June 2002. Nucleo Ricerca Desertificazione, University of Sassari, Sassari.

NRD (2009) DIS Database: A Database Application Developed by NRD in Collaboration with the DESERTLINKS and LADA Projects. Available at: http://dis-nrd.uniss.it/index.php?_mod=view&_section=intro (accessed 15 April 2013).

OECD (2001) *Environmental Indicators for Agriculture, Volume 3: Methods and Results.* OECD, Paris.

Portugal National Focal Point for Combating Desertification (1999) National Action Programme to Combat Desertification. Available at: www.unccd.int/ActionProgrammes/portugal-eng1999.pdf (accessed 15 April 2013).

Sepehr, A., Hassanli, A. M., Ekhtesasi, M. R., Jamali, J. B. (2007) Quantitative assessment of desertification in south of Iran using MEDALUS method. *Environmental Monitoring and Assessment* **134**, 243–254.

CHAPTER 7

Geobrowser-based simulation models for land degradation policy support

Mark Mulligan

Department of Geography, King's College London, London, UK

Introduction

In this chapter, I examine some of the scientific benefits of modelling and the role of environmental modelling in supporting decisions around the implementation of policy. I examine the roots of current policy support systems (PSSs) to be found in some of the early models developed for the MEDALUS projects. I trace the history of desertification modelling in the EU from MEDALUS onwards and highlight the increasing emphasis on the role of models in policy support. Finally, I discuss some of the characteristics of the current breed of models targeted at policy support and highlight their relationship with the early Thornes models before defining some of the key ongoing challenges for policy-relevant modelling.

PSSs

What is policy support?

PSSs are extensions of the ubiquitous and highly variable decision support systems (DSS) that can range from simple flowcharts to sophisticated Geographic Information Systems (GIS)-based simulation tools. DSS are usually intended to assist decision-making around a specific issue, such as whether or not to implement a specific land management intervention (e.g. where to install check dams, where to permit irrigation, where to build terraces). PSSs, on the other hand, assist decision-making around the design of (often much broader) policies such as adaptation of agriculture to climate change, land-use planning or land-use incentive schemes, all of which may include a range of individual management actions.

DSS and PSS are usually targeted at technical assistants to policymakers and form only part of the information input to the policymaking process. They are not designed to 'advise' on which policy to adopt but rather to act as digital test beds to understand better the likely implications of adopting various policies and, thus, add to the weight of evidence in favour of, or against, a particular policy. The availability of a digital test bed is particularly important if the policy involves a variety of landscapes, ecosystems, socio-economic activities and stakeholders, which may prevent the usual expert evaluations and conceptual scenario analyses from identifying all positive and negative outcomes of a particular policy. The policy may thus yield unintended and unhelpful surprises upon implementation. Where landscapes are spatially heterogeneous and/or temporally variable, as is

Monitoring and Modelling Dynamic Environments, First Edition. Edited by Alan P. Dykes, Mark Mulligan and John Wainwright.
© 2015 John Wiley & Sons, Ltd. Published 2015 by John Wiley & Sons, Ltd.

very much the case for the Mediterranean, the implications of policy implementation can be very difficult to trace conceptually as the observed variability yields complexity and scale dependence in outcomes. Spatially explicit, data-based simulation tools can help the handling and communication of the outcomes of such complexity over particular administrative or biophysical regions. Such tools combine relatively generic rules for the operation of biophysical and socio-economic processes with highly specific, spatially explicit data on biophysical and socio-economic properties. PSSs thus enable individual learning and co-learning of stakeholder groups and can also provide project-specific advice for the implementation of more robust and better-tested policy.

What policies?

Careful land-use and land management policy is fundamental to sustainable agriculture and development in marginal environments such as those found in the Mediterranean. The environment of the Mediterranean results from its semi-arid to sub-humid climate that is highly seasonal for both solar radiation (and thus temperature) and for rainfall. The long, hot and dry summer 'drought' followed by heavy October rainfall provides both a constraint to vegetation production and conditions that favour soil erosion and degradation. Over the long term, the Mediterranean climate has combined with geology to create rugged, highly dissected landscapes with sparse vegetation and poor, stony soils. Traditional agriculture is well adapted to this environment through sparse cropping of hardy species (almond, *Prunus dulcis*; olive, *Olea europaea*; fig, *Ficus carica*) combined with careful management of water through runoff capture, mulching of soils, terracing and occasional irrigation. Modern methods which include larger fields, mechanised ploughing and soil movements (Ramos and Mulligan 2005; van Wesemael et al. 2006) and intensively irrigated crops including greenhouse

vegetables, rice, wheat and citrus may not be so well adjusted to both the seasonal water shortages and inter-annual and longer-term climate variability of the Mediterranean.

Agricultural change in the Mediterranean is driven by (i) complex changes in population distribution (especially urbanisation); (ii) demography and socio-economic change, including the development of the service sector and of tourism; (iii) changes to transportation costs; (iv) subsidies and markets in Europe and beyond; and (v) technological change. Change in agricultural policy is not often tested for long-term sustainability in the face of soil degradation and long-term climate variability, and this has the potential to lead to severe degradation that may see short-term economic gains give rise to long-term losses. To be of real benefit, land-use strategies and incentives for the Mediterranean must be sustainable and profitable in the long term, and land management options must also be effective for the long term. The impacts of both land-use and land management strategies must also be understood for the range of socio-economic and biophysical conditions that exist in the region of influence of the proposed policy, since the response will not be uniform across variable landscapes and those applying policy should not assume that it will be.

Policies that need to be tested in this way include (i) investment or incentivisation of particular crop choices or land management techniques including irrigation, terracing, contour ploughing, ploughing and slope reforming; (ii) infrastructural investments such as dams, water transfers and desalinisation facilities; and (iii) conservation and protection schemes such as designation of protected areas, deforestation, buffer strips and check dams. All of these have complex interactions with biophysical processes and, thus, stores and fluxes of water, sediment, soil and plant productivity and on human behaviour such as crop choice and land management. Over spatially and temporally explicit landscapes, the only effective way to manage such complexities is through the application of data and process-based, spatially explicit modelling (Mulligan 2009).

What types of support?

A number of well-defined mechanisms exist to better understand the likely impacts of interventions. These mechanisms include environmental impact assessments (EIA; Holder 2004), environmental and social impact assessments (ESIA) and strategic environmental assessments (SEA), which are routinely employed at corporate or governmental levels in advance of major infrastructure projects. Processes such as integrated watershed management (Heathcote 2009), schemes such as the EU environmentally sensitive area (ESA) scheme and other such mechanisms are also directed towards the careful management of resources in complex circumstances. These all contribute to understanding and minimising negative impacts while making the most of opportunities. None provide data and understanding for scenario analysis of the proposed intervention, but all provide a legal, institutional or technical framework to facilitate such analysis to take place. The key role of PSS is to provide scientific support and to help bridge science with its application to policy. As such, PSS can contribute to many of these mechanisms and other policy-development or intervention-assessment processes. The support provided by the types of process-based, spatially explicit PSS discussed here may include:

(a) *Spatial targeting* of suitability for particular land-use or land management strategies

(b) *Analysis of long-term* land, water or production sustainability for particular crops or land management strategies

(c) *Analysis of* land, water or production *sustainability* given impacts of *climate change*

(d) *Integrating* the impacts of a range of *concurrent interventions* or understanding *runaway adoption* of a particular policy or intervention

(e) *Choosing* between particular interventions such as conservation versus land management approaches

(f) *Quantifying and negotiating* the sharing of common resources such as water

The role of modelling in policy support

Modelling is an important basis for policy support since it makes explicit our understanding of processes (represented as model equations) and couples them with spatio-temporal data sets representing the system state. Modelling is thus potentially a robust, negotiable and explicit abstraction and representation of the system under study and hence a potentially excellent framework for communication and analysis. Modelling can help bridge the gap between science and policy by providing summaries of complex processes and communicating their outcomes as maps or charts. It can also reach beyond simple mapping and GIS overlay to provide dynamic scenario analyses indicating system change over time and space in response to the operation of processes over a landscape and population. Modelling offers at least the following benefits to understanding the environment in policy-relevant situations:

Simplification – Models are simplifications of complex systems. Careful abstraction and conceptualisation can capture the important elements of a dynamic system while ignoring the less relevant details. Making the right assumptions here can simplify a policy problem without ignoring important elements.

Quantification – Models produce numbers. The precision and objectivity of modelling can be of great value in complex policy problems where expert opinion disagrees or where understanding magnitudes is critical (e.g. the potential of check dams to reduce sedimentation into a reservoir and the cost–benefit comparison of check dams vs. sediment dredging). Uncertainties can be rendered explicit and thus communicated.

Integration – Models can integrate across heterogeneous spaces, across variable time horizons, across processes and between disciplines and institutions. They are thus an important tool for bringing highly reductionist science into a more holistic realm that can be used for improved understanding of complex and interdependent social–environmental systems.

Communication – Science is difficult to communicate, even between scientists. Communications between scientists and policy advisors, who may or may not be scientists themselves, can be very difficult – in both directions. To design models, the problem being modelled needs to be very clearly specified. Modelling can therefore act as a means of producing a very clear specification for the problem being addressed. Moreover, modelling outputs can be highly graphical either as charts or as maps, and these can be very effective tools for communicating outcomes of scenarios for change or impacts of policy interventions.

Unfortunately, despite these positive characteristics, most models seem to be regarded as 'black boxes' that actually hide the basis for their results in mathematical mystery and technological magic. This attitude arises in part because most models have been developed for use only by modellers and sometimes very crudely interfaced for use by others largely in order to attract funding as policy-relevant research. In part, this may also result from the highly specialist nature of both mathematical modelling and computer programming, coupled with poor or vague documentation and a lack of training. In many cases, models are not 'user-friendly', they are very data demanding, they do not tackle the problems of interest to potential users or they are not validated nor trusted by users. Taken together, these factors have led to a slow and limited uptake of simulation models in policy support, despite their clear potential. The issue of land degradation and desertification is a case in point.

Models have been developed to improve understanding of this policy-relevant issue since at least the early 1990s, but there is still relatively little uptake of these in policy support today. This lack of uptake reflects the fact that from the early 1990s to around 2000, the models built were intended as science models, not as policy models (see discussion on the difference in Mulligan 2009), and were not meant for application by anyone apart from the model developers. Rather, most models were originally considered as tools for the scientists themselves to better understand the systems being modelled. However, since the 1990s, modelling of Mediterranean land degradation has become more and more focused on providing policy support, as outlined in the succeeding text.

A timeline of models for understanding Mediterranean land degradation

Science models

John Thornes was instrumental in leading the development of modelling of desertification both through his own conceptual modelling (e.g. Thornes 1990; Thornes and Brandt 1993) and through his leadership of a number of projects which developed some of the first detailed, physically based models for the Mediterranean (see Kirkby et al., 2015). These projects included the MEDALUS series of projects 1991–1999. MEDALUS I developed and applied models for soil erosion at the hillslope scale (Thornes et al. 1996), MEDALUS II combined these soil erosion models with models for vegetation growth and development as well as soil hydrology and MEDALUS III culminated in the development of the MEDRUSH model for forecasting hydrology, soil erosion and sediment yield in catchments of up to $2000\,km^2$ in a spatially explicit fashion using the GRASS GIS (Kirkby et al. 2002). This series of projects provided data sets, experimental results, maps and some of the first sophisticated models to deal with these issues. The resulting science base was fundamental to the next series of projects which built upon these efforts to interface and adapt models for use in policy. This was driven by increasing pressure from the European Commission to connect its science with its policy agendas.

Moving towards policy models

The MODMED project (1995–1999: Mazzoleni and Legg 1998) built predictive models for understanding natural vegetation dynamics and their response to changes in grazing and burning pressure resulting from developments in agricultural policy. The project was led by Stefano Mazzoleni, Istituto di Botanica, Università di Napoli, and continued the effort to develop and apply policy-relevant modelling that started in the MEDALUS projects. The MODULUS project (1998–1999), led by the Research Institute for Knowledge Systems (RIKS), took models and data sets developed in a range of previous EC projects (in particular EFEDA 1988–1991 (Bolle 1995), MEDALUS and MODMED 1995–1999) and combined them in an integrated model for climate, hydrology, plant growth and vegetation development, soil erosion and socio-economic processes as an operational DSS aimed at regional policymakers and applied in the Marina Baixa, SE Spain, and the Argolid of Greece (Oxley et al. 2004). The hydrology, crop growth, climate and erosion models in MODULUS were based on the PATTERN model (Mulligan 1996, 1998) developed under the PhD supervision of John Thornes alongside the EFEDA project.

MODULUS was largely a technical experiment that took existing models that had been developed in previous projects, left them in their native code and wrapped them with a software 'layer' to permit their integration as a single model capable of running simulations for scenarios of land use and climate change and provided results as graphics and maps. This approach was indeed closer to the description of 'policy model' described by Mulligan (2009) but was very cumbersome because the models came

from different backgrounds and philosophies, worked at different time resolutions and spatial scales and thus interacted inefficiently. Passing parameters correctly between the modules was a technical integration challenge, and the MODULUS simulations were slow to perform and difficult to apply. Nevertheless, this project showed that models developed for scientific purposes could be integrated to tackle policy-relevant questions and then, supported by a sophisticated but user-friendly interface, approach the requirements for policy application.

MedAction (2001–2004) further developed the MODULUS DSS into a PSS capable of advising policy. This development required the engagement of end users at the start of the project, the definition of clear policy goals with them and the careful re-engineering of models from the bottom up to produce an integrated model. The models were re-conceptualised and recoded for integrated operation and for operation without the need for intensive scientific field-parameterisation campaigns such as were available to EFEDA and MEDALUS field sites for example. The PSS was applied to land-use and water management issues in the Guadalentín basin (SE, Spain) (van Delden et al. 2007). The MedAction PSS thus became a very sophisticated spatio-temporal PSS that incorporated an intuitive and responsive interface using RIKS' *Geonamica* system. Since end users were engaged from the start of the project, the system better linked the science with the policy needs than previous efforts were able to achieve. A range of users were trained in the use of MedAction, but its uptake and use remain small, which is attributed by van Delden et al. (2007) to potential failings in one or more of the following requirements for successful uptake of PSS:

(a) Strategy: Is the system useful?
(b) Availability: Are the system and the data needed to apply it available?
(c) Credibility: Is the output verified and trustworthy?

(d) Language: Does the system fit with the users' information needs and the available data?

(e) Culture: Is there willingness to adopt PSS in the decision-making process?

(f) Structure: Who will work with the system and what is their role in the organisation?

In the DESURVEY project (2005–2010), the MedAction PSS was further developed for application across Europe and North Africa as the DESURVEY surveillance system. It was enhanced through the incorporation of multi-scale modelling with a simpler model, based on PESERA (Kirkby et al. 2003), being applied for application at the pan-European scale (1 km resolution) and a more complex and data-demanding model based on PATTERN (Mulligan 1998), as implemented in MedAction, used for application for local-scale analyses (at 1 ha resolution). New process descriptions, policy options and data sets necessary for application to North Africa were also developed alongside much more extensive training and application throughout Europe.

Barriers to uptake in policy support

The period 1991–1999 clearly saw significant advances in land degradation science and in the development of spatial environmental modelling within the context of Mediterranean land degradation. Subsequently, 1998–2010 saw significant scientific, technical and outreach developments in the application of models as fledgling PSSs. It is therefore disappointing to note that uptake of these tools is still limited, despite these significant efforts. This limitation is widely acknowledged for other types of DSS in the literature (Jakeman et al. 2006; McIntosh et al. 2011; van Delden et al. 2011). Taking the headings of van Delden et al. (2007) and applying them to the DESURVEY surveillance system, there follows an examination of the barriers that still remain to the uptake of these PSS in Mediterranean land degradation policy:

(a) *Is the system useful?* Yes it is but it can only be applied for the catchments in which the project has worked and provided data. At the coarse scale, the project worked across all of Europe, but for finer scales, data for only a few catchments are available.

(b) *Are the system and the data needed to apply it available?* The system is available for download and installation on application for a free licence from RIKS (www.riks.nl), and though the system has been designed to work with generally available data, providing such data in the form in which the PSS can use it is a considerable technical task.

(c) *Is the output verified and trustworthy?* Both MedAction and DESURVEY are sophisticated multi-component models. Their general operation and outputs have been verified as reasonable through testing, but only some of the model components have been validated against measured data (as is the case for most models) and then only for the catchments in which the systems have been applied and where such validation data exist.

(d) *Does the system relate to the users' information needs and available data?* They are intended to do so and have been designed with end-user input but only from the limited number of end users that could be engaged in the MedAction and DESURVEY research projects.

(e) *Is there willingness to adopt PSS in the decision-making process?* There was certainly willingness from certain end users to engage in the process of development of such systems. Whether this translates to adoption depends upon the willingness of those users to move the systems forward, the level of post-project support available from system developers and the users' role in the policymaking institution.

(f) *Who will work with the system and what is their role in the organisation?* For MedAction and DESURVEY, the users will tend to be scientists based at national institutes or GIS technicians supporting policy development at local to national scales in governmental and non-governmental contexts.

It is clear that many of the remaining barriers to the uptake of PSS are difficult to remove.

These include the issue of uncertainty (point (c)) which will only become more important by further uptake, application and development of the tools and the data sets upon which they rely. Questions of willingness to adopt (e) and institutional role (f) are dependent upon the individual and institutional incentive to use the system, which is itself a function of the quality of the system as expressed in (a)–(d) and other incentives or benefits of use, for example, for personal development or employability. Those barriers which can more readily be removed – at least partially – include (a) strategy or utility and (b) availability. By removing these barriers and thus extensively increasing the potential user base, model developers should achieve improvements in credibility (c) and language (d) as well as a user-driven increase in willingness to adopt (e) and clarification of PSS role in the policymaking process (f).

The State of the Art (2011)

Since 2009, members of the Environmental Monitoring and Modelling Group at King's College London (a research group originally set up by John Thornes) have worked hard to remove barriers related to (i) strategy and (ii) availability. Improving availability means making both the software and the data that it requires more accessible. For many users, the negotiation of licences, the download and installation of software and the parameterisation of models with local data are subject to very significant technical, institutional and capacity overheads and barriers. Solutions to these include (i) making models available as web-based models that can be run through a WWW browser and thus available to anyone irrespective of technical capacity and (ii) building self-parameterising models, that is, developing models that can be run with existing data sets that are provided globally, thus providing with the model all the data required for model application anywhere globally while giving the user

the capacity to add their own data if they have better or different data to those provided by the developer.

These two changes have the potential to dramatically increase the user base of a model and facilitate model-user communication in the long term since (i) modellers can examine web-usage logs to understand how their web-based model is being used; (ii) modellers know who is using the system and can interact with and support them; and (iii) any updates to the system or data are immediately available to all users without needing to re-download and reinstall them. Having a large number of users and keeping in contact with them for the long term, as well as being able to update software frequently and easily, can help in the process of making PSS more useful since users – rather than modellers – can continue to influence the development trajectory of the model.

WaterWorld

WaterWorld (Mulligan and Burke 2005; Mulligan 2013; WaterWorld 2014) is a spatial PSS focused on better understanding of the hydrological and soil erosion baseline of an area and the impacts of land use, land management and climate change on that baseline. Like most models, *WaterWorld* is not an entirely new model but rather has a long heritage. Some of the equations are based on the PATTERN model developed alongside the EFEDA project under the PhD supervision of John Thornes (Mulligan 1998) and later incarnations in the MODULUS (www.ambiotek.com/modulus) (Oxley et al. 2004), MedAction (www.ambiotek.com/medaction) (van Delden et al. 2007) and DESURVEY (www.ambiotek.com/desurvey) (van Delden et al. 2008) PSSs, but many of the equations were brought together for the FIESTA model (Mulligan and Burke 2005; Bruijnzeel et al. 2011) then modified for the AguaAndes model (Mulligan et al. 2010a, b) and finally connected with global databases for *WaterWorld*.

WaterWorld is applicable in the Mediterranean but also anywhere else and has been built as a self-parameterising, web-based model in order to overcome the availability and utility barriers discussed. To facilitate the building of self-parameterising models, a significant effort had to be invested in the collection, processing, integration and harmonisation of the required input data, globally. The resulting database of raster data sets at standardised 1 km and 1 ha spatial resolutions (referred to as SIMTERRA) is the foundation for *WaterWorld* and a number of other decision, policy and negotiation support systems developed within our web-based PSS framework (referred to as ECOENGINE).

WaterWorld was developed to reduce further the barriers to the use of spatial modelling in policy support. It is intended to meet the need for spatial hydrological baselines and understanding of the impact of land use and climate change in environments throughout the world. It works by spatial application of a series of process-based sub-models for key hydrological and soil erosion processes (see Figure 7.1) to supplied data sets describing climate, landscape, vegetation and hydrology. The system can be run at national scales at 1 km spatial resolution and at local scales at 1 ha resolution. For the 1 ha resolution runs, only the terrain data sets are native 1 ha resolution and all other data sets are interpolated from data sets at resolutions varying between 1 ha and 1 km. *WaterWorld* uses a raster spatial data model and a monthly time step. It represents the current environmental baseline using mean climate data from 1950 to 2000 and terrain, land cover and socio-economic data from around the year 2000. All data required to run simulations are supplied globally, and the PSS produces a range of outputs that can be visualised online using ubiquitous and user-friendly geobrowsers such as Google Earth and Google Maps or can be downloaded for GIS analysis. Geobrowsers have become a familiar and useful tool in policy support, especially in data-poor situations. The mapped data, high-resolution imagery (which can be better than locally available data) and ease of use of these systems mean that they are heavily used

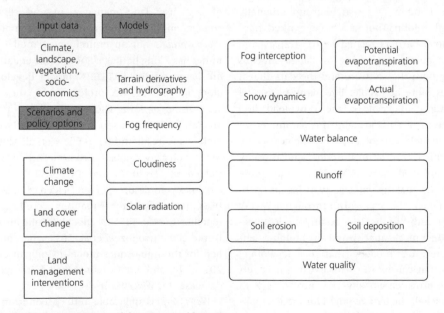

Figure 7.1 Main components of the WaterWorld system.

and well known. Using geobrowsers as the visualisation platform for sophisticated PSSs was thus an obvious choice because of their familiarity to a broad audience, their well-developed Application Protocol Interfaces (APIs) and their wealth of inbuilt map functionality.

WaterWorld has some 1000 registered users as of August 2013, around 60 of which have used the system more than 10 times. Users are from all over the world but particularly Latin America, North America, United Kingdom and Europe. The system is often used simply to assess the current baseline (water balance, soil erosion) but is also often used in scenario analysis for understanding the implications of climate change, land-use change or land management practices. Three case study applications are examined later, using *WaterWorld* Version 2.2, as a means of introducing the capabilities of the PSS.

There are five stages to using *WaterWorld*. Stage 1 involves defining the analysis area. Currently, users can choose 10° tiles at 1 km resolution or 1° tiles at 1 ha resolution. Stage 2 involves preparing the data. During this stage, *WaterWorld* interrogates the SIMTERRA databases and extracts and then prepares the data required for the simulation in the chosen tile. Stage 3 is to run the baseline simulation. Stage 4 is to apply any policy options or scenarios and then rerun the simulation with these changes so that their impact can be analysed relative to the baseline. The final stage is to examine the results as maps, charts or a model-generated narrative of outcomes. Full model documentation can be found online with the model (Mulligan 2013; WaterWorld 2014).

The baseline

Here, I run a baseline simulation at 1 km spatial resolution for the 10° tile centred on coordinates 35.0°N, −5.0°W (covering southern Spain, Portugal, Morocco and parts of Algeria). Figures 7.2 and 7.3 show typical map outputs of water balance and erosion and give an indication of the highly visual and dynamic nature of the system. The system outputs some

32 variables as mean annual values and a further 14 variables monthly.

Scenario analysis (impacts of climate change)

Here, I use the previous simulation as the baseline for a climate change scenario. A scenario is an outside influence on the system that is not under the direct control of the decision-maker. Climate change scenarios can either be chosen (i) from a range of around 175 IPCC assessment, scenario and GCM combinations (all downscaled to 1 km resolution), (ii) by choosing the mean of all GCMs for a given IPCC scenario (the ensemble mean method), (iii) by uploading user-defined scenario maps or (iv) by specifying a scenario simply as arbitrary seasonal changes in precipitation and temperature. All scenarios are applied in equilibrium rather than transient mode, with the 1950–2000 period used as the baseline against which the scenario is applied using the delta method at 1 km resolution. The high-resolution downscaling is necessitated by the spatial variability of climates in mountain regions. For this example, I selected IPCC AR4 A2a scenario, 2050s, ensemble mean of all 17 GCMs available. Figure 7.4 is the system output for the change in mean annual water balance by administrative region and indicates significant drying expected for southern Spain and northern Morocco but wetting for other areas including southern Algeria. The figure also shows the seasonal progression of the water balance for the baseline (right) and the scenario (left) and indicates approximately uniform change in the water balance seasonally on a tile-average basis – although there may be greater differences regionally if regions within the tile were to be examined.

Policy options

A policy option is a policy or intervention that is under the control of the decision-maker and can thus be applied to change the socio-environmental system for the better. *WaterWorld* allows users to examine the impacts of such

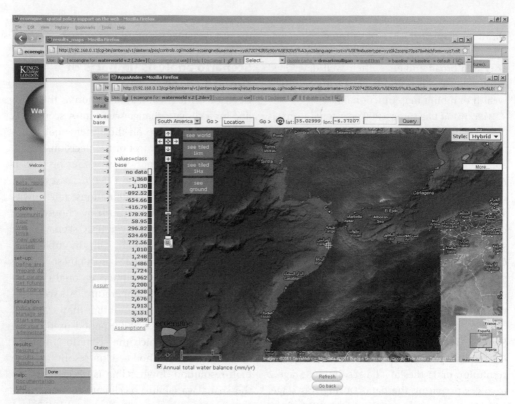

Figure 7.2 Annual total water balance (mm/yr), indicating areas of positive balance in the northern Mediterranean and mountainous regions and areas of negative water balance (local actual evapotranspiration maintained by flows from upstream and/or groundwater in southern Morocco and Algeria). Background map from Google Maps. Image shown within the context of the PSS interface.

policy options relative to the baseline situation. Here, I zoom in to the local scale of the 1° tile centred on 37.5°N, –2.5°W covering the Guadalentín basin (although a *WaterWorld* user could choose anywhere) and run a 1 ha resolution baseline simulation against which two policy options will be examined. The first policy option examines deforestation in the Sierra de Baza protected area (protected area number 20946 in the World Database of Protected Areas) near to Guadix. In the PSS, I convert a randomly assigned 75% of land within the protected area and where the slope gradient <30° from its original cover to agriculture (represented as per pixel: tree cover 10%, herbaceous cover 50%, bare soil 40%) and examine the implications. This scenario leads to an increase

in tree cover of up to 10% in some parts of the reserve and a decrease of up to 40% in others, while herbaceous cover increases by up to 17% in a few areas but decreases by up to 50% elsewhere in the reserve. Overall, bare soil increases throughout the reserve by up to 40%. These changes lead to complex changes in water balance within the reserve, depending on the change in vegetation cover and the topographic context. Water balance at a point is rainfall plus fog plus snowmelt inputs minus actual evapotranspiration. Some areas showed increased water balance, and some showed decreased water balance (because the balance between changes in evapotranspiration and fog interception on agriculturalisation varies over space). Figure 7.5 shows the resulting impact on flows

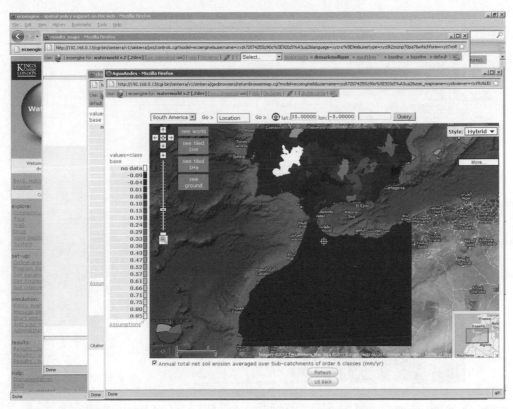

Figure 7.3 Annual total net soil erosion (erosion minus deposition, hillslope and channel) averaged over sub-catchments of Strahler stream order 6 (mm/yr), indicating high erosion in some basins in Spain and Portugal compared with Morocco and Algeria. Background map from Google Maps. Image shown within the context of the PSS interface.

in which some tributaries of the Guadalentín show increases in flows, while other neighbouring tributaries show decreases. The overall outcome depends very much on the original vegetation complex, the nature of vegetation change and the landscape characteristics and is thus not simple to predict. Such deforestation also leads to an increase in soil erosion in catchments draining the protected area with the consequent risk of sedimentation in downstream reservoirs.

The final scenario examines the impact on erosion of bench terracing all land with slope gradient greater than 10° in the Sierra de Baza protected area. The implementation of bench terracing reduces the local slope and thus

decreases erosive power and reduces sediment transport capacity, leading to soil deposition. Net soil erosion may thus decrease and even become negative (net deposition). Figure 7.6 shows the difference between baseline and terraced net soil erosion expressed as a percentage of baseline erosion. Where values are negative, soil erosion has decreased, where values are positive it has increased. Clearly, terracing in this area leads to spatially variable decreases in erosion.

The examples are clearly highly simplistic compared with most policy problems in which multiple interventions are targeted at multiple (sometimes conflicting) objectives and the outcomes need to be viewed from a number

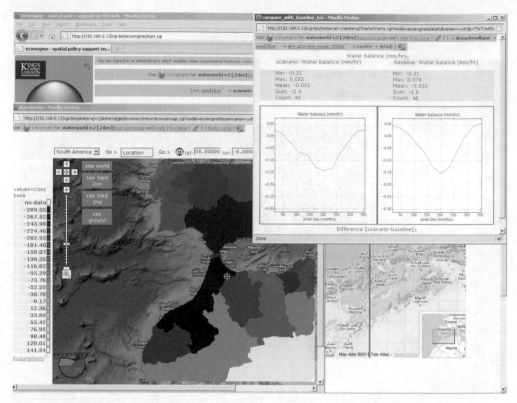

Figure 7.4 Change in annual total water balance (mm/yr) under AR4 A2a 17GCM ensemble simulation expressed as a map and through comparison of seasonal patterns of change. Background map from Google Maps. Image shown within the context of the PSS interface.

of perspectives (e.g. environmental, socio-economic, political). Nevertheless, they do give an indication of the functionality of *WaterWorld*. The interested reader is encouraged to work with the PSS themselves in order to better understand its potential and limitations for their own field of enquiry.

Addressing the remaining barriers

A number of barriers to uptake of PSS like *WaterWorld* remain. Perhaps the most substantial is the relative simplicity of such systems compared with the policy problems that they could support. I have shown simple climate change, land cover change and land management interventions applied singularly, but in reality, they will occur concurrently and may need to be applied concurrently in complex

spatial patterns. This complexity leads to difficulties in building simple interfaces for the application of such scenario/policy option combinations and also difficulties in interpreting the complex impacts of such combinations on the system under study. While interventions can be 'stacked' in *WaterWorld*, understanding their impacts and their interactions becomes very complex. Aside from issues of model complexity, communication, validation and trust which develop incrementally with use in a system like *WaterWorld*, the key remaining barriers to use are (e) willingness to adopt PSS in the policy process and (f) who will use these systems and what is their role in decision-making organisations.

Willingness to adopt depends upon finding an appropriate place for PSS in decision-making

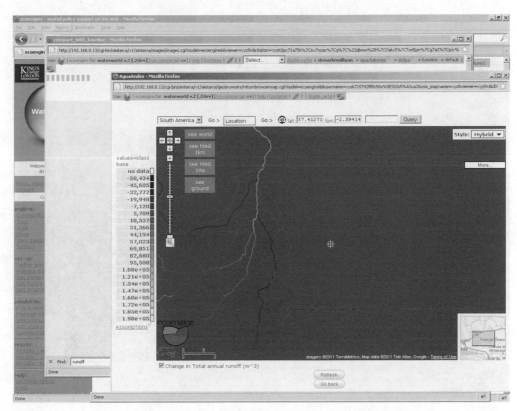

Figure 7.5 Change in total annual runoff (m³) after deforestation in Sierra de Baza protected area. Some channels show a net increase in flow while others a net decrease. Background map from Google Maps. Image shown within the context of the PSS interface.

processes. Projects such as the current CGIAR COMPANDES project (see www.benefitsharing.net) are working with a variety of stakeholders in the Andes to build confidence and competence in the use of *WaterWorld*. Greatest willingness seems to be within the NGO and advocacy group communities for whom systems like this can provide scientific decision support and potentially lend scientific legitimacy to policy stances or proposed interventions. Uptake of *WaterWorld* is thus currently greatest outside of policymaking organisations, that is, in those groups advocating support for or against particular policies or interventions that may be made by those with power and influence. As the user base grows, systems like *WaterWorld* should become more influential in providing a scientific

evidence base for or against the utility of particular interventions. Therefore, those developing PSS, and those using them, will need to be very careful that the data and models are robust enough and that their outputs are trustworthy and at least better than any alternative sources of evidence that would otherwise be applied.

Conclusion

John Thornes initiated a more nuanced understanding of Mediterranean desertification and supported the development of some of the simplest and some of the most sophisticated models for better understanding it. The process-based models that he championed have evolved

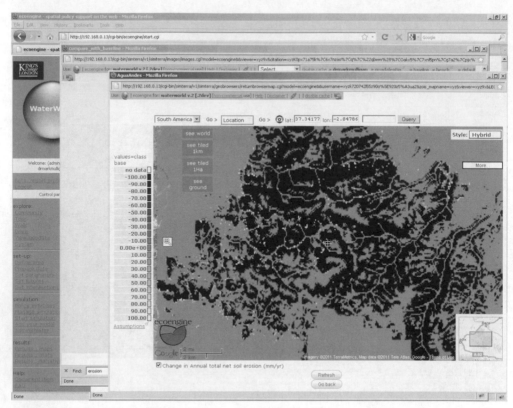

Figure 7.6 Change in annual total net soil erosion (mm/yr) on implementing bench terraces on land with gradient >10° in the Sierra de Baza protected area. Changes are expressed as a percentage of the baseline value. Most areas show a decrease in net erosion (erosion–deposition). Background map from Google Maps. Image shown within the context of the PSS interface.

over subsequent decades into software DSS and PSSs whose use is increasing both within and outside the Mediterranean. While such tools are still not routinely used to advise those making policy, they certainly are used by the multitude of interested parties who advocate particular stances for or against such policies.

Environmental science needs to link better to – and have impact upon – real-world policy applications. Simulation-based PSSs provide an appropriate way to do this. Unfortunately, time spent building, testing and interfacing models with policymakers is time not available for publication and more career-centric activities. Moreover, policy-relevant modelling is often viewed as not robust-enough science to be funded through the normal academic routes

and insufficiently policy focused (too technical and science focused) to be funded through social science, policy or development routes. This activity thus falls between, for example, the Natural Environment Research Council (NERC) and the Economic and Social Research Council in the United Kingdom, so who will fund such activity? It is clear that most policy problems applied to heterogeneous landscapes are far too complex to keep track of with conceptual and analytical approaches, so numerical, computer-based approaches need to be better understood and valued by the donor and policy communities so that their development is championed and supported. This need seems to be clear at least to the policy-driven EU research offices.

John Thornes had an outstanding career and left an important legacy for many reasons but not least because he worked on what he thought was important both academically and practically, not on what he thought would lead to the most rapid career progression. He made sure his science was robust but useful, and he engaged the talents of many colleagues to build projects in which the sum was greater than the parts. He took time away from activities of personal academic benefit towards those that benefited others both inside and outside of academia, 'in the service of society' as his final academic home, King's College London, would describe it. For environmental science to become more policy focused, more scientists need to work that way and the scientific establishment needs to reconsider which metrics are really important to the role of science in modern society. Some headway is being made in this area with respect to the UK research 'impact' agenda, but there is a long way to go.

References

Bolle, H. J. (1995) Identification and observation of desertification processes with the aid of measurements from space: results from the European Field Experiment in Desertification-threatened Areas (EFEDA). *Environmental Monitoring and Assessment* **37**, 93–101.

Bruijnzeel, L. A., Mulligan, M., Scatena, F. S. (2011) Hydrometeorology of tropical montane cloud forests: emerging patterns. *Hydrological Processes* **25**, 465–498.

Heathcote, I. W. (2009) *Integrated Watershed Management: Principles and Practice* (2nd Ed.). John Wiley & Sons, Inc., Hoboken, NJ.

Holder, J. (2004) *Environmental Assessment: The Regulation of Decision Making.* Oxford University Press, New York.

Jakeman, A. J., Letcher, R. A., Norton, J. P. (2006) Ten iterative steps in development and evaluation of environmental models. *Environmental Modelling and Software* **21**, 602–614.

Kirkby, M. J., Abrahart, R. J., Bathurst, J. C., Kilsby, C. G., McMahon, M. L., Osborne, C. P., Thornes, J. B., Woodward, F. I. (2002) MEDRUSH: a basin-scale, physically based model for forecasting runoff and sediment yield. In: Geeson, N. A., Brandt, C. J., Thornes, J. B. (eds.) *Mediterranean Desertification: A Mosaic of Processes and Responses.* John Wiley & Sons, Ltd, Chichester, pp. 203–229.

Kirkby, M. J., Jones, R. J. A., Irvine, B., Gobin, A., Govers, G., Cerdan, O., Van Rompaey, A. J. J., Le Bissonnais, Y., Daroussin, J., King, D., Montanarella, L., Grimm, M., Vieillefont, V., Puigdefabregas, J., Boer, M., Kosmas, C., Yassoglou, N., Tsara, M., Mantel, S., Van Lynden, G. (2003) Pan-European Soil Erosion Risk Assessment: The PESERA Map, Version 1 (October 2003). Explanation of Special Publication Ispra 2004 No.73 (S.P.I.04.73). European Soil Bureau Research Report No.16, EUR 21176. Office for Official Publications of the European Communities, Luxembourg.

Kirkby, M. J., Bracken, L., Brandt, C. J. (2015) John Thornes and desertification research in Europe. In: Dykes, A. P., Mulligan, M., Wainwright, J. (eds.) *Monitoring and Modelling Geomorphological Environments.* John Wiley & Sons, Ltd, Chichester, pp. 317–326.

Mazzoleni, S., Legg, C. (1998) Modmed: modelling vegetation dynamics and degradation in Mediterranean ecosystems. In: Mairota, P., Thornes, J. B., Geeson, N. (eds.) *Atlas of Mediterranean Environments in Europe – The Desertification Context.* John Wiley & Sons, Ltd, Chichester, pp. 14–18.

McIntosh, B. S., Ascough, J. C., Twery, M., Chew, J., Elmahdi, A., Haase, D., Harou, J. J., Hepting, D., Cuddy, S., Jakeman, A. J., Chen, S., Kassahun, A., Lautenbach, S., Matthews, K., Merritt, W., Quinn, N. W. T., Rodriguez-Roda, I., Sieber, S., Stavenga, M., Sulis, A., Ticehurst, J., Volk, M., Wrobel, M., van Delden, H., El-Sawah, S., Rizzoli, A., Voinov, A. (2011) Environmental decision support systems (EDSS) development – challenges and best practices. *Environmental Modelling and Software* **26**, 1389–1402.

Mulligan, M. (1996) Modelling Hydrology and Vegetation Change in a Degraded Semi-arid Area. Unpublished PhD thesis. King's College, University of London, London.

Mulligan, M. (1998) Modelling the geomorphological impact of climatic variability and extreme events in a semi-arid environment. *Geomorphology* **24**, 59–89.

Mulligan, M. (2009) Integrated environmental modelling to characterise processes of land degradation and desertification for policy support. In: Hill, J., Roeder, A. (eds.) *Remote Sensing and Geoinformation Processing in the Assessment and Monitoring of Land Degradation and Desertification.* Taylor & Francis, London, pp. 45–72.

Mulligan, M. (2013) WaterWorld: a self-parameterising, physically-based model for application in data-poor but problem-rich environments globally. *Hydrology Research* **44**, 748–769.

Mulligan, M., Burke, S. M. (2005) FIESTA: Fog Interception for the Enhancement of Streamflow in Tropical Areas. Report to UK DfID. Available at: www.ambiotek.com/fiesta (accessed 14 July 2014).

Mulligan, M., Rubiano, J., Hyman, G., White, D., Garcia, J., Saravia, M., Leon, J. G., Selvaraj, J. J., Gutierrez, T., Saenz-Cruz, L. (2010a) The Andes 'basin': biophysical and developmental diversity in a climate of change. *Water International* **35**, 472–492.

Mulligan, M., Rubiano, J., Rincon-Romero, M. (2010b) Hydrology and land cover change in tropical montane environments: the impact of pattern on process. In: Bruijnzeel, L. A., Scatena, F. N., Hamilton, L. S. (eds.) *Tropical Montane Cloud Forests: Science for Conservation and Management*. Cambridge University Press, Cambridge, UK, pp. 516–525.

Oxley, T., Winder, N., McIntosh, B. S., Mulligan, M., Engelen, G. (2004) Integrated catchment modelling and decision support tools: a Mediterranean example. *Environmental Modelling and Software* **19**, 999–1010.

Ramos, M. C., Mulligan, M. (2005) Spatial modelling of the impact of climate variability on the annual soil moisture regime in a mechanized Mediterranean vineyard. *Journal of Hydrology* **306**, 287–301.

Thornes, J. B. (1990) The interaction of erosional and vegetational dynamics in land degradation: spatial outcomes. In: Thornes, J. B. (ed.) *Vegetation and Erosion: Processes and Environments*. John Wiley & Sons, Ltd, Chichester.

Thornes, J. B., Brandt, C. J. (1993) Erosion-vegetation competition in and environment undergoing climate change with stochastic rainfall variations. In: Millington, A. C., Pye, K. T. (eds.) *Environmental Change in the Drylands: Biogeographical and Geomorphological Responses*. John Wiley & Sons, Ltd, Chichester, pp. 305–320.

Thornes, J. B., Shao, J. X., Diaz, E., Roldan, A., Hawkes, C., McMahon, M. (1996) Testing the MEDALUS hillslope model. *Catena* **26**, 137–160.

van Delden, H., Luja, P., Engelen, G. (2007) Integration of multi-scale dynamic spatial models of socio-economic and physical processes for river basin management. *Environmental Modelling and Software* **22**, 223–238.

van Delden, H., Ouessar, M., Sghaier, M., Ouled Belgacem, A., Mulligan, M., Luja, P., de Jonga, J., Fetoui, M., Ben Zaied, M. (2008) User driven application and adaptation of an existing Policy Support System to a new region. iEMSs 2008. International Congress on Environmental Modelling and Software Integrating Sciences and Information Technology for Environmental Assessment and Decision Making, 7–10 July 2008, Barcelona, Catalonia, Spain.

van Delden, H., Seppelt, R., White, R., Jakeman, A. J. (2011) A methodology for the design and development of integrated models for policy support. *Environmental Modelling and Software* **26**, 266–279.

van Wesemael, B., Rambaud, X., Poesen, J., Mulligan, M., Cammeraat, E., Stevens, A. (2006) Spatial patterns of land degradation and their impacts on the water balance of rainfed tree crops: a case study in South East Spain. *Geoderma* **133**, 43–56.

WaterWorld (2014) The WaterWorld Policy Support System. Available at: www.policysupport.org/waterworld (accessed 14 July 2014).

CHAPTER 8

Application of strategic environmental assessment to the Rift Valley Lakes Basin master plan

Carolyn F. Francis and Andrew T. Lowe

Halcrow Group Ltd., a CH2M HILL Company, UK

Introduction

Halcrow Group Ltd in association with an Ethiopian Consultancy Generation Integrated Rural Development Consultants (GIRDC) was appointed by the Ministry of Water Resources (now the Ministry of Water and Energy) of the Federal Democratic Republic of Ethiopia to undertake an integrated natural resources master plan for the Rift Valley Lakes Basin (RVLB) (Halcrow Group Ltd in association with GIRDC 2008, 2009, 2010). The overall objective of the master plan was to provide sustainable development over a 25-year planning horizon. The commission also included a strategic environmental assessment (SEA) of the master plan, the preparation of a portfolio of development projects and engineering and environmental feasibility studies for four development projects in preparation for seeking international funding for project implementation. The selected development projects were an irrigation scheme, a town water supply scheme and two integrated watershed development projects. The contract was undertaken between December 2006 and January 2010.

The RVLB forms part of the East African Rift Valley which extends through Tanzania, Kenya and Ethiopia. The RVLB lies in southern Ethiopia, extending from the Kenyan border to the watershed with the Awash Valley, which also forms part of the East African Rift Valley. Highlands separate the RVLB from the Omo river basin to the west and the Wabi Shebele and Genale Dawa river basins to the east.

The economy of the RVLB is highly dependent on agriculture, which provided an estimated 66.6% of regional gross domestic product (RGDP) in 2005 (combining crops, livestock, apiculture, forestry and fisheries), compared with industry 9.7% and services 23.7%, respectively.[1] The greater part of the agricultural RGDP was provided by crops (37.64%), with dryland

---

[1] These are Halcrow's calculations based on various reports from the Central Statistical Agency of Ethiopia including CSA/EDRI/IFPRI (2006) and the Ethiopian Agricultural Sample Enumeration (EASE) returns for 2001/2002 in CSA (2002).

Monitoring and Modelling Dynamic Environments, First Edition. Edited by Alan P. Dykes, Mark Mulligan and John Wainwright.
© 2015 John Wiley & Sons, Ltd. Published 2015 by John Wiley & Sons, Ltd.

farming contributing 36.72% and irrigation only 0.92%. In monetary terms, RGDP in 2005 was estimated at 7.96 billion Ethiopian Birr, or about USD105 per capita, which is very low even by Ethiopian standards. The future economic development in the RVLB very much rests on agriculture and agro-processing.

The concept of sustainable development has evolved over recent decades and has come to be interpreted in different ways; indeed, according to Adams, there is 'no simple, single meaning of "sustainable development"' (Adams 2009, p. 23). The phrase comprises two seemingly contradictory ideas – sustainability, which implies at least the maintenance of existing environmental conditions, and development, which implies an improvement on current usually economic conditions, often by exploiting the natural environment.

An early definition comes from Brundtland (1987, paragraph 27), where sustainable development meets 'the needs of the present without compromising the ability of future generations to meet their own needs'. The definition of sustainable development has since broadened from a narrow view of environmental sustainability whereby the natural capital remains intact to include social cohesion and economic feasibility (Gilbert et al. 1996; Adams 2009).

Within the context of the preparation of an integrated natural resources master plan with a 25-year planning horizon, sustainable development implies the implementation of interventions or projects using the natural resource base, which lead to a material improvement in the economy, or at least people's living conditions, without prejudicing the very natural resource base on which the development depends. One area studied in detail for the RVLB master plan was the potential development of water resources. Future water demand was projected to increase to meet drinking water supply for a growing population, plans for irrigation expansion and industrial development particularly in agro-processing. Water resources are also required to meet what may

be called environmental demands, support wetland habitats and the fauna that depend upon them, provide the natural resources used by local communities, dilute pollutants, attenuate flooding and so on. The impact of future water resource demands on the environment, in particular the terminal lakes, was evaluated as part of the SEA of the master plan.

The approach to SEA has been evolving over the last two decades, and it is clear from the review literature that different approaches are being developed in different parts of the world. For example, Thérivel and Partidário (1996) mostly review case studies in Europe and North America. Kjorven and Lindhjem (2002) reviewed SEAs conducted in developed and developing countries, illustrating emerging approaches and lessons learnt. Dalal-Clayton and Sadler (2005) reviewed case studies in developing countries including numerous examples in Africa. In Ethiopia, the Federal Environmental Protection Authority (EPA) has prepared draft guidelines on the application of SEA, which draw upon the approach developed in South Africa (Council for Scientific and Industrial Research 2000). A central plank of this approach is to consider fully the constraints imposed by the environment on scenarios for development. This involves preparing an inventory of the bio-physical, economic and social resources of the plan area and ongoing trends that may influence the value of those resources and how they are used over the plan horizon; identifying environmental opportunities and constraints in the plan area which are incorporated into development scenarios; and identifying sustainability objectives, criteria and indicators which are used during plan development to select preferred scenarios and then monitor the implementation of the plan. The overriding objective of the SEA process is to develop and promote sustainable policies, plans and programmes.

A detailed assessment of the surface water resources of the RVLB has been undertaken. However, what are the implications for promoting sustainable development if the water resources inventory of the RVLB shows that (i) the existing resources are degrading; (ii) this trend is likely to continue for the foreseeable future due to population pressure, economic growth, climate change or some other factor; and (iii) the proposed development projects are likely to exacerbate the rate of environmental degradation? Advocates of 'strong' sustainability, which requires the maintenance of the natural and human condition, may argue for interventions to reduce or even reverse the growth in water demand. However, such an approach is impractical given the growing population and desire to alleviate poverty through development. An alternative approach, known as 'weak sustainability', would be to seek trade-offs between further environmental degradation and human capital (Adams 2009, p. 144).

In our studies, irrigation expansion was initially considered to be an important component in the economic development of the RVLB, but it became clear that expansion is water constrained and that even small increases in abstraction of water for irrigation have significant impacts on lake levels. Consequently, it was expedient to consider how much environmental degradation would be acceptable in return for irrigation expansion and whether there were other means to achieve the economic development goals of the master plan without overexploiting the water resources.

Approach

The approach to the SEA involved a multidisciplinary diagnostic study of the environmental, ecological and human situation of the RVLB; an assessment of the impact of development scenarios on the water resources; and the environmental and social implications for long-term economic development in the RVLB based on agriculture and agro-processing. The potential effects of climate change were incorporated into the water resources modelling.

Current land use (for 2007) was evaluated using satellite imagery including Landsat MSS 1970, Landsat TM for 1986–1990, Landsat ETM for 2000, MrSID composite imagers for 1987–1990, MrSID composite images for 2000 and Aster Imagery for 2006. Extensive field surveys were undertaken between February and May 2007 to verify the interpretation of land use from the satellite imagery.

A land evaluation exercise was undertaken to identify land suitable for agricultural uses following the methodology outlined in the Food and Agriculture Organisation Soils Bulletin No. 32 (FAO 1976). The evaluation process involved matching areas characterised by different environmental and social conditions with possible land-use options. Land suitability is then defined as the fitness of a particular area for a specific land use or land utilisation type (LUT) under a stated system of management. In other words, this approach seeks to ensure that the LUT proposed is suitable for the land in question, embedding long-term sustainability in the planning process.

Soil surveys were undertaken with a coverage of soil sampling at approximately 1 per 2,500 ha, and soil classes were mapped at a scale of 1:250,000. Agro-ecological classes were mapped at a scale of 1:2,000,000. Slope classes were mapped from a digital elevation model with a resolution of 57 m, re-sampled from an initial 90 m resolution data at about 1:250,000.

The mapping of natural resources is inherently imprecise due to the fact that the natural resources (soil, slope, climate, etc.) exist as a continuum. Consequently, the mapping of natural resources should ideally be undertaken at one scale and presented at a smaller scale to reduce the imprecision. Here, the land suitability classes were mapped at 1:250,000, which is the same scale as the soil mapping and a larger scale than some of the other input data. It is recognised that given the resolution of some of the input data, which reflects the lack of spatial data coverage in the study area, there is a great deal of imprecision in the land suitability mapping.

Land is considered to be suitable (S) or not suitable (N) for given LUTs. The suitable land is further subdivided into highly suitable (S1), moderately suitable (S2) and marginally suitable (S3), based on limiting factors such as climate (rainfall and duration of growing season), topography and slope, soil moisture availability and retention, drainage, soil depth, nutrient levels, erosion hazard and soil toxicity (salinity and sodicity). Five main LUTs were identified and defined: smallholder rainfed agriculture, large-scale mechanised rainfed agriculture, large-scale irrigated agriculture, production forestry and rangeland.

Climatic and hydrological data were collected from monitoring stations in the RVLB and analysed to obtain monthly and annual statistics for rainfall, temperature, evapotranspiration, river flows and lake levels. A water balance model using the Water Evaluation and Planning System (WEAP) software was built and calibrated to observed river flows and lake levels over a 30-year historical period from 1975 to 2004. The 30-year historical time series was then used as a surrogate for a future time series in order to model the future response of the surface water balance of the lakes to increased abstractions that will occur as the basin develops, with particular reference to irrigation expansion, and as a response to climate change.

A water quality monitoring programme was undertaken in the dry (July 2007) and wet (August 2007) seasons. A portable Hach probe was used to monitor water temperature, pH, electrical conductivity (EC) and total dissolved solids (TDS) in the field. Water samples were collected, transported to Addis and analysed in the laboratory for the remaining parameters.

The impact of climate change on the water resources of the RVLB was assessed using a hypothetical increase in mean annual temperature of +2°C, resulting in a 10% increase of evapotranspiration and a 10% decrease of rainfall over the modelling period of 30 years (taken to be 2006–2036). A uniform change in evapotranspiration and rainfall in response to climate change is assumed across the RVLB, and no account has been made for seasonal variations in climate change impacts.

This is a high climate change impact scenario within the range of predictions for temperature and rainfall in the published literature for Ethiopia. While there is some consistency in predictions of temperature increases by 2050, there are significant differences in predictions for rainfall. Hulme et al. (2001) compared the results of 10 GCMs (including four simulations for HadCM2), each one modelled for four global climate scenarios ranging from low to high climate sensitivities, to examine forecast changes in temperature and rainfall. The results for Ethiopia indicated a warming of between 1 and 4°C by 2050 over 1961–1990 values. The results for rainfall were less consistent, with three models indicating increased aridity of between 5 and 10% and six indicating increased rainfall by 5–25%. Furthermore, most of the changes in rainfall forecast using HadCM2 GCM were within the natural rainfall variability. The National Meteorological Agency (NMA) of Ethiopia predicted an increase in mean annual temperature of 1.7–2.1°C and a small increase in rainfall of about 3–8% by 2050 over 1961–1990 values, based on the IPCC mid-range (A1B) scenario modelled using the software Model for the Assessment of Greenhouse Gas Induced Climate Change/Regional and Global Climate SCENario GENerator (MAGICC/SCENGEN; Tadege 2007).

Hulme et al. (2001) discuss some of the inherent difficulties in forecasting climate change, resulting from the large inter-annual variations in climate masking small long-term changes, the absence of the effect of land cover changes in the GCMs and the poor representation of some aspects of climate variability in GCMs which are important in Africa.

The climate change scenario adopted here exacerbates future aridity through a combination of increased in temperature and reduced rainfall, superimposed on a rainfall record which included a prolonged period of dry years towards the end of the record. If this does not represent a 'worse-case' scenario, then it is at least a challenging one. However, our approach is not to make predictions of environmental degradation within a given time frame but rather to use the modelling as a basis for decision-making to manage a scarce resource – water. As such, we consider that a conservative climate change scenario is acceptable.

Data on current and proposed irrigation schemes were obtained from the Ministry of Water Resources.

Environmental and social characteristics of the RVLB

Water resources of the RVLB

The RVLB covers an area of about 53,000 km^2 (Figure 8.1). The topography ranges from over 3000 m above sea level (masl) in the Ethiopian highlands on the western and eastern flanks of the RVLB to about 1650 masl at the northern end of the Rift Valley and to about 500 masl at the southern end of the Rift Valley on the Kenyan border.

The climate of the RVLB is tropical monsoon with a bimodal rainfall distribution caused by the passage of the Intertropical Convergence Zone creating the 'small' or spring rains (the *belg*) and the main summer rains (*meher*). In the southern part of the RVLB, the small rains start earlier in the year and are less dependable, while the main rains occur later with a more pronounced intervening dry period, compared to the northern part. There is widespread variation in rainfall induced by topography, with mean annual rainfall varying from about 2000 mm in the highlands to 700 mm in the northern part of the central valley and 400 mm at the southern end of the central valley. Inter-annual variability in rainfall is high and shows sequences of years with above and below average rainfall, with a run of years with below average rainfall causing widespread drought. Mean annual temperatures range from 13°C in

the highlands to about 27°C at the southern end of the Rift Valley.

The RVLB is subdivided into four interdependent hydrological sub-basins, each one characterised by numerous mountain streams and rivers rising in the highlands and draining to the lakes in the central valley, several of which are terminal lakes (Figure 8.1). The RVLB as a whole is considered to be a hydrologically closed basin. The total annual average river flow into the lakes systems is estimated to be 5200 Million m^{3}year^{-1}. The groundwater resources, although important for rural water supply, are small in relation to the surface water resources and are not considered further here.

The Ziway–Abijata–Langano–Shala sub-basin covers 14,479 km^2 in the northern part of the RVLB. Lakes Ziway, Abijata and Langano lie in broad, shallow depressions and are the remnants of a large freshwater lake that existed several times during the Early–Mid-Holocene and Late Pleistocene wet periods (Grasse and Street 1978). Lakes Abijata, Shala and Chitu are all terminal lakes, and the last two named are crater lakes. Morphometric data are presented in Table 8.1.

Lake Ziway is fed by two major tributaries, the Meki and the Katar Rivers, with mean annual flows of about 180 and 300 Million m^{3}year^{-1}, respectively. Lake Ziway is drained via the Bulbula River which flows to Lake Abijata. The outflow from Lake Ziway to the Bulbula River is controlled geologically by a sill. Lake Langano is fed by several streams draining the Eastern Highlands and overflows via the River Horakelo to Lake Abijata.

The Awasa–Cheleleka sub-basin covers an area of 1404 km^2 and is located in the central eastern flank of the RVLB. Lake Awasa is a large, relatively shallow terminal lake, lying in the remains of the Corbetti Caldera, the outer edges of which form moderate to steeply sloping sides around the lake. The lake was once much larger, probably incorporating the Cheleleka wetland to the east. The main streams to Lake Awasa rise in the Eastern Highlands and descend to the Cheleleka plains before discharging via the

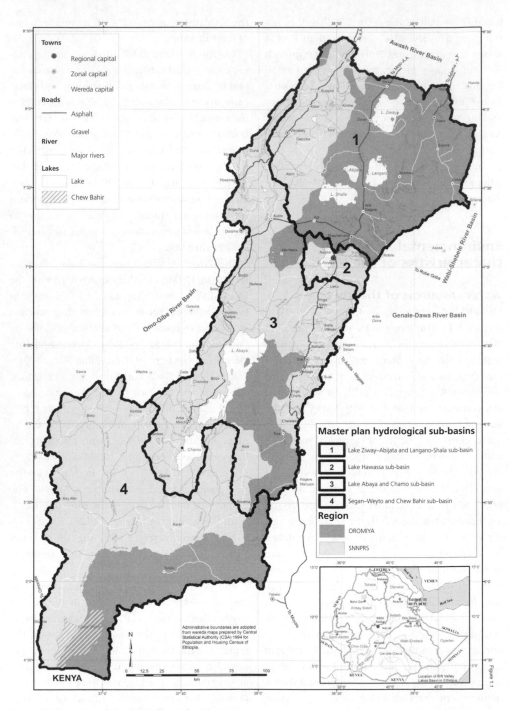

Figure 8.1 Location of the Rift Valley Lakes Basin.

Table 8.1 Characteristics of the central valley lakes.

Lake	Lake level (masl)	Area (km²)	Max. depth (m)	Mean depth (m)	Volume (Million m³)
Ziway	1636	423	9	2.5	1,158
Langano*	1590	247	42	23.0	5,553
Abiyata[†]	1581	132	12	6.0	744
Shala[‡]	1558	302	252	121.0	36,472
Awasa	1680	94	23.2	13.6	182
Abaya	1169	1140	24.5	8.6	9,819
Chamo[§]	1110	317	14.2	10.2	3,242

From Halcrow Group Ltd in association with GIRDC (2009).
*4–11 April 2008.
[†]30 April–12 May 2008.
[‡]14–27 May 2008.
[§]May 1998 by Seleshi (2006).

Tikur Wuha to Lake Awasa near Awasa Town. The mean annual flow of these streams is estimated to be 143 Million m³ year⁻¹.

The Abaya–Chamo sub-basin covers an area of 18,120 km² in the southern part of the RVLB. Numerous rivers flow into Lakes Abaya and Chamo, the largest ones being the Bilate, Gidabo and Gelana Rivers with mean annual flows of 830, 550 and 290 Million m³ year⁻¹. In the recent past, Lake Abaya overflowed to Lake Chamo, and Lake Chamo overflowed via the Haro Shet to the Segen River and the Chew Bahir sub-catchment. Both links are believed to have been severed in the last two to three decades, converting both to terminal lakes, but evidence in the hydrological records has not been found.

The Chew Bahir sub-basin covers an area of 19,031 km² in the southernmost part of the RVLB. The main rivers are the Weyto and the Segen Rivers, which have a combined mean annual flow of about 700 Million m³ year⁻¹ and drain to the salt flats and seasonal wetlands of Chew Bahir on the Ethiopian–Kenyan border.

The rivers and Lake Ziway are freshwater bodies, which are important for drinking water supply, irrigation and watering cattle (Table 8.2). The remaining lakes are highly alkaline–saline which are unsuitable for drinking water supply and irrigation, but do support wildlife and fisheries. Lake Abijata also supports a soda ash factory. There are marked spatial variations in water quality on Lake Abijata due to the freshwater influence of the Bulbula River compared with the middle of the lake. The limited historic data obtained from Wood and Talling (1988), the Addis Ababa Water and Sewerage Authority monitoring programme in 2004 and Halcrow's water quality sampling programme in 2007, indicate a trend of increasing salinity on Lake Abijata as measured by EC, TDS and chloride, while the results for other ions such as sodium and potassium are highly variable (Table 8.3). The increase in salinity is attributed to the reduction in lake volume in recent decades. The seasonal variations indicated by the data for all the lakes monitored in July and August 2007 are modest and likely to be within the sampling error resulting from the small number of samples.

Nature conservation value and habitat types

The Rift Valley Lakes are internationally important wetlands. There are two national parks: the Abijata Shala Lakes National Park, which is also a candidate Ramsar Site,[2] and the Nech Sar National Park, which includes parts of Lakes Abaya and Chamo. Lakes Ziway, Langano, Abijata, Shala, Awasa, Abaya and Chamo have all been identified as Important Bird Areas by BirdLife International and the Ethiopian Wildlife and Natural History Society (Edwards 1996). All of the lakes, except for Abijata where fishing is banned but believed to

---

[2] Ethiopia is not yet a signatory to the Ramsar Convention on Wetlands.

Table 8.2 Selected water quality parameters of the Rift Valley Lakes in April, 2007.

Lake	pH	Dissolved oxygen (mg l^{-1})	Electrical conductivity (μS cm^{-1})	Total dissolved solids (mg l^{-1})	Sodium (mg l^{-1})	Chloride (mg l^{-1})	Fluoride (mg l^{-1})
Ziway	8.7	6.0	460	220	63.5	12.5	1.51
	8.7	5.9	459	219	66.0	13.0	1.49
Langano	8.9	6.7	1,932	923	405	182	7.9
	8.9	5.8	1,967	932	216	183	9.6
Abijata	9.2	4.9	11,580	5,520	1,940	774	70
	10.5	2.9	83,580	41,520	12,940	10,778	270
Shala	9.8	2.8	48,150	23,160	6,000	3,250	156
Awasa	9.0	6.7	886	424	162	27	7.7
	9.0	6.8	887	424	162	27	6.1
Abaya	9.1	6.1	1,319	628	246	67	8.2
	9.0	5.9	1,308	626	244	65	7.6
Chamo	9.6	6.3	2,104	1,006	430	130	9.4

From Halcrow Group Ltd in association with GIRDC (2008).

Table 8.3 Physico-chemical water quality of Lake Abijata.

Parameter	Year		
	1964	2004	2007
Electrical conductivity (μS cm^{-1})	15,800	51,950	83,580
Total dissolved solids (mg l^{-1})		25,100	41,520
Sodium (mg l^{-1})	5,108	11,251	12,940
Potassium (mg l^{-1})	252	16.9	6,284
Calcium (mg l^{-1})		3.2	4.0
Magnesium (mg l^{-1})		0.49	0
Chloride (mg l^{-1})	1,830	3,530	10,778
Fluoride (mg l^{-1})		228	270
HCO$_3$+ CO$_3$ (mg l^{-1})	7,330		26,000

From (Wood and Talling 1988; Addis Ababa Water and Sewerage Authority water quality database for a monitoring programme in 2004; and Halcrow Group Ltd in association with GIRDC 2008).

occur illegally, support commercial fisheries which are important within the region but are showing signs of over-exploitation through falling landings, smaller size of fish and use of smaller sized nets. In addition to the lakes, the RVLB also supports the Senkelle Swayne's Hartebeest Sanctuary, the Chelbi Wildlife Reserve, several Controlled Hunting Areas and four Priority Forest Areas. Management of these various designated sites ranges from poor to non-existent.

Some 57% of the land cover of the RVLB comprises natural habitats (Table 8.4). The main natural habitat is shrubland with 19,494 km^2 and smaller areas of forest and woodland (1,717 and 3,858 km^2). Five broad vegetation groups are found:

1 Afro-alpine and sub-Afro-alpine vegetation which is found in the highlands at elevations about 3200 masl.
2 Dry Evergreen Montane (high) Forest occurs in remnants in the eastern and western escarpments between 1500 and 3200 masl.
3 Acacia–Commiphora–Boswellia (small-leaved deciduous) Woodland is widely distributed between 900 and 1900 masl and is found in a narrow strip through the central valley and in the south-western lowlands.
4 Combretum–Terminalia (broad-leaved deciduous) Woodland is found in higher rainfall

areas (>700 mm year⁻¹) in the south-western lowlands.

5 Shrublands which are found in the south-western lowlands.

Human environment

The population of the RVLB was estimated to be 8.89 million in 2005, calculated from the Statistical Abstract 2005. Based on the national

Table 8.4 RVLB land cover in 2007.

Land cover	Area (km²)	% of basin	% of land area
Annual cultivation (intensive and moderate)	14,803	27.9	30.1
Perennial cultivation	4,976	9.4	10.1
Mechanised farms	161	0.3	0.3
Total cultivation	**19,940**	**37.6**	**39.6**
Urban	73	0.1	0.1
Afro-Alpine vegetation	421	0.8	0.9
Forest	1,717	3.2	3.5
Woodland	3,858	7.3	7.7
Riparian vegetation	549	1.0	1.1
Shrubland	19,494	36.8	38.8
Grassland	1,291	2.4	2.6
Marsh	1,185	2.2	2.4
Bare land	1,854	3.5	3.7
Total land	**50,381**	**57.3**	**100.0**
Water	2,653	5.0	
RVLB total	**53,034**	**100.0**	

population growth rate of about 2.9% per annum, the population of the basin would double within 25 years (Table 8.5). The population is predominantly rural, and while this will continue to be the case for the plan period, an appreciable rate of urbanisation is expected.

The average population density for the RVLB is 167 persons km⁻², which is almost three times the national average of 65 persons km⁻², and the population is distributed unevenly over the basin. The low population density areas are mostly located in the southernmost part of the RVLB in Teltele Zone, where there is low and unpredictable rainfall and the main agricultural system is pastoralism, as well as areas with poor access in the midlands and highlands. The rural population density in Teltele Zone is only about 5 persons km⁻². The highest population densities are found in the eastern and the western highlands, where the higher rainfall supports intense agricultural systems, including the coffee production areas of Gedeo Zone where rural population densities exceed 1000 persons km⁻². The largest towns in the RVLB range in population from about 50,000 to 100,000 persons.

Based on the land cover mapping, some 19,940 km² of the RVLB is cultivated, comprising 14,803 km² of moderate to intensive annual cultivation, 4,976 km² of perennial cultivation and 161 km² of mechanised farms (Table 8.4). These areas actually include five main types of farming

Table 8.5 Current and projected population in the RVLB (millions) based on a medium growth rate.

Year	Total			Urban			Rural		
	Both	Male	Female	Both	Male	Female	Both	Male	Female
2005	8.89	4.43	4.46	1.12	0.56	0.56	7.77	3.87	3.90
2010	10.77	5.38	5.39	1.50	0.75	0.75	9.28	4.64	4.64
2015	12.82	6.41	6.40	1.96	0.98	0.98	10.86	5.43	5.42
2020	14.98	7.50	7.48	2.52	1.26	1.26	12.47	6.24	6.22
2025	17.34	8.69	8.65	3.20	1.60	1.59	14.14	7.09	7.05
2030	20.00	10.03	9.96	4.05	2.03	2.02	15.95	8.00	7.95
2035	22.92	11.51	11.42	5.10	2.56	2.54	17.82	8.95	8.88

From CSA (2002).

Table 8.6 Estimated land tenure in woredas associated with RVLB.

	Holders <1 ha	Holders >1 ha	State farms	Commercial farms	Total
Numbers	1,749,973	638,925	13	70	2,388,981
Average farm size	0.4	1.51			
Area farmed (ha)	699,989	964,134	28,095	193,168	1,885,386
% of land held	37	51	1	10	100.0
% numbers	73.3	26.7	0.0	0.0	100.0

From Halcrow Group Ltd in association with GIRDC (2008).

systems. Three of these are subsistence-level mixed farming systems under dryland conditions combining crop cultivation and animal husbandry. These are, namely, enset-based mixed farming systems with perennial crops, cereals and root crops; highland cereal (barley and wheat) mixed farming; and lowland cereal (maize, teff, sorghum, wheat and pulses) mixed farming. The fourth farming system is commercial farming comprising smallholders and medium- to large-scale commercial farms. Smallholders grow cereals, coffee, vegetables and fruit crops. Medium- and large-scale commercial farms, including three state farms, grow a range of cash crops such as wheat, barley, rapeseed, linseed, faba and haricot beans, coffee, corn and banana. Cotton is grown in the Segen River valley. Private enterprises have developed horticulture around Lake Ziway. Cash crops are mostly marketed within the RVLB, with some private investors exporting high-value crops such as coffee, flowers, vegetables, fruits and herbs. The fifth farming system is agro-pastoral and pastoral farming, which is practised in the arid, southern part of the RVLB, sometimes supported with supplementary cultivation of drought-resistant short-season crops such as sorghum, maize, millet and beans using spate irrigation along ephemeral streams. Pastoralists also supplement their diet through the collection of wild plants, fishing and hunting.

There are about 7,534 ha of commercial and large-scale farms under irrigation run by government and private investors and a further 21,242 ha of community-based schemes

supporting an estimated 56,500 smallholders. The total area under irrigation is less than 2% of all cultivated land. Most irrigation schemes are gravity fed, with traditional or modern river diversions and furrow or field flooding. Pumped schemes supply about 3000 ha around Lake Ziway. Some private investors have invested in spray and drip irrigation.

EASE agricultural data for 2001/2002 was used to characterise land tenure (Table 8.6). These data provide slightly different estimates of the area of cultivation than the land cover mapping described earlier. An estimated 7000 km^2 are farmed by some 1.75 million smallholder families with less than 1 ha and an average farm size of 0.4 ha. A further 9650 km^2 are farmed by about 0.64 million smallholder families with more than 1 ha and an average farm size of 1.51 ha.

With households typically comprising five members, a large rural population lives at or below subsistence level. Assuming an adult requires 2100 kcal day^{-1}, which is supplied by 1.25 quintals year^{-1} of wheat equivalent, a family of five requires about 6.25 quintals of wheat equivalent in a year. In the RVLB, wheat yields vary between 12 and 30 quintals ha^{-1}, with the highest yields in the north-east, the lowest yields in the south and typical yields of about 15 quintals ha^{-1}. For a family of five to be self-sufficient in food, they need about 0.5 ha, and to have enough surplus to meet their basic needs such as seed, fertiliser and household goods and services, they need about 1.0 ha. On this basis, about three-quarters of smallholder families in the RVLB live on landholdings which are too small to be self-sufficient in food, and the remaining quarter are living at or just above the level to meet their basic needs.

This leads to three important points: (i) given the large numbers of households living at or below subsistence levels, many households require food aid even in average and good years; (ii) relatively small 'shocks' such as poor rainfall or harvest affect large numbers of people; and (iii) the carrying capacity for subsistence farming RVLB in the existing cultivated areas and under current methods of farming is at or close to the maximum as further land subdivision between family members cannot support their needs.

Environmental and social trends

An understanding of recent and ongoing environmental and social trends is beneficial in formulating development policy. However, as Adams (2009) pointed out in relation to dryland political economy, all too often there are insufficient data and a lack of understanding of the complex interactions between the environment and local communities to identify environmental trends, and he provides ample warning against assuming too readily that environmental degradation is occurring. In the RVLB, long-term records are largely confined to the census, agricultural returns and river and lake hydrology. These may be supplemented with a snapshot of data from individual development projects which are spatially and temporally confined and a mix of quantitative and qualitative data, including anecdotal reports of environmental change from community consultation.

It is usually assumed that environmental degradation is a problem in much of Ethiopia and that the primary driving force is the large and growing population which is causing pressure on the land and natural resources. There is evidence to support this such as changes in land use, the subdivision of land to small and fragmented plots, the expansion of agriculture mostly into unsuitable areas and the widespread degradation of common resources such

as grazing land, woody biomass collection for fuelwood and overfishing.

A comparison of satellite imagery from 1986 to 2006 provides statistics on changing land cover at the basin scale (Halcrow Group Ltd in association with GIRDC 2008). The area of cultivated land rose from 30.4 to 37.6% and shrublands from 26.5 to 36.8%, but the area of grasslands fell from 18.0 to 2.4%. There were relatively modest increases in the areas of forests (from 2.8 to 3.2%) and woodlands (5.4 to 7.3%), but these figures mask deforestation of natural habitats, afforestation in previously clearly forest and plantations. Field surveys indicate that the main land uses are grazing and browsing with some woody biomass collection (39%), intensive cultivation (17%) and moderate cultivation (12%). Some 16 classes of land-use change were identified, the largest being a shift from large-scale (often state farm) to smallholder cultivation (30,180 ha), an increase in small-scale irrigated agriculture from rainfed cultivation or grassland/shrubland (18,324 ha), an increase in urban areas (12,816 ha) and a reduction in lake size leading to increased areas of bare soil (7,708 ha). These statistics alone do not tell us about the changing quality of the remaining natural resources, as the degradation of land cover is very difficult to discern from remotely sensed data or visual observations on the ground.

The Southern Nations, Nationalities and Peoples' Regional States Livelihood Zones Study (USAID 2005) of 38 Livelihood Zones, of which 25 occur in the RVLB, showed widespread evidence of pressure on agricultural production, as evidenced by (i) increasing population leading to the shrinking landholding size per household, fragmentation of landholdings and increased risk of food insecurity in previously food secure areas; (ii) land degradation and soil erosion contributing to loss of soil fertility and declining yields; (iii) widespread reporting of the shortage of grazing land and a shift to a 'zero grazing' system where cattle are kept and fed close to the homestead; and (iv) development of sedentary farming systems in valley bottoms near water in the southern agro-pastoral zone.

Time series data are difficult to acquire for the RVLB, but comparing the EASE 2001 enumeration data with the agricultural statistics collected for the period 2005–2007 during the master plan phase 1 studies (Table 8.7) shows that the cultivated area is expanding extremely fast, at about 10% per annum, while overall yields are increasing more slowly at about 1% per annum.

Most of the best land for irrigation is already developed, but irrigation expansion continues, particularly along the Meki and Katar Rivers, Lake Ziway and the Bulbula River. Some 64 potential irrigation schemes of various sizes have been identified by the Ministry of Water Resources, with a total planned area of new irrigation of some 151,050 ha (Table 8.8). If implemented, the area under irrigation would increase sixfold with the area of irrigated land rising from about 1.5 to 9% of the current cultivated land.

Table 8.7 Changes in annual field crop area, production and yield, 2001–2007.

All annual field crops	EASE 2001	Average 2004/2005–2006/2007	% change per annum during period
Cultivated area (ha)	774,149	1,275,690	10.47
Production (tonnes)	1,621,374	2,825,276	11.60
Yield (tonnes ha^{-1})	2.09	2.21	1.03

From (EASE 2001 in CSA (2002) and Wereda statistics, average of 2004/2005–2006/2007 reported in Halcrow Group Ltd in association with GIRDC (2008)).

There is evidence of long-term climatic trends superimposed on strong inter-annual variability. Hulme et al. (2001) reviewed observed (1900–2000) changes in temperature and rainfall across Africa. They concluded that while the climate has warmed over the last century, there were areas of cooling including parts of Ethiopia, Sudan and Egypt, where the mean annual temperature was reported to have decreased by about 0.5°C. The long-term change in rainfall was more mixed, with increases of rainfall of 10% along the Red Sea and in parts of the Horn of Africa but a 10–20% decrease in rainfall of the eastern Sahel. The NMA examined more recent climate data for Ethiopia. They noted strong inter-annual variations in mean annual rainfall and temperature, with trend analysis indicating that rainfall had remained constant in the period 1971–2000 over the country, while temperature had increased by an average of 0.37°C every 10 years between 1961 and 2006 (Tadege 2007).

Data on lake water levels in the RVLB show short- and long-term variations over the last 30 years (Figure 8.2), with fairly consistent seasonal variations in water levels of about 1 m, with maximum lake levels occurring towards the end of the wet season in September or October, superimposed on longer-term trends of rising and falling lake levels. Several lakes show higher water levels in the mid- to late 1990s followed by a period of lower lake levels in the early 2000s, reflecting a period of above average rainfall followed by several years of below average rainfall. The terminal lakes Abijata and Chamo show marked falls in lake levels over the

Table 8.8 Existing and potential irrigation in the RVLB.

Sub-basin	Existing area (ha)	Percentage of total	Total planned (ha)	Cumulative (ha)
Ziway–Abijata–Langano–Shala	7,156	24.1	49,250	56,406
Awasa	1,900	6.4	10,000	11,900
Abaya–Chamo	13,591	45.8	44,800	58,391
Segen–Weyto	7,025	23.7	47,000	54,025
Total	**29,672**		**151,050**	**180,722**

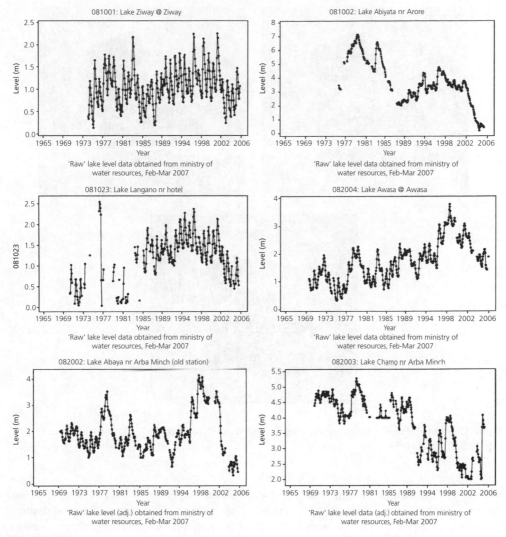

Figure 8.2 Historical water levels on Lakes Ziway, Abijata, Langano, Awasa, Abaya and Chamo.

record. The picture for Lake Chamo is confused by possible errors in adequately recording the datum as the stage level has been relocated more than once. The most interesting example is Lake Abijata, which has decreased in area from a maximum of 225 km² in 1970–1972 to 166 km² in 2000 and to 87 km² in 2006 (Figure 8.3). The decline in water levels is not explicable through the rainfall record alone, with water abstractions for irrigation, drinking water supply and soda ash production playing a major role.

Various authors have considered the implications of future climate change on the water resources in Ethiopia. Zeray et al. (2006) used the HadCM3 GCM to assess the impact of climate change on Lake Ziway basin hydrology over the coming century, predicting a decrease in the total average inflow to the lake of between 19 and 27% caused by decreasing inflows in the rainy season (June–September). Kinfe (1999) assessed the impact of climate change on the water resources of the Awash River, showing

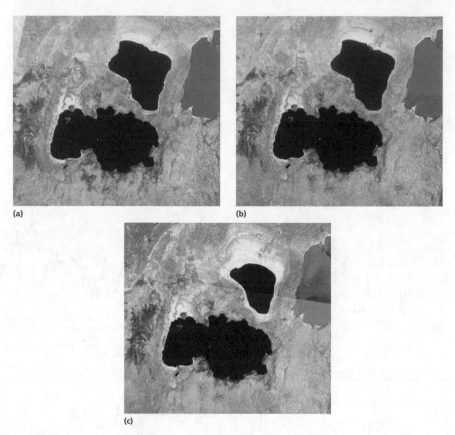

(a)

(b)

(c)

Figure 8.3 Satellite images of Lakes Abijata, Langano and Shala, showing changes in lake areas between (a) 1986, (b) 2000 and (c) 2006.

that for a +2°C increase in temperature and a decrease of 10% in rainfall, there is a 25% change in total annual runoff for the Awash River basin. Hassan (2006) assessed the impact of climate change on the water resources of the Lake Tana sub-basin and showed that if rainfall decreases by 10%, the mean annual runoff falls by 29%.

As indicated by Stern (2006) and the NMA (2007), among others, the potential impacts of climate change may include:

- Decreases in the effective runoff and water storage in the lakes
- Increases in the variability of weather such as the failure of the small rains and delays to the onset of the main rains

- Reductions in crop yields and increases in food insecurity, hunger, malnutrition and death
- Degradation of aquatic habitats, fisheries, rangelands and biodiversity
- Expansion of desertification (human-induced land degradation)

The over-abstraction of water for domestic consumption and irrigation as well as the impact of climate change are most clearly going to be observed in the changing water levels of the terminal lakes of the RVLB. The effects of climate change would be most felt in poor communities characterised by low income levels, poor health, limited adaptive or coping mechanisms and high reliance on natural resources for their livelihoods.

Constraints on agricultural development in the RVLB

Land suitability in the RVLB

The results of the land suitability evaluation for the RVLB for all five land uses are shown in Table 8.9 for the whole RVLB. Not surprisingly, given the climate, terrain and soil characteristics, there is very little highly suitable land for any land use. In terms of moderately suitable land, forestry (37%) and smallholder rainfed agriculture (36%) occupy the greatest area with an appreciable amount also suitable for mechanised rainfed agriculture (17%) and a far lesser area (3%) moderately suitable for irrigation. The highest proportion of marginally suitable land is taken by irrigation (41%) followed by smallholder rainfed agriculture (34%), mechanised rainfed agriculture (26%) and forestry (8%). The land use with the highest proportion of not suitable land is rangeland at 58%, essentially land above 1500 masl, followed by irrigation at 51% due largely to slope constraints and mechanised rainfed agriculture at 50% due both to slope and climatic constraints. Production forestry is not suitable in 47% of the area mainly due to climatic constraints, and smallholder rainfed agriculture is not suitable in

22% of the basin again largely due to climatic constraints but with some erosion hazard constraints. Overall, the land evaluation study identified about 170,000 ha of highly but mostly moderately suitable land for irrigation, which is not dissimilar to the areas of planned irrigation identified by the Ministry (Table 8.8).

Response of the lakes of the RVLB to future planned irrigation

WEAP was used to model the impact of varying levels of irrigation development on lake water levels and areas. The results of irrigation expansion on the Ziway–Abijata Lakes are described by Lowe and Francis (2010). Full development of 49,250 ha of new irrigation identified in this sub-catchment leads to year-on-year decreases in lake levels on Lake Ziway (Figure 8.4a and b). The long-term average (LTA) lake level falls by 1.3 m and the lake area falls by 24% over the 30-year simulation period. More importantly, the lake level falls below the geological sill height which controls overflows to the Bulbula River and Lake Abijata. Lake Ziway becomes a terminal lake and inflows to Lake Abijata are much reduced.

The potential impact on Lake Abijata is catastrophic (Figure 8.4b). For example, when the flow in the Bulbula River is halved, the LTA of

Table 8.9 Results of the land suitability evaluation.

Land use type (LUT)	Highly suitable (S1)		Moderately suitable (S2)		Marginally suitable (S3)		Not suitable (N)	
	Area (km²)	%	Area (km²)	%	Area (km²)	%	Area (km²)	%
Smallholder rainfed	1,457	2.7	18,947	35.7	18,158	34.2	11,807	22.3
Mechanised rainfed	1,286	2.4	8,732	16.5	13,867	26.1	26,483	49.9
Large-scale irrigated	289	0.5	1,437	2.7	21,719	40.9	26,923	50.7
Production forestry	1,639	3.1	19,850	37.4	4,023	7.6	24,856	46.8
Rangeland			19,726	37.2			30,643	57.8

The LUTs are not mutually exclusive.

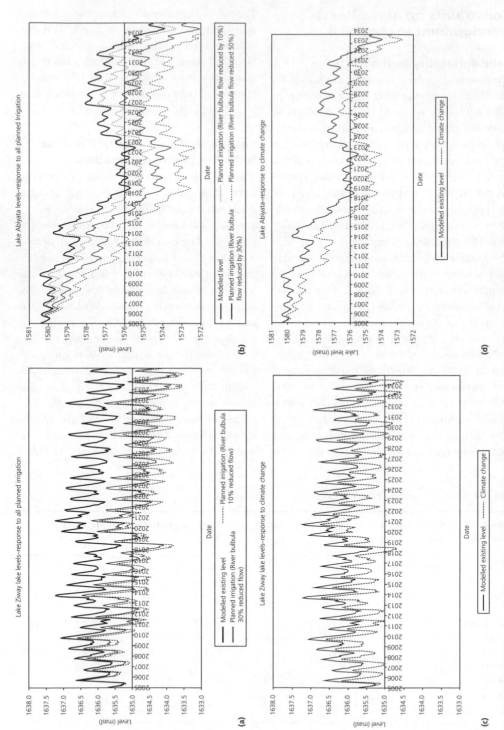

Figure 8.4 Modelled water levels on Lakes Ziway and Abijata for different areas of irrigation development (a and b) and climate change (c and d).

the lake level on Abijata decreases by 2.64m and the lake area reduces by 35%. By the end of the modelled time series in year 2034, lake levels have fallen by 2.84m and the lake area has declined from the current area of 82.3 to 28.6km^2, shrinking by a further 65.3%. If the Bulbula River flow were to be depleted by more than 70%, the model shows that the reduction in flows together with the evaporative losses would result in the disappearance of Lake Abijata.

Changes in lake morphology significantly affect the water quality of the lakes due to changes in dilution capability. Lake Ziway is the largest freshwater lake in the RVLB, and by reducing the outflow from Lake Ziway to the Bulbula River, this would reduce the flushing ability of Lake Ziway, causing an increase in salinity and pollution levels (Jansen et al. 2007). Because of its shallow depth, there is real danger that Lake Ziway is turned into a saline terminal lake over the few next decades due to over-abstraction of water for irrigation. This would have serious repercussions on the livelihoods of the many local communities who depend on water from Lake Ziway for

small-scale pumped irrigation and domestic water supply, as well as recent floriculture development (Jansen et al. 2007).

Capping irrigation expansion

The historic record shows that water levels on some of the terminal lakes are declining and this is mostly attributed to abstraction of water for irrigation. Further irrigation development will result in continued falls in lake levels. The questions then arising are, if irrigation development is important to the development of the economy of the RVLB, what would be an acceptable level of environmental degradation, as measured by changes in lake morphology, and how much new irrigation can be supported for a given environmental threshold?

Table 8.10 shows the approximate areas of potential irrigation that could be developed, were the irrigated area to be capped at an environmental impact threshold defined as a lowering of lake levels from the LTA of 10 and 25%. These thresholds are arbitrary, but they serve to illustrate the following points. Even modest increases in irrigation of between 13,250 and

Table 8.10 Area of future irrigation when capped by an impact threshold defined as a lowering of lake levels from the long-term average (LTA) of 10 and 25%.

Lake	Impact threshold of 10% of average lake levels				Impact threshold of 25% of average lake levels			
	10% fall in LTA lake level (m)	Proposed irrigation (ha)	Change in lake level for proposed irrigation (m)	Change in lake area for proposed irrigation (%)	25% fall in LTA lake level (m)	Proposed irrigation (ha)	Change in lake level for proposed irrigation (m)	Change in lake area for proposed irrigation (%)
Ziway	0.11	5,500	−0.05	−1.0	0.26	8,000	−0.17	−3.0
Abijata	0.64	0	0.64	−8.5	1.06	0	−1.06	−14.2
Langano	0.13	750	−0.10	−0.5	0.33	1,500	−0.15	−0.7
Shala	0.18	1,000	−0.13	−0.12	0.44	3,000	−0.40	−0.4
Awasa	0.19	500	−0.43	−1.7	0.48	500	−0.43	−1.7
Abaya*	0.20	5,500	−0.13	−0.7	0.44	9,000	−0.22	−1.2
Chamo*	1.00	0	−1.38	−8.8	1.01	0	−1.86	−9.6
Total		13,250				21,000		

*Planned irrigation is permitted in the Abaya basin, but not in the Chamo basin.

21,000 ha for the whole of the RVLB will result in appreciable changes in lake morphology. For example, the 10% impact threshold results in reductions in lake areas of 8.5 and 8.8% for Lakes Abijata and Chamo, rising to 14.2 and 9.6% for the 25% impact threshold. The capped levels of irrigation development would increase existing irrigated lands by between 50 and 70%, well short of the sixfold increases indicated by the land suitability assessment.

The effects of climate change on water resources

WEAP was used to model climate change during the 30-year simulations using current levels of water use for irrigation and domestic water supply requirements. The response of water levels on Lakes Ziway and Abijata is shown in Figure 8.4c and d. Both lakes show a continual decline in lake levels over the 30-year period. On Lake Ziway, the LTA lake water level falls by 0.57 m with a reduction in lake area of 11%, while the impact on Lake Abijata is an overall fall in the LTA water level of 1.3 m and a 17% decrease in lake area which is equivalent to a 25% reduction in flows in Bulbula River (Table 8.11).

The impact of climate change in the RVLB is high which is expected as most of the lakes are terminal. However, it is worth noting that the impact of climate change on lake levels is less than that modelled for full irrigation development of 49,250 ha over the same period. The superimposition of irrigation development and climate change has not been modelled, but clearly this would accelerate the degradation of the lakes.

Discussion

Environmentally acceptable levels of degradation

Given that increasing abstraction from the rivers impacts on the levels and water quality of the Rift Valley Lakes, an important area for discussion for policymakers, environmental conservation bodies and local communities alike is how much degradation of lake morphology, wetland habitat, wildlife and water quality would be acceptable as a trade-off to develop irrigation to improve agricultural yields, stimulate agro-industrial development and improve the income of farmers. This is a difficult question to answer.

Table 8.11 Long-term average (LTA) lake level response to climate change.

Lake	Level threshold 10% of average lake level	All planned irrigation	Capped irrigation expansion	Climate change	Climate change
			LTA lake level change (m)		Change in area (%)
Ziway	−0.11	−1.30	−0.05	−0.57	−10.7
Abijata	−0.64	−1.57	−0.64	−1.28	−17.0
Langano	−0.13	−0.10	−0.10	−1.08	−3.1
Shala	−0.18	−0.46	−0.07	−1.74	−0.6
Awasa*	−0.19	−11.05	−0.26	−1.39	−5.5
Abaya†	−0.20	−1.66	−0.03 (−0.15)	−1.73	−9.5
Chamo†	−1.00	−6.50	−1.38 (−1.28)	−6.87	−18.5

*All planned irrigation is for 10,000 ha in Awasa.
†First number is for an additional 1500 ha in Abaya basin and 4000 ha in Chamo; second number in parentheses is for an additional 5500 ha in Abaya basin only.

There is a need for further scientific studies on the impact of lake level change on lake morphology, water quality and wetland ecology in order to understand better the effects of falling lake levels on the natural environment. There is also a socio-economic dimension to the discussion, as the lakes and wetlands are used by local communities for other uses such as fishing, collecting reeds for thatch, cut and carry of grasses for animal fodder, dry season grazing and hunting, in addition to any aesthetic value the lakes and wetlands may have for local communities.

In a household survey undertaken by the Halcrow Group Ltd in association with GIRDC of 183 household heads currently farming in a proposed new irrigation scheme on the western short of Lake Chamo, 96% of respondents were very positive towards maintaining water levels on Lake Chamo, and over 80% said that they were willing to receive less irrigation water supply with project in order to achieve this, with the average response prepared to accept a 15% reduction in irrigation water to conserve the lake. While it is difficult to know whether the respondents understood the meaning of such a reduction, clearly the conservation of Lake Chamo is important to them.

Given the above, it is worth asking who should be making the decision on what the acceptable level of change should be. Inevitably, the Ministry is responsible for approving and implementing major irrigation schemes, but the impacts on lake levels would be felt by local communities and not necessarily those benefiting from the new irrigation systems. The wider environmental lobby also has interests in the conservation of the Rift Valley Lakes, which support migratory bird species.

At present, much irrigation follows traditional methods, so a move to more efficient water use through drip and sprinkler methods has potential to expand the area under irrigation for the same level of water abstraction. While private investors have adopted these methods, there has been widespread reluctance to invest in the necessary on-farm equipment for smallholder schemes, given factors such as dependence on traditional farming methods, low educational levels and lack of operation and maintenance capacity among smallholders. Given that large numbers of smallholders have taken up small-scale pumped irrigation schemes, for example, around Lake Ziway, there may be a case to trade off investment in new irrigation schemes with investment in rehabilitation of existing irrigation schemes, the institutional capacity of farmer extension programmes and farmer training.

A final point to consider is that the investment required to construct new irrigation schemes is substantial and benefits a relatively small number of beneficiaries and consideration should be given to whether similar sums could be spent to benefit larger numbers of recipients.

Implications for agricultural development in the RVLB

The scope for improving agricultural yields by conversion to irrigated farmland is constrained by the lack of water resources in the RVLB rather than the lack of suitable land. There is scope to improve irrigation efficiencies as described earlier. However, economic growth based on agriculture requires improved production in the dryland sector which dominates the agricultural sector accounting for 98% of the RGDP from crops in 2005 through increased yields and the expansion of cultivation into new areas. Agricultural expansion is initially easier than raising yields, as evidenced by current trends (Table 8.7).

Based on the land suitability evaluation, 3.86 million ha is suitable for smallholder cultivation, of which 4% is highly suitable (S1), 49% is suitable (S2) and 47% is marginally suitable (S3). Given that the present cultivated area of 1.98 million ha is a subset of this, then it could be assumed that 95% of S1 and S2 land is already committed, leaving about an additional 2 million additional hectares that is available for cultivation, mostly comprising S3 land.

The area of land suitable for agriculture but not currently being cultivated has been estimated as

(i) some 0.45 million ha of farmland which is not being cultivated based on agricultural statistical returns and (ii) 1.5 million ha of natural and semi-natural habitats suitable for agriculture from the land cover mapping. This additional land would approximately double the area of land currently under cultivation, but bringing this land into cultivation poses major difficulties. The land on farmland which is not being cultivated includes land which is left fallow, field boundaries, patches of steeper slopes or poorer soils within the farm unit. To bring this land into cultivation requires increased use of inputs (fertiliser), land levelling and other on-farm improvements.

The 1.5 million ha of moderately and poorly suitable land (S2 and S3) which could be converted to agriculture mostly lies in the lowlands where unreliable rainfall and poor soils already provide constraints to agriculture (Figure 8.5). Many of these areas have intensive seasonal and, near permanent waterbodies, all-year risks of malaria transmission which has a debilitating impact on the health of smallholder farming communities. Tsetse fly infestation is common below 1500 masl, although it is also found at higher altitude, which causes trypanosomiasis (sleeping sickness) in cattle on which small-holders are reliant for land preparation. Without mechanisation on subsistence and smallholder

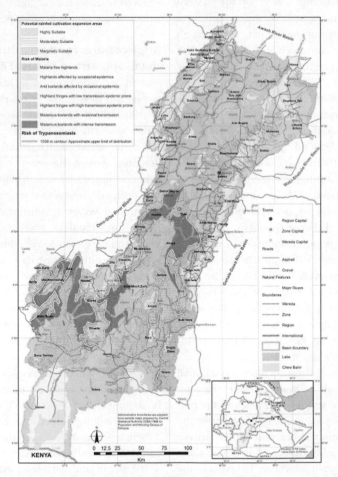

Figure 8.5 Environmental health risks in potential agricultural expansion areas.

farms, the risk of trypanosomiasis on cattle alone could inhibit much of the agricultural expansion into the lowlands. Agricultural expansion into these areas would need to be supported by considerable investment in public and environmental health.

The conversion of 1.5 million ha (15,000 km²) of natural habitats to smallholder agriculture would also impact severely on the natural habitats and fauna of the southern lowlands. Shrublands cover some 19,494 km² of the RVLB, much of it found in the more arid, lower-lying central valley (Table 8.3). Agricultural development of suitable agricultural land may see the loss of up to three-quarters of this habitat.

Conclusions

Economic development in the RVLB is dependent on improvements in agricultural production and industrial development based on agro-processing. Improvement in agriculture through irrigation expansion is constrained due to water scarcity, although there is scope for improvements through greater water use efficiency and a modest expansion of the area under irrigation, provided further lowering of water levels in the lakes is acceptable to stakeholders. Failure to control the ongoing expansion of irrigation to within acceptable limits will lead to major adverse impacts on the Rift Valley Lakes, including the reduction in the lakes, increasing salinity and the disappearance of one or more lakes, with Lake Abijata being the most fragile in the basin.

Agricultural production is dominated by dryland smallholder mixed farming systems. Current yields and a growth rate in yields of 1% per annum are very low. Raising the growth rates in yields to 3% per annum would double production within the plan period, but even this is a poor rate of return given that in many cases, yields could be improved fourfold. Improving agricultural yields requires considerable investment. On the supply side, interventions are needed for soil

and water conservation; improvements in the productivity of crops, livestock and forestry; and rural services including education, health and agricultural extension services. On the demand side, yield improvements will be stimulated by the increasing and changing demands from the urbanising population and through investment on infrastructure, especially for transportation and communications to access markets.

Without improvements in yields on existing agricultural land, current levels of agricultural production could only be maintained over the plan period through rural-to-rural migrations, largely due to agricultural expansion into areas which are marginally suitable for smallholder cultivation. However, agricultural expansion in these areas would need to be coupled with investment in disease vector control, public health and veterinary services and the mechanisation of smallholder farms to address the problems of vector-related diseases.

A major environmental consequence of smallholder expansion would be the loss of extensive areas of natural habitats, particularly shrublands, and the wildlife dependent on them. Like-for-like mitigation of the loss of such extensive areas is not possible, although some compensation could be achieved through the delineation of the best habitats for nature conservation, including linking corridors, and investment in the habitat and wildlife management of these areas.

If the environment of the potential agricultural expansion in the southern lowlands proves too hostile for smallholders, who are mostly highlanders, then rural-to-urban migrations would be more likely than rural-to-rural migrations. This raises the prospect of rapidly growing urban centres, environmental and health problems associated with poor urban planning and the lack of built sanitation and high rates of under- and unemployment and associated socio-economic problems.

On a final note, much of the previous discussion is based on the existing levels of exploitation of natural resources and future trends given predicted population increases, without taking account of climate change. The carrying capacity of the RVLB for the current smallholder population is at or near its limits, and any change towards greater aridity would exacerbate the overlying trends. There is an urgent need for a step change in agricultural production to raise yields, promote off-farm employment through agro-processing and control the rate of population growth.

Acknowledgements

This paper arises from a consultancy contract commissioned by the Ministry of Water and Energy of the Government of Ethiopia from Halcrow Group Ltd in associated with GIRDC of Ethiopia. The content of this paper draws upon the work of a multidisciplinary team of over 30 specialists. We would like to thank our colleagues for their contributions and companionship over the course of three years and the Ministry for their support.

Personal note by Dr Carolyn Francis. Professor John Thornes was my PhD supervisor in the mid-1980s on a research project on gully growth and hillslope erosion in semi-arid Spain. I subsequently worked for him as a postdoctoral researcher on the interrelation between soil erosion, vegetation cover and hillslope hydrology, again in Spain. These studies integrated extensive field work, laboratory soil testing and mathematical modelling of soil erosion. Although very much a process geomorphologist, Professor Thornes was also interested in the influence of people's activities on soil erosion. I have now spent 20 years in environmental consultancy assessing the environmental impact of development projects. I have taken from Professor Thornes an enthusiasm for multidisciplinary environmental studies and a love of fieldwork. While the work described in this paper is a collective effort, rather than an example of my own work, I hope that Professor Thornes would have found it interesting. Professor Thornes was not only an inspirational academic but, with the help of his family, also very generous in his support towards his students. He is sadly missed.

References

Adams, W. M. (2009) *Green Development: Environment and Sustainability in a Developing World* (3rd Ed.). Routledge, London.

Brundtland, G. H. (1987) *Our Common Future. Report of the World Commission on Environment and Development.* United Nations, New York.

Council for Scientific and Industrial Research (CSIR) (2000) *Guideline Document: Strategic Environmental Assessment in South Africa.* CSIR, Pretoria.

CSA/EDRI/IFPRI (2006) Atlas of the Ethiopian Rural Economy. Available at: http://www.ifpri.org/node/3763 (accessed 19 March 2015).

CSA (2002) *Ethiopia Agricultural Sample Enumeration (EASE) 2001/02.* Central Statistical Agency, Addis Ababa

Dalal-Clayton, B., Sadler, B. (2005) *Strategic Environmental Assessment: A Sourcebook and Reference Guide to International Experience.* Earthscan, London.

Edwards, S. (1996) *Important Bird Areas of Ethiopia: A First Inventory.* Ethiopian Wildlife and Natural History Society, Addis Ababa.

Food and Agriculture Organisation (1976) *A Framework for Land Evaluation.* Soils Bulletin No. 32. FAO of the United Nations, Rome.

Gilbert, R., Stevenson, D., Girardet, H., Stren, R. E. (1996) *Making Cities Work.* Earthscan, London.

Grasse, F., Street, F. A. (1978) Late Quaternary lake level fluctuations and environments of the northern Rift Valley and Afar Region (Ethiopia and Djibouti). *Palaeogeography, Palaeoclimatology, Palaeoecology* **24**, 279–325.

Halcrow Group Ltd in association with GIRDC (2008) Rift Valley Lakes Basin Integrated Resources Development Master Plan Study Project. Phase 1 Final Report. Unpublished report to the Ministry of Water Resources, Ethiopia.

Halcrow Group Ltd in association with GIRDC (2009) Rift Valley Lakes Basin Integrated Resources Development Master Plan Study Project. Phase 2 Final Report. Unpublished report to the Ministry of Water Resources, Ethiopia.

Halcrow Group Ltd in association with GIRDC (2010) Rift Valley Lakes Basin Integrated Resources Development Master Plan Study Project. Phase 3 Final Report. Unpublished report to the Ministry of Water Resources, Ethiopia.

Hassan, R. (2006) Vulnerability of Water Resources of Lake Tana, Ethiopia to Climate Change. CEEPA Discussion Paper No. 30. CEEPA, University of Pretoria, South Africa.

Hulme, M., Doherty, R., Ngara, T., New, M., Lister, D. (2001) African climate change: 1900–2100. *Climate Research* **17**, 145–168.

Jansen, H., Hengsdijk, H., Legesse, D., Ayenew, T., Hellegers, P., Spliethoff, P. (2007) Land and Water Resources Assessment in Ethiopian Central Rift Valley. Alterra-rapport 1587. Alterra, Wageningen. Available at: http://edepot.wur.nl/19397 (accessed 19 March 2015).

Kinfe, H. (1999) Impact of climate change on the water resources of the Awash River Basin, Ethiopia. *Climate Research* **12**, 91–96.

Kjorven, O., Lindhjem, H. (2002) *Strategic environmental assessment in World Bank operations – experience to date – future potential*. Environment Strategy Papers Series No. 4. The World Bank, Washington, DC. Available at: http://documents.worldbank.org/curated/en/2002/05/1993711/strategic-environmental-assessment-world-bank-operations-experience-date-future-potential (accessed 19 March 2015).

Lowe, A. T., Francis, C. F. (2010) Water Resources Modelling in the Rift Valley Lakes Basin, Ethiopia – a Tool to Define Acceptable Limits of Environmental Degradation to Underpin Economic Development Strategy. IWA World Water Congress and Exhibition, 19–24 September 2010, Montreal, Canada.

Tadege, A. (ed.) (2007) Climate Change National Adaptation Programme of Action (NAPA) of Ethiopia. National Meteorological Agency, Addis Ababa. Available at: http://unfccc.int/resource/docs/napa/eth01.pdf (accessed 19 March 2015).

Seleshi, B. (2006) Modelling natural conditions and impacts of consumptive water use and sedimentation of Lake Abaya and Lake Chamo, Ethiopia. *Lakes & Reservoirs: Research and Management* **11**, 73–82.

Stern (2006) Stern Review: The Economics of Climate Change. HM Treasury, London. Available at: http://webarchive.nationalarchives.gov.uk/20130129110402/http://www.hm-treasury.gov.uk/stern_review_report.htm (accessed 19 March 2015).

Thérivel, R., Partidário, M. R. (1996) *The Practice of Strategic Environmental Assessment*. Earthscan, London.

USAID (2005) *Southern Nations Nationalities and Peoples' Regional State Livelihood Profiles*. United States Agency for International Development, Washington, DC. Available at: http://www.dppc.gov.et/Livelihoods/Downloadable/Regional%20Overview.pdf (accessed 19 March 2015).

Wood, R. B., Talling, J. F. (1988) Chemical and Algal Relationships in a Salinity Series of Ethiopian Waters. *Hydrobiologia* **158**, 29–67.

Zeray, L., Roehrig, J., Checkol, D. A. (2006) Climate Change Impact on Lake Ziway Watershed Water Availability. Conference on International Agriculture Research for Development, 11–13 October 2006. University of Bonn, Bonn.

CHAPTER 9

Modelling hydrological processes in long-term water supply planning: Current methods and future needs

Glenn Watts

Evidence Directorate, Environment Agency, Bristol, UK

Department of Geography, King's College London, London, UK

Introduction

This paper explores the way that hydrological and geomorphological processes are represented and modelled in planning for the provision of water supply in England and Wales. Water supply is one of the most important practical applications of hydrological theory, directly affecting the lives of the whole population every day and underpinning the entire economy (NAO 2005).

Clean reliable water supply is a familiar and essential part of life. To meet people's water requirements, water companies in England and Wales take around 15 million cubic metres of water from the environment every day, or just under $6\,km^3$ a year (Ofwat 2009). This is about 8% of average runoff (Environment Agency 2001). About 70% of public water supply is from the surface water (rivers, reservoirs and lakes) and 30% comes from groundwater (Defra 2009). About 58% of this water is used in homes, 23% is in business, commerce, industry and the public sector, and the remaining 19% is lost through leakage or used in system operations (Figure 9.1).

In England and Wales, water supply is provided by around twenty private water companies. These companies have statutory duties that include making and publishing plans for the long-term management of water supplies. The preparation of such plans on a consistent basis was first proposed in 1996 (Department of the Environment 1996), and 2 years later, all water companies contributed voluntarily to a major survey of the yield of all public water supply sources in England and Wales (Environment Agency 1998). Subsequently, water companies produced voluntary water resources plans in 1999 (Environment Agency 1999) and 2004 (Environment Agency 2003) before the first plans prepared on a statutory basis in 2009. These plans, of course, build on a long tradition of public water supply planning over the 20th century, including the strategic assessments of water supply options by the Water Resources Board in the early 1970s (e.g. Water Resources Board 1973).

Conceptually, the water resources planning process is simple. A forecast of available water is compared with a forecast of demand. As long as supply exceeds demand, no action is required. When demand approaches supply, actions must be taken to restore supply to a safe margin above demand (Figure 9.2). The conceptual simplicity is, of course illusory – an assessment of the available supply requires a detailed

Monitoring and Modelling Dynamic Environments, First Edition. Edited by Alan P. Dykes, Mark Mulligan and John Wainwright.

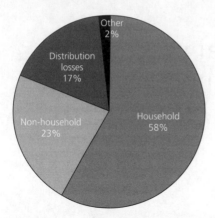

Figure 9.1 Water use from public water supply in England and Wales, 2008–2009 (data from Ofwat 2009).

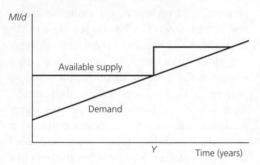

Figure 9.2 Conceptual approach to the supply–demand balance. While available supply is greater than demand, no intervention is necessary. In year Y, demand equals supply; as a result, additional supply is made available so that supply continues to exceed demand.

Table 9.1 Stages in the promotion and construction of a public water supply reservoir.

Reservoir stage	Duration (years)
Clarify scale of deficit	1
Identify possible schemes and carry out initial investigations	2–3
Select preferred scheme and develop a detailed plan	2
Public inquiry including preparation	3–5
Construction including necessary infrastructure	5–10
Filling and commissioning	1–2
Total	**14–23**

understanding of the natural variability of precipitation and evapotranspiration, of catchment hydrology and of the behaviour of the physical water supply system. In particular, it is important to understand how the system responds to droughts. This is difficult because in most cases there is very limited experience of drought and little relevant measured hydrological data.

Typically, water supply plans look 25 years ahead. This reflects the maximum time needed to construct the most difficult water resources schemes from inception to operation. This may seem excessively long but becomes more comprehensible with a breakdown of the individual

stages (summarised in Table 9.1). Suppose the initial analysis demonstrates that a large new scheme will be needed. Further work will be needed to clarify the scale of the deficit (1 year). A long list of possible schemes is developed together with initial feasibility studies, for example, looking at site geology (2–3 years). A decision is made on a preferred scheme, and detailed plans are drawn up (perhaps 2 years). As the scheme is large, a public inquiry may be necessary (up to 5 years). Ten years has now elapsed before construction can start. A major civil engineering scheme like a reservoir may take 5–10 years to construct and another 18 months to 2 years to fill and bring into operation. Carsington Water in Derbyshire and Roadford Reservoir in Cornwall both took over 20 years from pre-feasibility to operation (Wilby and Davies 1997). There is a strong case for looking several decades further ahead in plans: any scheme that is constructed will be in use for many decades, so it may make sense to consider water demand over the lifetime of the system.

Climate change adds a further dimension of complexity to the water supply problem. Climate change affects both supply and demand (Watts 2010) and has been examined in water supply plans in England and Wales since 1999 (see Charlton and Arnell 2011 for a review of the approaches taken in draft plans in 2008).

Climate change presents a particular challenge in estimating the available water supply, as many of the methods used were developed assuming a stationary climate (Milly et al. 2008), and the volume and variety of information on climate change make engagement difficult for the practitioner with little spare time.

Much of the focus of water supply planning falls on the demand for water and any proposed new infrastructure that is needed for meeting this demand. The basic raw material for water companies, though, is water in the environment. A good understanding of catchment hydrology is essential to maintaining secure water supply. This paper explores the way that hydrological processes are represented in water supply planning in England and Wales. Specific hydrological requirements are discussed in the next section. Subsequent sections examine (i) the hydrological models currently used by water companies, (ii) areas where further understanding of hydrological processes would contribute to improved water supply planning and (iii) the benefits of improving the representation of hydrological processes in water supply planning. This chapter concentrates on surface water hydrology, as this is the source of more than two-thirds of water supply in England and Wales. Much of the discussion of principles is relevant also to the assessment of the performance of groundwater sources, although groundwater yield is normally constrained not only by regional groundwater flow but also by local hydrogeological conditions around the source.

Hydrological requirements for water supply planning

Understanding the hydrological requirements for water supply planning requires an examination of the main sources of water used for public water supply, the way that supply systems work and current approaches and possible developments in water supply planning.

Sources of water and supply system function

Surface water abstraction points fall into three broad categories:

- Impounding reservoirs, where the dam wall blocks a valley and captures the entire flow of the upstream catchment. These are typically in upland areas.
- Off-line or pumped storage reservoirs, where some or most of the stored water is pumped from a river. Such reservoirs are generally in lowland areas.
- Direct abstraction from rivers and lakes. These are sometimes small abstractions serving rural communities. More often, direct abstractions are used in conjunction with other sources.

In each case, to forecast source performance, it is necessary to understand the hydrological response of the catchment to the abstraction point and how this may change over time. All catchments in England and Wales are influenced by human activity to some extent: even for small upland catchments, hydrological response is conditioned by managed activities such as grazing. Most public water supply abstractions come from medium to large catchments that are affected by a wide range of anthropogenic activities (Weatherhead and Howden 2009; O'Driscoll et al. 2010). Most obvious are abstractions and discharges, modifying flows directly in a way that may change from hour to hour and day to day. Channel modification, for example for flood management, and impoundments such as weirs also alter the timing and magnitude of flows. Land-use and cropping patterns affect evapotranspiration and runoff. Urban surfaces and drainage systems collect water and deliver it to river channels very quickly.

Water supply systems are designed to smooth the natural variability of climate to provide secure water supplies through all but the worst droughts. The ultimate aim of a water supply system is to transform the highly non-linear hydrological system to a linear, predictable water supply system that can provide water through any period of dry weather. In practice, this aim is tempered by practicality and cost: it is not possible to provide unlimited supplies through very rare and unusual droughts, so supply systems are designed to meet some sort of standard. This design standard is often expressed as a frequency of failure. For example, hosepipe bans or other restrictions on outdoor water use may be planned to occur no more than once every 10 years, and further restrictions on water use may be expected once every 30 or 40 years. It is clear that in assessing water supply system performance, it is necessary to understand the types of drought that may occur very rarely, which means that very long flow sequences are necessary. On the other hand, it is rarely necessary to understand the precise timing and magnitude of high flows and floods. Upland impounding reservoirs store all the water that arrives until they reach a target level (typically close to full, unless levels are being held low for downstream flood protection). Detailed flood models can be important operationally, for example, to manage reservoir levels to avoid damaging spills, but are not needed for long-term planning. In the lowlands, the maximum volume of abstraction during floods is limited by pump capacity, so a detailed understanding of peak flows is normally superfluous. For both upland and lowland reservoirs, an assessment of mean daily flows usually provides sufficient precision for planning purposes. Even using monthly flows may not introduce significant error, especially for large schemes. However, as most abstraction rules are set with daily conditions, it is usually most convenient to use sequences of daily mean flow.

Approaches to water resources planning

The conventional approach to water resources planning simplifies the complex problem to a comparison of a single, annual measure of average demand with a measure of reliable supply (Environment Agency 2008; Watts 2010; Hall et al. 2012). In the United Kingdom, the measure of reliable supply is known as the yield or deployable output (Environment Agency 2008). For stand-alone sources, the deployable output is simply the lowest volume of water available over a defined period or through a drought with a specified probability or return period. There are very few stand-alone sources in England and Wales: most are used in conjunction with other groundwater or reservoir sources. For this reason, the rest of this section concentrates on the calculation of reservoir deployable output.

For reservoirs, deployable output is usually calculated by simulation modelling. A simple model calculates daily reservoir level using a water balance approach, where inflows come from the natural catchment, and, if appropriate, by pumping from a river. Daily demand is taken from the reservoir. Deployable output is calculated by iterating demand to find the maximum demand that can be met within predefined failure conditions. The failure condition can be simply that the reservoir must not empty. In more sophisticated simulations, restrictions on water use can be introduced, and the failure conditions can then include a target frequency for restrictions. The simulation can be over a specified record length: for example, the 1998 review of water company yields (Environment Agency 1998) specified a period from 1920 to date to include the very dry year of 1921 and the serious drought of 1933–1934, as well as more familiar recent droughts such as of 1976 (Marsh et al. 2007b reviewed the drought record of the 20th century). Alternatively, the system may be expected to perform adequately through a drought with a specified return period, such as 1 in 50 years.

In reservoir simulation modelling, a long historic record is used to try to characterise natural variability, recognising both that reservoirs smooth climatic variability and that the most testing droughts tend to be of long duration. It is important to note that reservoir performance in a specific year such as 1921 is of little interest from a water resources planning perspective.

Indeed, in many cases, the reservoir did not actually exist during the most testing historic periods. Instead, the planner needs to understand how the reservoir will perform in future droughts. The historic (or reconstructed) record is used as a surrogate for future conditions. Implicit in this is an assumption of climatic stationarity: the climate of the last century is used as a guide to future droughts. Climate change means that stationarity can no longer be assumed (Milly et al. 2008), but outputs from global climate models (GCMs) can be downscaled to provide factors to scale historic rainfall and temperature data to represent future climatic conditions (Vidal and Wade 2006; Fowler et al. 2007; Murphy et al. 2009; Wilby and Fowler 2010; Christierson et al. 2012).

One important problem with the use of historic droughts as a guide to future water resources performance is that even with the United Kingdom's extensive rainfall data, the historic record provides only a small sample of feasible droughts. Future developments in water resources planning may see further use of modelled climate data to provide multiple feasible weather time series both for the historic period and for the future. Studies using multiple climate simulations have demonstrated the opportunities that such an approach may give (New et al. 2007; Lopez et al. 2009; Fung et al. 2013), and it is thought that probabilistic approaches should lead to improvements in water supply planning (Hall et al. 2012). In evaluating future hydrological requirements, this need for efficient and effective hydrological modelling of large climate ensembles must be considered.

Summary of hydrological requirements

Water resources planning, at least when using historical or reconstructed records, appears to have relatively parsimonious data requirements: all that is needed is a long time series of daily mean flow at each intake point. However, the scale of these data requirements is significant.

In England and Wales, there are about 2000 separate abstraction points for public water supply (NAO 2005). Many of the larger intakes are close to permanent river gauging structures, but only a few have records that start earlier than the 1960s. In any case, water resources planners are interested in the catchment's current and future response to drought, rather than past responses.

It is clear that the main hydrological requirement – long time series of river flows at discrete locations using historical or synthetic climate records – can best be met by hydrological modelling. The next section explores the approaches used by water companies in England and Wales to address these requirements.

Current approaches to hydrological modelling in water supply planning

Simplicity is often regarded as a virtue in hydrological modelling (Anderson and Burt 1985; Watts 1997). With a large number of sites to model, it is no surprise that water companies tend to use modelling approaches that require little observed data beyond measured river flows. There is little published or peer-reviewed literature on water companies' hydrological modelling approaches. The information used in this section comes mainly from a survey of water companies conducted as part of an Environment Agency research project in 2009 and presents only a snapshot of the approaches used in the 2009 water resources plans. Just over half of water companies answered questions on approaches to modelling. For water resources planning, hydrological modelling approaches used by water companies fall broadly into two categories: empirical models and conceptual models.

Empirical models
Empirical models make no attempt to understand physical processes and have no explanatory powers (Watts 1997). Water companies make

considerable use of empirical relationships to derive flow sequences at abstraction points. Half of the water companies that responded only use empirical approaches. Many of the others use empirical methods for parts of their supply areas. The methods used are usually based on scaling long flow records from other locations. Where short gauged records exist, scaling factors are often based on linear regression. Where there is no gauged record, the ratio of catchment areas is often used. In both cases, this means that simulated record will exhibit the same pattern as the gauged record, with flows only differing in magnitude.

Conceptual models

Around half of the companies that responded make use of conceptual rainfall–runoff models for some or all of their surface water sources. The main models used are the Stanford Watershed Model (SWM) (Crawford and Linsley 1966) (one company), HYSIM (Manley 1975) (six companies) and Catchmod (Wilby et al. 1994) (one company). Other models may be in use by other companies, and some companies use a wider range of hydrological models for other applications such as site-specific investigations.

These models are conceptually similar (Figure 9.3), though they differ in detail. Used for water resources planning, each takes as input daily precipitation and potential evapotranspiration (PE). Water is routed through a series of subsurface stores that conceptually represent the soil and groundwater zones. HYSIM has four such stores: two represent soil and two represent groundwater. The SWM and Catchmod each have three stores, with one for the soil, one representing deep soil or shallow groundwater and a final store for deep groundwater. All three models have channel routing components but these are seldom used in water resources planning applications. Channel routing allows the timing of flows to be assessed more accurately, but for water supply planning in UK catchments, there is usually little benefit from understanding flow timing on anything

other than a daily average. The SWM and HYSIM have snowmelt modules, but these are not normally used for water resources planning in England and Wales as snow occurs rarely and is not usually important in water supply.

Each model has a large number of parameters. Without snowmelt, the SWM has 25 parameters (Shaw 1988). HYSIM uses over 20 parameters, though some are derived from hydrograph analysis (Manley 2006). Catchmod has around 13 parameters for each zone (Wilby 2005) and is often used with two or three zones representing different parts of the catchment. For both HYSIM (Manley 2006) and Catchmod, the manuals make recommendations about some parameter values but many need to be derived by calibration. The most commonly used version of Catchmod is built in a spreadsheet and has to be calibrated manually. HYSIM can be calibrated manually but there is also an optimisation routine that can search automatically for values for up to six pre-defined parameters (the user cannot change the parameters that can be optimised).

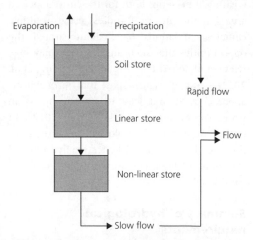

Figure 9.3 Typical conceptual rainfall–runoff models. Rainfall and evaporation act on the soil store. Excess water becomes rapid flow. Water also flows from this store into lower stores and ultimately becomes slow flow. Models may have more than three stores and may generate flow directly from more than one subsurface store.

Parameter estimation in rainfall–runoff models involves comparing the modelled flow with the recorded flow. One commonly used approach splits the data into two, using one period for calibration and the other for validation. It is also necessary to identify one or more objective functions that can be used to test model fit. All objective functions are to some extent a compromise, in that because they summarise model performance, they are biased towards achieving a good fit in some parts of the hydrograph, potentially at the expense of others (ASCE 1993). Objective functions aid in the assessment of model fit, but they do not find optimal parameter sets. With so many parameters, all of these models may suffer from equifinality (Beven 2001; Ivanović and Freer 2009): multiple sets of parameters may lead to an equally good, but different, model fit. This is not usually a problem for infilling data, but for hindcasting and forecasting, there may be a range of possible feasible results. It is clear that calibration of each of these models is to some extent subjective: different users would achieve different results for the same catchment. These models are a means to an end: once calibrated, the model is used to produce flows that can be used in further modelling. It is unlikely that model calibration will be revisited unless the results are very clearly erroneous. This means that any information about possible errors is likely to be lost. Modelled time series will be treated as accurate forecasts rather than the approximations that they really are. This is a particular problem because some of the errors are likely to be systematic rather than random. For example, low flows may be consistently over- or underestimated, and this may lead to bias in modelled deployable output.

It is worth noting that the gauged river flow record itself will also contain errors. Errors may be consistent (e.g. a gauging station may always underestimate high flows), may change over time (e.g. with the accumulation of sediment) or may be effectively random (e.g. a log may be stuck on a weir for a few hours). Gauging authorities try to correct obvious errors and usually provide valuable information about the quality of stations and even individual days' flow records (see the UK National River Flow Archive for examples). In model calibration, much of this information is lost or ignored by practitioners partly because it is difficult to make sense of such detail but also because practitioners rarely have experience or expertise in flow measurement. Gauging is usually carried out by a different organisation, which in itself is a barrier to understanding and accommodating measurement error in hydrological models.

Modelling the impact of climate change

In recent years, much attention has been paid to the impact of climate change on water supply. The importance of maintaining water supply was underlined by the South East England drought of 2004–2006 (Marsh et al. 2007a) and then the exceptionally dry winters of 2010–2011 and 2011–2012 (Kendon et al. 2013), causing some water users to question the extent to which suppliers are considering the full range of possible future conditions, including the impact of climate change (e.g. POST 2012). Water companies in England and Wales first looked at climate change in their 1999 water supply plans. Conventional methods for the calculation of deployable output implicitly assume that the climate is stationary and that historic droughts provide a useful guide to future droughts (Hall et al. 2012). This approach is well suited to perturbation using the delta change or scale factor approach (Wilby and Fowler 2010) where historic climate series are scaled to represent future conditions. This has the advantage of simplicity but retains the temporal structure of the historic record. For water resources planning, it is particularly important to note that this means that drought sequencing will remain the same: droughts will not become longer, and there will be no change in the number or frequency of dry winters.

Even simple scale factor approaches present some problems for water companies. Until the latest generation of UK climate change projections (UKCP09, Murphy et al. 2009), it was not easy for non-specialists to acquire appropriate change factors. As discussed earlier in this chapter, for many sources, water companies do not have models that can turn climate time series into river flows. Even where rainfall–runoff models exist, they use PE as an input. GCMs simulate most of the components of the common evapotranspiration models such as Penman (1948) and Penman–Monteith (1965), but preparation of scaling factors for PE is an additional data processing problem. To simplify these tasks, water companies and the Environment Agency commissioned a series of research projects, first to provide flow scaling factors (UKWIR 1997) and then to rework the flow factors and add rainfall and evaporation scaling factors (UKWIR 2003, 2006). The 2006 work also used a number of different GCMs and recognised the uncertainty around climate change modelling by providing a range of values around a central estimate.

In practice, most water companies continue to use flow scaling factors to give a central estimate of the impact of climate change on river flows. An observed flow time series is multiplied by monthly factors to provide a new flow sequence that represents different climatic conditions. The flow scaling factors are themselves derived from rainfall–runoff model results. In all three UKWIR studies, rainfall–runoff models were run for 60–70 catchments. Once calibrated, the models were run twice: once with baseline climate data (1961–1990) and again with the climate data perturbed by the monthly scaling factors. Monthly flow factors were calculated from the ratio of 30-year monthly mean flow with and without climate change. The catchments and rainfall–runoff models used changed between the 2003 and 2006 studies. In the earlier studies (UKWIR 1997, 2003), the flow factors were used without adjustment by selecting factors from nearby catchments with similar characteristics. The later study (UKWIR 2006) provided a method to adjust flow factors based on catchment characteristics. All three studies and the associated guidance from the Environment Agency (2008) emphasised the many shortcomings in the use of flow scaling factors and urged the use of appropriate climate data in well-calibrated hydrological models, though unhelpfully there was no explanation of the characteristics of a well-calibrated model.

The representation of hydrological processes

The previous sections of this chapter have concentrated on the approaches used to generate long river flow sequences. Where empirical relationships have been used to calculate flows, there is clearly no representation of hydrological processes. All three rainfall–runoff models make some claim towards representing hydrological processes. Each considers the catchment as a series of stores. All claim that at least some parameters are physical representations of catchment properties. In practice, it is difficult to think of a single value representing a whole catchment and virtually impossible to measure such values meaningfully in the field (Beven 2009). In any case, practitioners are generally happy to adjust any parameter to give a good fit to a measured hydrograph, potentially negating any claim to physical reality other than that imposed by model structure.

There is a second, less obvious, area where hydrological processes are modelled. All the rainfall–runoff models use two input time series. Rainfall is usually measured directly by rain gauges, but PE is modelled from meteorological data. In the United Kingdom, the most readily available PE data is provided by the Meteorological Office's MORECS system. MORECS uses a modified Penman–Monteith formulation calculated weekly with a 40×40 km resolution for the whole of the United Kingdom and is available from 1961 onwards (Hough and Jones 1997). Commonly used MORECS outputs are PE and actual evapotranspiration (AE) for grass and 'real land use', which is a static assessment of the land cover mix in each grid square.

For PE, land use is relevant because it determines the bulk canopy resistance of vegetation (Hough and Jones 1997) and the proportion of bare soil. AE is calculated using a soil moisture accounting scheme to modify evapotranspiration rates. This is used operationally but is not normally part of water companies' hydrological modelling.

PE is an abstract concept that cannot be measured directly and is highly dependent on the formula used for its calculation (Kay and Davies 2008). Evapotranspiration is an important part of the water balance, especially during low-flow periods (Oudin et al. 2005b). Practitioners tend to view PE time series as measured data just like rainfall, but the formulation of the PE model itself is important in determining rainfall–runoff model response. During calibration, rainfall–runoff model parameters will be adjusted so that the modelled flow matches the gauged flow. The parameters that represent soil characteristics determine the relationship between PE and AE. The calibrated parameters will be to some extent a function of the model used to calculate PE. If a different PE formulation is used, the rainfall–runoff model should be recalibrated. There will be particular problems if an extended PE series is derived from different PE models. Many climate change studies do not use Penman–Monteith to estimate PE because it requires some meteorological data not readily available from GCMs. For example, the first release of UKCP09 did not provide probabilistic estimates of wind speed (Murphy et al. 2009). Climate change studies often use simpler, temperature-based models of PE (e.g. Chun et al. 2009) which may be quite effective in representing catchment behaviour (Oudin et al. 2005a, b). In principle, there may be no problem in using temperature-based PE models, provided that they are used consistently throughout model calibration and simulations. In practice, it will be tempting to mix data sources, with unpredictable consequences.

In considering hydrological processes in water supply plans, it is essential to note the scale of the assessments that water companies must carry out. It is not surprising that many simplifications are made and that standard models are used. The next section examines how enhanced representation of hydrological processes could improve water supply planning.

Improving hydrological representation

The previous sections have shown how water companies tend to use simple approaches to hydrological modelling. The youngest of the three models, Catchmod, is over 25 years old (Greenfield 1984); HYSIM is 10 years older (Manley 1975) and the SWM is not far from its 50th birthday (Crawford and Linsley 1966). The main approach to the calculation of PE has its origins in the 1940s (Penman 1948; Monteith 1965). Durability and simplicity may, of course, be beneficial in many ways, but there are also developments in hydrological science that could improve the understanding and estimation of the availability of water. It is worth remembering why accurate estimates of water availability are important. Overestimation could lead to unnecessarily frequent restrictions on supply, while underestimation may lead to wasted investment and unnecessary environmental impact.

Consideration of the needs of water companies and recent hydrological literature reveals five areas where improvements in hydrological representation would be of benefit to water companies:

- Exploration of alternative model structures
- Improved understanding of parameter uncertainty
- Better understanding of catchment change
- Further work on important catchment-scale processes
- Guidance and protocols for consistency

This section explores each of these in more detail, before considering the barriers to adoption of relevant changes.

Alternative models and model structures

There are many different hydrological models. Many were developed for specific research needs and are in limited use (e.g. Baird et al. 1992). Limited distribution allows models to be modified and improved quickly and easily, but such models are unlikely to be used by practitioners. Other models that started as research applications have been taken up more widely. For example, one version of PDM (Moore 1985) became part of the operational MOSES evaporation model (Blyth 2002) which in turn is now part of the Joint UK Land Environment Simulator (JULES) global modelling scheme (Blyth et al. 2006). Some models that used to be available to end users are no longer supported. For example, HYRROM (Blackie and Eeles 1985) is another lumped model conceptually similar to Catchmod and HYSIM, in addition incorporating a relatively sophisticated parameter optimisation facility. This model was available commercially in the 1990s but has fallen out of use. The way that the available models are packaged has a significant influence on the capacity of water companies to explore new approaches and take on new ideas. For example, the latest UK climate projections allow users to explore climate uncertainty by running hundreds or even thousands of different sets of change factors through hydrological models (Murphy et al. 2009). This is effectively impossible with many versions of models like Catchmod because each data set will have to be specified manually, taking a very long time and with a significant risk of error.

The transfer of models from research to practitioners is far from trivial. Practitioners need models that are proven, reliable and stable. There must be clear documentation and effective user manuals. There is little incentive or reward for researchers to package hydrological models for a wider audience, especially as the number of users will be small. There is, though, a need for other models to be available, both to represent some catchments better than the

current models can and also to improve the understanding of uncertainty in model simulations. Many studies of parameter variability use a fixed model structure and vary parameter values (e.g. Wilby 2005; New et al. 2007), but exploring a range of different model structures may be particularly valuable in examining catchment response to changing climate. There may be a role here for regulators to work with water companies to develop a range of effective models to a consistent standard. These would be packaged in a way that allows different calibrations to be explored and permits batch runs. Researchers would have to allow their models to be used by other people without supervision but could benefit from insights resulting from the wider use of their approaches. The transparency of such a modelling approach would allow others to replicate results and understand the importance of different assumptions.

Improved understanding of parameter uncertainty

Water resources planners tend to seek a single, optimal parameter set for a hydrological model. Beven (2009, p. 145) suggests that 'optimisation of environmental models cannot be considered a good strategy when the optimum model found may depend on input and model structural errors'. The 'best' parameter set is not only a function of catchment hydrological response but also changes depending on the period chosen for optimisation (Wilby (2005) illustrates this for Catchmod) and also the metrics used to evaluate model fit. This becomes particularly important where models are to be used to project flows using climate data outside historic experience. Some parameter sets that appear to perform similarly in the calibration period may give completely different results with perturbed climate data (Figure 9.4). This may occur because some sort of hydrological threshold is crossed, but it could also be caused by a problem with the parameter values or the model structure. The challenge for the modeller is to recognise these circumstances and to understand

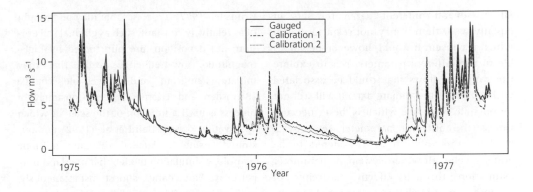

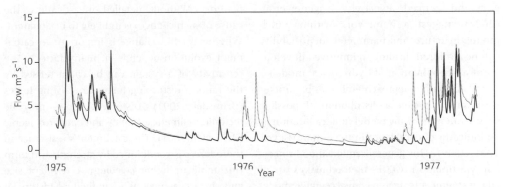

Figure 9.4 Different parameter estimates may give different results outside the calibration period. In this example, two sets of parameters represent the gauged flow equally well (upper graph). With scaled rainfall and evapotranspiration, representing climate change, the results differ, especially at low flows.

what is happening. This is very unlikely to be possible if only a single calibration of the model is available. Indeed, in these circumstances, the modeller is unlikely to realise that anything unusual has occurred.

One way to deal with this problem is to improve the estimation of model parameters. This has often been the preferred solution (Beven 2009), and many different approaches to seeking optimal parameter sets have been developed. Most involve identifying an objective function that describes the difference between modelled and observed flows (e.g. ASCE 1993) and then searching the parameter space for the point at which this objective function is at a minimum. When computing power was limited, efficient search routines were essential (Blackie and Eeles 1985). Where there

are large numbers of parameters, this remains important. There are particular problems if parameters are correlated (Kirkby et al. 1987; Watts 1997), which is often the case in hydrological models: the apparently optimal parameter set found may actually depend on the order in which the parameters were searched by the optimisation routine. With powerful computers, it is possible to search the parameter space randomly or systematically with thousands of different simulations. This is most likely to reveal that a number of different parameter sets are equally good at reproducing observed flows, as far as can be seen using the chosen objective function and within the accuracy tolerances of the data – that is, the problem of equifinality (Beven 2007; Ivanović and Freer 2009). For the practitioner, this may not seem like a problem at all: if several different parameter sets fit equally well, then it may not seem to matter which is chosen. It would, however, be valuable to use different parameter sets to explore the range of flows that could be associated with a given set of climate data. It will still not be possible to decide which is 'best', because unless the different parameter sets reveal model errors, all can be considered to be equally valid. However, systematic parameter estimation through effective searches of parameter space will lead to improved consistency and reduced subjectivity. Wagener et al. (2001) suggest a framework within which model structure and parameter identifiability can be explored, leading to improved development and application of hydrological models.

An alternative approach embraces parameter uncertainty to give a distribution of possible flow sequences, using model fit in calibration as an indicator of confidence in the output (Beven 2001, 2009). This generalised likelihood uncertainty estimation (GLUE) method makes use of all parameter sets that give an acceptable model fit, rather than concentrating only on the best. A probability is assigned to each flow prediction based on the quality of fit in the calibration period. There is an underlying assumption here that the model will perform equally well within and outside calibration periods (Beven 2009). A probability distribution of flow predictions may help with decision-making, though it does not readily fall into the essentially deterministic framework currently used to estimate the supply–demand balance. Decision-making will be considered further later in this chapter.

Better understanding of catchment change

Catchment change is perhaps the biggest challenge facing water resources planners, but it is rarely considered in water resources planning except indirectly as part of a water demand forecast. Geomorphological processes act to change catchments over timescales ranging from a few seconds to many millennia (Thornes and

Brunsden 1977). Processes that may be thought of as relatively constant, such as gradual soil erosion and deposition, are punctuated by high-magnitude, low-frequency events that may initiate significant and irreversible change (Brunsden and Thornes 1979). Constancy of process is itself a function of the scale at which observation occurs: Baird et al. (1992) demonstrated, using a detailed dynamic hillslope hydrology simulation model, that overland flow pathways can change almost instantaneously, with an irreversible impact on erosion rates and patterns. Without detailed investigation, the cause of such changes is unlikely to be apparent. Whether such a change is important in catchment evolution depends on many factors. The sensitivity of a system can be characterised by the ratio of disturbing forces to resisting forces (Brunsden 2001). Disturbing forces may be tectonic, climatic, biotic and anthropogenic (Brunsden 2001). The catchment's resistance to these forces is conditioned by strength and form (including underlying geology, catchment size and shape), as well as its individual history of change (Brunsden 2001; Phillips 2006). For example, the elapsed time since the last significant change event plays a role in determining how the system will respond to the next event. This effect is obvious in flood generation: rain falling on an already wet catchment gives a different response compared to the same volume of rain on a dry catchment. It is also apparent in water pollution events: when a high-flow event follows a drought summer, nutrients and pollutants are flushed into rivers in high concentrations (Whitehead et al. 2009). Geomorphological catchment evolution may be dominated by non-linearity because there are many thresholds in the processes that determine response (Phillips 2006). As the catchment changes, its performance through time will also change (Brunsden 2001).

It is apparent that the water resources planner's simplifying assumption of time-invariant catchment hydrological response must be questioned. While tectonic change is unlikely to be a driver of change in UK catchments in water resources planning timescales, the other drivers of change identified by Brunsden (2001) – climatic, biotic and anthropogenic – are indubitably important. For the water resources planner, the problem is both identifying drivers of change and, perhaps more importantly, understanding the relative importance of the different drivers and the way that these will change catchment hydrological response.

Much – indeed most – attention has been paid to *climate change* (Watts et al. 2015). Climate change will directly alter rainfall patterns and timing (Murphy et al. 2009) and may modify the characteristics and frequency of droughts in the United Kingdom (Blenkinsop and Fowler 2007; Burke et al. 2010). Increases in temperature may enhance evaporation rates (Kay and Davies 2008), while increased CO_2 concentrations may increase plant growth but reduce transpiration (Ficklin et al. 2009) and change recharge to groundwater (Ficklin et al. 2010). Sea-level rise could affect some water sources in lowland catchments and may induce saline intrusion in some coastal aquifers (Ranjan et al. 2006; Bates et al. 2008). Climate change may also increase water temperatures (Webb et al. 2008): combined with lower summer flows, this may lead to increased spread of fish diseases and a loss of some ecosystem services (Johnson et al. 2009). More intense rainfall may drive enhanced erosion and mobilisation of sediment (Orr et al. 2008).

Biotic change may be driven by many factors. In the absence of other disturbance or intervention, plants will grow, die and be replaced by other individuals or species. Human interventions change this: agricultural crops change from year to year, grazing alters vegetation, and apparently natural areas even in the uplands are managed for amenity or other activities (Orr et al. 2008). Unplanned activity may also affect plant distribution: during the 1976 drought summer, there were 10 times as many outdoor fires as in 1974 (Doornkamp et al. 1980).

Anthropogenic changes present the biggest pressures on the UK water environment and perhaps the greatest unknown in projecting future water availability. People take water from the environment for a variety of reasons at tens of thousands of different locations across England and Wales. In 2004, there were nearly 48,000 abstraction licences (NAO 2005); deregulation has now reduced the number of licences but not the number of abstraction points, and some uses of water have never needed legal permission and are unrecorded. Some of this water is returned to the environment later or in a different location. Other human activities also affect the water environment directly. Channels are modified to reduce flood risk, for navigation, for amenity, or to make water easier to abstract. Drainage is improved in agricultural and urban areas to reduce flood risk. Indirect hydrological impacts include land-use change and urbanisation.

Faced with this diverse range of pressures, how should the water resources planner allow for hydrological change in forecasting water availability? The simplest approach is to continue to ignore these changes and assume that past hydrological response is a good indicator of future response. This is becoming increasingly untenable. The reference period for climate change factors of 1961–1990, centred on 1975, is already nearly 40 years in the past. Water supply plans look at least 25 years ahead. Over this period, some geomorphological changes could be important in modifying catchment response. Whole forests could be planted and harvested, and new crops will be introduced. For example, new biofuels are now grown to help to reduce greenhouse gas emissions: these may affect hydrological response (Vanloocke et al. 2010), but there have been few assessments of hydrological impact.

One way to address catchment change in hydrological modelling is to adjust model parameters to reflect future conditions (Wagener 2007). This implies an understanding of the relationship between catchment characteristics and model parameters, as well as an ability to predict the way that important catchment characteristics will change in the future. In a typical rainfall–runoff model such as those discussed earlier in this chapter, this is very difficult. Parameters are adjusted to achieve a good model fit, with little or no reference to catchment conditions. One possible approach would be to extend the use of multiple feasible model calibrations discussed previously. It may be possible to identify areas of the feasible parameter space that are more likely to represent future catchment conditions, though it is not immediately obvious how this would be done. Nonetheless, it may be worth investigating this approach further.

There is a class of rainfall–runoff model that may provide a better way to deal with catchment change. The models are conventional rainfall–runoff models, but their parameters are derived from regionalisation studies (e.g. Hundecha and Bárdossy 2004; Lee et al. 2005; Young 2006; Kay et al. 2007; Pechlivanidis et al. 2010). A large number of hydrological records are used. For example, Young (2006) used 260 separate locations across England, Wales and Scotland. The rainfall–runoff model is calibrated for each location, and then the parameters are related to a range of catchment characteristics. This regionalisation approach allows hydrological models to be applied to ungauged catchments, needing only an understanding of catchment characteristics. In principle, this approach should be able to cope with some degree of catchment change, by changing model parameters in line with the expected change in catchment characteristics. Hundecha and Bárdossy (2004) illustrated this approach for the Rhine in Germany, altering model parameters to examine the impact of urbanisation and afforestation. Bulygina et al. (2009) adjusted soil classification parameters to look at catchment change in mid-Wales. There are limitations: the hydrological models are untested outside their range of calibration, even though this range represents a wider

range of conditions than a single calibration on one catchment. The approach also depends on finding significant relationships between model parameters and catchment characteristics that are expected to change: it is quite possible that the catchment characteristics used may be effectively static and not readily related to catchment change. However, regionalisation of model parameters may provide an objective approach to the assessment of the impact of catchment change on hydrological response.

Modifying model parameters to represent catchment change is one way to improve hydrological representation of changing conditions, but more work is needed before practitioners would be able to adopt this approach. Water resources planners would find it enormously valuable to have a qualitative or semi-quantitative approach of the relative hydrological impact of different catchment changes. Drawing this together is far from trivial: there is much information in the geomorphological and hydrological literature but little work that synthesises this. In the United States, O'Driscoll et al. (2010) looked at the impact of urbanisation and other changes on the hydrology of the southern states. In the United Kingdom, Weatherhead and Howden (2009) reviewed a range of hydrological effects of land-use management, but for most, only a direction of change was reported, with little quantification of the rate of change. Most studies looked only at hydrological response to a single change, such as new crops (Vanloocke et al. 2010) or afforestation (Johnson 1995). Studies looking at the interaction between multiple factors such as river flow and water quality often assumed a fixed hydrological response (e.g. Wilby et al. 2006; Johnson et al. 2009). Table 9.2 is a preliminary attempt to consider the possible direction of change in river flow regimes caused by a range of possible catchment changes. In passing, it is worth noting that many of the cited studies modelled these changes and that there are few field-based comparative studies. Correcting and expanding this table and moving towards a broader quantification

Table 9.2 The impact of catchment change on hydrological response.

Catchment change	Hydrological impact	Source
Afforestation	Reduced peak runoff in floods (12% one catchment, mid-Wales)	Bulygina et al. (2009)
	Reduced annual recharge to groundwater (30–45% in English Midlands), with winter recharge reduced most	Zhang and Hiscock (2010)
	Reduced water yield and reduced low flows but may lead to increased recharge and low flows where soils are degraded. Little impact on big floods but may reduce flash floods	Van Dijk and Keenan (2008)
	Increased actual evaporation and reduced runoff, especially in dry areas	Trabucco et al. (2008)
	Increased infiltration and reduced overland flow compared to pasture, upland mid-Wales	Marshall et al. (2009)
Grassland improvement	Increased runoff from grazed improved grassland compared to unimproved, upland mid-Wales	McIntyre and Marshall (2010)
Intensified grazing	Increased runoff through soil poaching and soil degradation	Weatherhead and Howden (2009)
Urbanisation	Increased direct runoff; decreased actual evapotranspiration (Leipzig, Germany)	Haase (2009)
	Increased recharge from leaking pipes and sewers in some cities; reduction in recharge and low flows because of reduced infiltration in others. Southern United States	O'Driscoll et al. (2010)
	Increased summer peak runoff; little change in winter peak runoff, Germany	Hundecha and Bárdossy (2004)
Soil degradation	Increased peak runoff in floods (8% one catchment, mid-Wales)	Bulygina et al. (2009)
Soil improvement	Reduced flood runoff but <5% in most catchments across England and Wales	Hess et al. (2010)
New crops and cropping patterns	Increased water use from Miscanthus, Midwest United States	Vanloocke et al. (2010)

of these changes would provide an exceptionally useful resource for those involved in practical catchment planning.

Improved understanding of catchment processes

There is, of course, an enormous body of work that addresses catchment processes in England and Wales. Process-based studies usually examine small areas in detail, in contrast to catchment-scale or river basin assessments, which often make statistical inferences about the link between input (e.g. rainfall) and response (e.g. flow). Bridging the gap between small-scale process and whole catchment response is very difficult but is clearly of great importance in the application of hydrological theory to

water supply planning, which tends to be most interested in large catchments. One of the major concerns in climate change research is the possibility of passing the so-called tipping points or thresholds beyond which system response alters in some way (Molina et al. 2009; Russill and Nyssa 2009). Phillips (2006) observed that, contrary to popular belief, real-world geomorphological systems frequently exhibit complex non-linear dynamics. Water supply systems are designed to smooth non-linear catchment hydrological responses: in long-term water resources planning, it is important to understand where changes in processes will lead to a non-linear response that disturbs the capacity of supply systems to smooth hydrological variability.

Droughts present the most testing conditions for water supply systems, but because they occur very rarely in the hydrological record (Marsh et al. 2007b; Hannaford et al. 2011; Watts et al. 2012) and may change in character in the future (Burke et al. 2010), further work on catchment response to drought would be of great value. In the uplands, severe droughts are often associated with events such as drying of organic soils or fires: these change soil and vegetation, sometimes irreversibly (Doornkamp et al. 1980; Orr et al. 2008). The recovery of catchments after prolonged dry periods is particularly important in water supply (Watts et al. 2012), so an improved understanding of the way that drought may alter short- and long-term catchment response would be of particular interest in planning for future conditions. Further analysis of historical drought response would be of particular value. For example, droughts of the 19th century were longer than those of the 20th century (Jones et al. 2006; Marsh et al. 2007b).

Over the 21st century, PE is expected to increase almost everywhere in the world (Bates et al. 2008), mainly as a response to increased temperatures. PE is a poorly defined concept that is hard to relate to measured meteorological parameters on a catchment scale (Oudin et al. 2005a, b), but evapotranspiration is an important part of the water balance, especially at low flows. Further work on evapotranspiration would also improve the understanding of the impact of vegetation and land-use changes on catchment hydrology. It may be possible to improve hydrological understanding by recasting rainfall–runoff models so that they calculate evaporation directly as part of an energy balance. This would add complexity to the models but would eliminate the conceptual problems associated with PE. Some studies take this integrated approach to model evaporation as a dynamic response to atmospheric and land surface conditions (e.g. Rosero et al. 2010). Applying such models on a catchment scale could prove challenging for practitioners: the data requirements are large

and some of the climatic variables may not be available for important historical droughts. However, such approaches may be particularly useful for climate change studies where the output from GCMs or RCMs could be used directly in hydrological models.

Guidance and protocols

Current water resources planning guidance (Environment Agency 2008) does not attempt to address questions of how hydrological modelling should be carried out for water resources assessments. This is perhaps understandable: hydrological modelling needs to be appropriate for the catchment in question, and there is no point in constraining new approaches with unnecessary guidance. However, there may be some benefit in developing guidance on how approaches to hydrological modelling should be reported. Such guidance need not be complicated. It may be sufficient to describe the model, document the parameter values used and provide some indication of the goodness of fit of the model during calibration and validation. ASCE (1993) attempted to introduce consistent reporting of model performance in such studies, and some performance indicators such as the Nash–Sutcliffe efficiency measure (Nash and Sutcliffe 1970) are reported routinely by many researchers. There may also be a case for extending such guidance to include protocols for calibration and validation as well as uncertainty analysis of hydrological models.

Conclusions: Improving decision-making in water resources planning

This paper has explored the representation of hydrological and geomorphological processes in water supply planning in England and Wales, identifying several areas where improvements could be made. These include:

• Exploration of alternative structures in hydrological models

- Improved understanding of hydrological model parameter uncertainty
- Better understanding of catchment change
- Further work on important catchment-scale processes
- Guidance and protocols for consistency

Some, such as improved approaches to model calibration, require the adoption of methods that are already widely used in research applications. Others, such as better understanding of catchment change and evapotranspiration, would benefit from further research. Water resources planning is, though, above all an immediate and practical problem that matters to the entire population of England and Wales and indeed most other countries. The importance of continued reliable water supply means that it should be relatively easy to justify expenditure on research and development, but it is also important to prioritise investment that will lead to improved decision-making.

The traditional approach to water resources planning balances available supply against demand: when supply is less than demand, an intervention is made to restore the intended balance. This approach is often referred to, sometimes pejoratively, as 'predict and provide', because the method itself encourages the planner to make supply and demand match precisely. There are, of course, many uncertainties in the supply–demand balance. These are grouped together into a planning margin called 'headroom' (UKWIR 2002). In more sophisticated approaches, the headroom calculation combines probability distributions of errors in the components of supply and demand using a Monte Carlo approach, but the result is a single number which is added to demand to make a new target for the planner to meet. For the decision-maker, this may seem like the perfect approach to uncertainty, as it allows this complex problem to be managed easily. Reduced to a single number, the perfect solution can be identified and implemented, and the planner's job is done.

Many of the improvements to hydrological assessment identified in this chapter will reveal

more uncertainty. For example, multiple model parameter sets would produce an envelope of possible flows for a given climate projection. Using multiple climate projections (e.g. Lopez et al. 2009; Fung et al. 2013) or probabilistic projections like UKCP09 (Murphy et al. 2009) with multiple hydrological model parameter sets could lead to a wide range of possible flow regimes, each of which would lead to a different supply–demand balance problem. Uncertainty in future emissions scenarios and in catchment change adds further to the possible range of supply–demand balances. Some studies suggest that climate model uncertainty outweighs hydrological model uncertainty (e.g. Wilby et al. 2006; New et al. 2007). However, such studies tend to compare different GCMs with the results from a single hydrological model, therefore perhaps missing the full range of plausible hydrological uncertainty. In all of this, there is, of course, no increase in total uncertainty: these methods simply allow us to understand uncertainty better, though not completely.

Can better understanding of uncertainty improve decision-making in water supply planning? The so-called end-to-end uncertainty studies may demonstrate a very wide range of uncertainty as it cascades through the different levels of assessment (Jones 2000; Dessai et al. 2009) and yet, paradoxically, still fail to accommodate all possible uncertainty (Hall 2008). This is because there are areas of each component that are understood very poorly or not at all: Dessai et al. (2009) call this 'deep uncertainty'. One solution to this problem is to pursue improved understanding and better models, but while this should characterise uncertainty better, uncertainty itself may not be reduced to an extent that helps decision-makers (Dessai et al. 2009; Wilby 2010). In any case, improved models take many years to produce. For example, it took 7 years between the last two UK climate projections (Hulme et al. 2002; Murphy et al. 2009). Water resources planners need to make decisions with the best information available *now*. Waiting for better models and new research is not an option, though knowing that improvements are likely to be made may help to shape decisions.

Water supply planning is a problem of adapting to future uncertain conditions. The 'predict and provide' approach is seductively simple but may lead to a solution that can only cope with one highly unlikely combination of future conditions. If conservative assumptions are made through the assessment, the solution itself may turn out to be quite robust. For example, Dessai and Hulme (2007) showed that Anglian Water's 2004 plan was robust to climate change, partly because many of the options selected were flexible but also because the single climate model used happened to be very dry and therefore presented particularly testing conditions for water supply.

There is an alternative to providing a single prediction of the future. Instead, planners can work to understand the performance of proposed strategies across a range of possible futures (Groves et al. 2008; Dessai et al. 2009; Hallegatte 2009; Gober et al. 2010; Ranger et al. 2010). In water resources planning, this is not new: the National Rivers Authority's 1994 water resources strategy for England and Wales (NRA 1994) considered three different scenarios of demand, using the scenarios to argue for demand growth to be limited. The Environment Agency's 2001 water resources strategy (Environment Agency 2001) examined supply and demand in four different scenarios driven by societal and political values, seeking a strategy that was robust to the range of scenarios and choosing staged actions in a way that would now be described as 'adaptive management'. In such an approach, the ability to model future impacts remains essential: the change is in the way that the future is considered, so that the planner no longer seeks perfect foresight but instead looks to understand how the system in question would respond to different future conditions. It is here that an improved understanding of hydrological processes and their associated uncertainties would play an enhanced role, by helping decision-makers to understand the conditions that would challenge the performance of water supply systems and also the conditions that would not prove particularly difficult.

Effective water supply planning depends on good hydrological analysis and interpretation. Central to any improvements in this are the skills and understanding of the hydrologists and planners working in water companies, engineering consultancies and regulators. The changes identified in this paper can be successful only if practitioners are given the time and resources to improve their knowledge. This also requires a commitment from academia to transfer knowledge effectively from research to users in a way that understands the pressures under which users work and addresses the most pressing problems that face water resources planners today. Assumptions of climatic and hydrological stationarity are no longer credible, and it seems inevitable that water supply planning will move from 'predict and provide' to the identification of solutions that are robust to a range of future conditions. Improvements in hydrological modelling and understanding will ensure that hydrological science continues to underpin this assessment, to allow people to continue to rely on their water supplies for decades to come.

Acknowledgements

I am grateful to the Evidence Directorate of the Environment Agency for permission to write this paper, but the opinions expressed are my own and not those of the Environment Agency. Two anonymous referees made helpful suggestions that improved the paper considerably.

I was fortunate to be one of John Thornes's postgraduate students when he was head of the Department of Geography at the University of Bristol. My research looked at soil piping in agricultural terraces in Murcia, Spain, combining extended fieldwork with physically based distributed mathematical models of terrace hydrology. I learnt from John the importance of questioning and testing assumptions, as well as the benefits of drawing from a wide range of research disciplines to investigate practical hydrological and geomorphological problems. It was in the field, though, that I learnt most: John's ability to read a landscape and explain complex ideas simply was inspirational, and he was, of course, always hugely entertaining. I hope that I have used this experience wisely through my work in water resources and water supply planning, and I find it increasingly important in my current role, leading the Environment Agency's research on climate change impacts and adaptation.

References

Anderson, M. G., Burt, T. P. (1985) *Hydrological Forecasting*. John Wiley & Sons, Ltd, Chichester.

ASCE (1993) Criteria for evaluation of watershed models. *Journal of Irrigation and Drainage Engineering* **119**, 429–442.

Baird, A. J, Thornes, J. B., Watts, G. P. (1992) Extending overland flow models to problems of slope evolution and the representation of complex slope topographies. In: Parsons, A. J., Abrahams, A. D. (eds.) *Overland Flow: Hydraulics and Erosion Mechanisms*. UCL Press, London, pp. 184–208.

Bates, B. C., Kundzewicz, Z. W., Wu, S., Palutikof, J. P. (2008) *Climate Change and Water. Technical Paper of the Intergovernmental Panel on Climate Change*. IPCC Secretariat, Geneva.

Beven, K. (2001) How far can we go in distributed hydrological modelling? *Hydrology and Earth System Sciences* **5**, 1–12.

Beven, K. (2007) Towards integrated environmental models of everywhere: Uncertainty, data and modelling as a learning process. *Hydrology and Earth System Sciences* **11**, 460–467.

Beven, K. (2009) *Environmental Modelling: An Uncertain Future?* Routledge, London.

Blackie, J. R., Eeles, C. W. O. (1985) Lumped catchment models. In: Anderson, M. G., Burt, T. P. (eds.) *Hydrological Forecasting*. John Wiley & Sons, Ltd, Chichester, pp. 311–345.

Blenkinsop, S., Fowler, H. J. (2007) Changes in drought frequency, severity and duration for the British Isles projected by the PRUDENCE regional climate models. *Journal of Hydrology* **342**, 50–71.

Blyth, E. (2002) Modelling soil moisture for a grassland and a woodland site in south-east England. *Hydrology and Earth System Sciences* **6**, 39–47.

Blyth, E., Best, M., Cox, P., Essery, R., Boucher, O., Harding, R., Prentice, C., Vidale, P. L., Woodward, I. (2006) JULES: A new community land surface model. *Global Change Newsletter* **66**, 9–11.

Brunsden, D. (2001) A critical assessment of the sensitivity concept in geomorphology. *Catena* **42**, 99–123.

Brunsden, D., Thornes, J. B. (1979) Landscape sensitivity and change. *Transactions of the Institute of British Geographers* **4**(4), 463–484.

Bulygina, N., McIntyre, N., Wheater, H. (2009) Conditioning rainfall-runoff model parameters for ungauged catchments and land management impacts analysis. *Hydrology and Earth System Sciences* **13**, 893–904.

Burke, E. J., Perry, R. H. J., Brown, S. J. (2010) An extreme value analysis of UK drought and projections of change in the future. *Journal of Hydrology* **388**, 131–143.

Charlton, M. B., Arnell, N. W. (2011) Adapting to climate change impacts on water resources in England – an assessment of draft Water Resources Management Plans. *Global Environmental Change* **21**, 238–248.

Christierson, B. V., Vidal, J-P., Wade, S. D. (2012) Using UKCP09 probabilistic climate information for UK water resource planning. *Journal of Hydrology* **424–425**, 48–67.

Chun, K. P, Wheater, H. S., Onof, C. J. (2009) Streamflow estimation for six UK catchments under future climate scenarios. *Hydrology Research* **40**, 96–112.

Crawford, N. H., Linsley, R. K. (1966) Digital simulation in hydrology: Stanford Watershed Model IV. Technical report no. 39. Department of Civil Engineering, Stanford University, Stanford, CA.

Defra (2009) *The Environment in Your Pocket 2009: Key Facts and Figures on the Environment of the United Kingdom*. Department for Environment, Food and Rural Affairs, London.

Department of the Environment (1996) *Water Resources and Supply: Agenda for Action*. Department of the Environment, London.

Dessai, S., Hulme, M. (2007) Assessing the robustness of adaptation decisions to climate change uncertainties: A case study on water resources management in the East of England. *Global Environmental Change* **17**, 59–72.

Dessai, S., Hulme, M., Lempert, R., Pielke Jr, R. (2009) Climate prediction: A limit to adaptation? In: Adger,

W. N., Lorenzoni, I., O'Brien, K. L. (eds.) *Adapting to Climate Change: Thresholds, Values, Governance.* Cambridge University Press, Cambridge, UK, pp. 64–78.

Doornkamp, J. C., Gregory, K. J., Burn, A. S. (eds.) (1980) *Atlas of Drought in Britain 1975–76.* Institute of British Geographers, London.

Environment Agency (1998) *Review of Water Company Yields.* Environment Agency, Bristol.

Environment Agency (1999) *Progress in Water Supply Planning.* Environment Agency, Bristol.

Environment Agency (2001) *Water Resources for the Future – A Strategy for England and Wales.* Environment Agency, Bristol.

Environment Agency (2003) *Securing Water Supply.* Environment Agency, Bristol.

Environment Agency (2008) Water resources planning guideline. Environment Agency, Bristol. Available at: www.environment-agency.gov.uk/business/sectors/39687.aspx (accessed 26 October 2010).

Ficklin, D. L., Luo, Y., Luedeling, E., Zhang, M. (2009) Climate change sensitivity assessment of a highly agricultural watershed using SWAT. *Journal of Hydrology* **374**, 16–29.

Ficklin, D. L., Luedeling, E., Zhang, M. (2010) Sensitivity of groundwater recharge under irrigated agriculture to changes in climate, CO_2 concentrations and canopy structure. *Agricultural Water Management* **97**, 1039–1050.

Fowler, H. J., Blenkinsop, S., Tebaldi, C. (2007) Linking climate change modelling to impacts studies: Recent advances in downscaling techniques for hydrological modelling. *International Journal of Climatology* **27**, 1547–1578.

Fung, F., Watts, G., Lopez, A., Orr, H. G., New, M., Extence, C. (2013) Using large climate ensembles to plan for the hydrological impact of climate change in the freshwater environment. *Water Resources Management* **27**, 1063–1084.

Gober, P., Kirkwood, C. W., Balling, R. C., Ellis, A. W., Deitrick, S. (2010) Water planning under climatic uncertainty in Phoenix: Why we need a new paradigm. *Annals of the Association of American Geographers* **100**, 356–372.

Greenfield, B. J. (1984) The Thames Water catchment model. Unpublished report. Technology and Development Division, Thames Water, Reading.

Groves, D. G., Yates, D., Tebaldi, C. (2008) Developing and applying uncertain global climate change projections for regional water management planning. *Water Resources Research* **44**, W12413.

Haase, D. (2009) Effects of urbanisation on the water balance – A long-term trajectory. *Environmental Impact Assessment Review* **29**, 211–219.

Hall, J. (2008) Probabilistic climate scenarios may misrepresent uncertainty and lead to bad adaptation decisions. *Hydrological Processes* **21**, 1127–1129.

Hall, J. W., Watts, G., Keil, M., de Vial, L., Street, R., Conlan, K., O'Connell, P. E., Beven, K. J., Kilsby, C. G. (2012) Towards risk-based water resources planning in England and Wales under a changing climate. *Water and Environment Journal* **26**, 118–129.

Hallegatte, S. (2009) Strategies to adapt to an uncertain climate change. *Global Environmental Change* **19**, 240–247.

Hannaford, J., Lloyd-Hughes, B., Keef, C., Parry, S., Prudhomme, C. (2011) Examining the large-scale spatial coherence of European drought using regional indicators of precipitation and streamflow deficit. *Hydrological Processes* **25**, 1146–1162.

Hess, T. M., Holman, I. P., Rose, S. C., Rosolova, Z., Parrott, A. (2010) Estimating the impact of rural land management changes on catchment runoff generation in England and Wales. *Hydrological Processes* **24**, 1357–1368.

Hough, M. N., Jones R. J. A. (1997) The United Kingdom Meteorological Office rainfall and evaporation calculation system version 2.0 – An overview. *Hydrology and Earth System Sciences* **1**, 227–239.

Hulme, M., Jenkins, G. J., Lu, X., Turnpenny, J. R., Mitchell, T. D., Jones, R. G., Lowe, J., Murphy, J. M., Hassell, D., Boorman, P., McDonald, R., Hill, S. (2002) *Climate Change Scenarios for the United Kingdom: The UKCIP02 Scientific Report.* Tyndall Centre for Climate Change Research, School of Environmental Sciences, University of East Anglia, Norwich.

Hundecha, Y., Bárdossy, A. (2004) Modeling of the effect of land use changes on the runoff generation of a river basin through parameter regionalization of a watershed model. *Journal of Hydrology* **292**, 281–295.

Ivanović, R. F., Freer, J. E. (2009) Science versus politics: Truth and uncertainty in predictive modeling. *Hydrological Processes* **23**, 2549–2554.

Johnson, R. C. (1995) Effects of upland afforestation on water resources: The Balquhidder Experiment 1981–1991 (2nd Edition). Report 116. Institute of Hydrology, Wallingford, CT.

Johnson, A. C., Acreman, M. C., Dunbar, M. J, Feist, S. W., Giacomello, A. M, Gozlan, R. E., Hinsley, S. A., Ibbotson, A. T., Jarvie, H. P., Jones, J. I., Longshaw, M., Maberly, S. C., Marsh, T. J., Neal, C., Newman,

J. R., Nunn, M. A., Pickup, R. W., Reynard, N. S, Sullivan, C. A., Sumpter, J. P., Williams, R. J. (2009) The British river of the future: How climate change and human activity might affect two contrasting river ecosystems in England. *Science of the Total Environment* **407**, 4787–4798.

Jones, R. N. (2000) Managing uncertainty in climate change projections – Issues for impact assessment: An editorial comment. *Climatic Change* **45**, 403–419.

Jones, P. D., Lister, D. H., Wilby, R. L., Kostopoulou, E. (2006) Extended river flow reconstructions for England and Wales, 1856–2002. *International Journal of Climatology* **26**, 219–231.

Kay, A. L., Davies, H. N. (2008) Calculating potential evaporation from climate model data: A source of uncertainty for hydrological climate change impacts. *Journal of Hydrology* **358**, 221–239.

Kay, A. L., Jones, D. A., Crooks, S. M., Kjeldsen, T. R., Fung, C. F. (2007) An investigation of site-similarity approaches to generalisation of a rainfall-runoff model. *Hydrology and Earth System Sciences* **11**, 500–515.

Kendon, M., Marsh, T., Parry, S. (2013) The 2010 drought in England and Wales. *Weather* **68**, 88–95.

Kirkby, M. J., Naden, P. S., Burt, T. P., Butcher, D. P. (1987) *Computer Simulation in Physical Geography*. John Wiley & Sons, Ltd, Chichester.

Lee, H., McIntyre, N., Wheater, H., Young, A. (2005) Selection of conceptual models for regionalisation of the rainfall-runoff relationship. *Journal of Hydrology* **312**, 125–147.

Lopez, A., Fung, F., New, M., Watts, G., Weston, A., Wilby, R. L. (2009) From climate model ensembles to climate change impacts and adaptation: A case study of water resource management in the southwest of England. *Water Resources Research* **45**, W08419.

Manley, R. E. (1975) A hydrological model with physically realistic parameters. In: *Application of Mathematical Models in Hydrology and Water Resources Systems*. IAHS Publication 115. IAHS, Rennes, pp. 154–161.

Manley, R. E. (2006) *A Guide to Using HYSIM*. R. E. Manley and Water Resource Associates, Ltd., Wallingford.

Marsh, T., Booker, D., Fry, M. (2007a) *The 2004–06 Drought*. National Hydrological Monitoring Programme, Centre for Ecology and Hydrology, Wallingford, CT.

Marsh, T., Cole, G., Wilby, R. (2007b) Major droughts in England and Wales, 1800–2006. *Weather* **62**, 87–93.

Marshall, M. R., Francis, O. J., Frogbrook, Z. L., Jackson, B. M., McIntyre, N., Reynolds, B.,

Solloway, I., Wheater, H. S., Chell, J. (2009) The impact of upland land management on flooding: Results from an improved pasture hillslope. *Hydrological Processes* **23**, 474–475.

McIntyre, N., Marshall, M. (2010) Identification of rural land management signals in runoff response. *Hydrological Processes* **24**, 3521–3534.

Milly, P. C. D., Betancourt, J., Falkenmark, M., Hirsch, R. M., Kundzewicz, Z. W., Lettenmaier, D. P., Stouffer, R. J. (2008) Climate change – stationarity is dead: Whither water management? *Science* **319**(5863), 573–574.

Molina, M., Zaelkeb, D., Sarmac, K. M., Andersend, S. O., Ramanathane, V., Kaniaruf, D. (2009) Reducing abrupt climate change risk using the Montreal Protocol and other regulatory actions to complement cuts in CO_2 emissions. Proceedings of the National Academy of Sciences, USA. Available at: http://www.pnas.org/content/106/49/20616.full (accessed 4 April 2010).

Monteith, J. L. (1965) Evaporation and environment. *Symposia of the Society for Experimental Biology* **19**, 205–234.

Moore, R. J. (1985) The probability-distributed principle and runoff production at point and basin scales. *Hydrological Sciences Journal* **30**, 273–297.

Murphy, J. M., Sexton, D. M. H., Jenkins, G. J., Booth, B. B. B., Brown, C. C., Clark, R. T., Collins, M., Harris, G. R., Kendon, E. J., Betts, R. A., Brown, S. J., Humphrey, K. A., McCarthy, M. P., McDonald, R. E., Stephens, A., Wallace, C., Warren, R., Wilby, R., Wood, R. A. (2009) *UK Climate Projections Science Report: Climate Change Projections*. Hadley Centre, UK Meteorological Office, Exeter.

NAO (2005) Environment Agency: Efficiency in water management. Report by the Comptroller and Auditor General, National Audit Office, London. Available at: www.nao.org.uk/wp-content/uploads/2005/06/050673.pdf (accessed 20 August 2013).

Nash, J. E., Sutcliffe, J. V. (1970) River flow forecasting through conceptual models. Part 1 – A discussion of principles. *Journal of Hydrology* **10**, 282–290.

New, M., Lopez, A., Dessai, S., Wilby, R. (2007) Challenges in using probabilistic climate change information for impact assessments: An example from the water sector. *Philosophical Transactions of the Royal Society A* **365**, 2117–2131.

NRA (1994) *Water: Nature's Precious Resource*. National Rivers Authority, Bristol.

O'Driscoll, M., Clinton, S., Jefferson, A., Manda, A., McMillan, S. (2010) Urbanization effects on watershed hydrology and in-stream processes in the southern United States. *Water* **2**, 605–648.

Ofwat (2009) *Service and Delivery – Performance of the Water Companies in England and Wales 2008–09.* Office of Water Services, Birmingham.

Orr, H. G., Wilby, R. L., McKenzie Hedger, M., Brown, I. (2008) Climate change in the uplands: A UK perspective on safeguarding regulatory ecosystem services. *Climate Research* **37**, 77–98.

Oudin, L., Hervieu, F., Michel, C., Perrin, C., Andréassian, V., Anctil, F., Loumagne, C. (2005a) Which potential evapotranspiration input for a lumped rainfall-runoff model? Part 2 – Towards a simple and efficient potential evapotranspiration model for rainfall-runoff modelling. *Journal of Hydrology* **303**, 290–306.

Oudin, L., Michel, C., Anctil, F. (2005b) Which potential evapotranspiration input for a lumped rainfall-runoff model? Part 1 – Can rainfall-runoff models effectively handle detailed potential evapotranspiration inputs? *Journal of Hydrology* **303**, 275–289.

Pechlivanidis, I. G., McIntyre, N. R., Wheater, H. S. (2010) Calibration of the semi-distributed PDM rainfall-runoff model in the Upper Lee catchment, UK. *Journal of Hydrology* **386**, 198–209.

Penman, H. L. (1948) Natural evaporation from open water, bare soil and grass. *Proceedings of the Royal Society of London A* **193**, 120–145.

Phillips, J. D. (2006) Evolutionary geomorphology: Thresholds and nonlinearity in landform response to environmental change. *Hydrology and Earth System Sciences* **10**, 731–742.

POST (2012) Water resource resilience. POST Note 419. Parliamentary Office of Science and Technology, London. Available at: www.parliament.uk/briefing-papers/POST-PN-419 (accessed 20 August 2013).

Ranger, N., Millner, A., Dietz, S., Fankhauser, S., Lopez, A., Ruta, G. (2010) *Adaptation in the UK: A Decision-making Process.* Grantham Research Institute on Climate Change and the Environment, London School of Economics and Political Science, London.

Ranjan, P., Kazama, S., Sawamoto, M. (2006) Effects of climate change on coastal fresh groundwater resources. *Global Environmental Change* **16**, 388–399.

Rosero, E., Yang, Z-L., Wagener, T., Gulden, L. E., Yatheendradas, S., Niu, G.-Y. (2010) Quantifying parameter sensitivity, interaction, and transferability in hydrologically enhanced versions of the Noah land surface model over transition zones during the warm season. *Journal of Geophysical Research: Atmospheres* **115**, D03106.

Russill, C., Nyssa, Z. (2009) The tipping point trend in climate change communication. *Global Environmental Change* **19**, 336–344.

Shaw, E. M. (1988) *Hydrology in Practice* (2nd Edition). Chapman and Hall, London.

Thornes, J. B., Brunsden, D. (1977) *Geomorphology and Time.* Methuen, London.

Trabucco, A., Zomer, R. J., Bossio, D. A., van Straaten, O., Verchot, L. V. (2008) Climate change mitigation through afforestation/reforestation: A global analysis of hydrologic impacts with four case studies. *Agriculture, Ecosystems and Environment* **126**, 81–97.

UK Water Industry Research/Environment Agency (UKWIR) (1997) Effects of climate change on river flows and groundwater recharge. Report 97/CL/04/1, UKWIR, London.

UKWIR (2002) An improved methodology for assessing headroom. Report WR-13, UK Water Industry Research Ltd., London.

UKWIR (2003) Effect of climate change on river flows and groundwater recharge: UKCIP02 Scenarios. Report 03/CL/04/2, UK Water Industry Research Ltd., London.

UKWIR (2006) Effects of climate change on river flows and groundwater recharge: Guidelines for resource assessment and UKWIR06 scenarios. Project CL04/C, UK Water Industry Research Ltd., London.

van Dijk, A. I. J. M., Keenan, R. J. (2008) Overview: Planted forests and water in perspective. *Forest Ecology and Management* **251**, 1–9.

Vanloocke, A., Bernacchi, C. J., Twine, T. E. (2010) The impacts of *Miscanthus giganteus* production on the Midwest US hydrologic cycle. *Global Change Biology Bioenergy* **2**, 180–191.

Vidal, J. P., Wade, S. (2006) Effects of climate change on river flows and groundwater recharge: Guidelines for resource assessment and UKWIR06 scenarios. UKWIR report, UK Water Industry Research Ltd., London.

Wagener, T. (2007) Can we model the hydrological impacts of environmental change? *Hydrological Processes* **21**, 3233–3236.

Wagener, T., Boyle, D. P., Lees, M. J., Wheater, H. S., Gupta, H. V., Sorooshian, S. (2001) A framework for development and application of hydrological models. *Hydrology and Earth System Sciences* **5**, 13–26.

Water Resources Board (1973) *Water Resources in England and Wales.* HMSO, London.

Watts, G. (1997) Hydrological modelling in practice. In: Wilby, R. L. (ed.) *Contemporary Hydrology: Towards Holistic Environmental Science.* John Wiley & Sons, Ltd, Chichester, pp. 151–194.

Watts, G. (2010) Water for people: Climate change and water availability. In: Fung, F., Lopez, A., New, M. (eds.) *Modelling the Impact of Climate Change on Water Resources*. John Wiley & Sons, Ltd, Chichester, pp. 86–127.

Watts, G., von Christiersen, B., Hannaford, J., Lonsdale, K. (2012) Testing the resilience of water supply systems to long droughts. *Journal of Hydrology* **414–415**, 255–267.

Watts, G., Battarbee, R. W., Bloomfield, J. P., Crossman, J., Daccache, A., Durance, I., Elliott, J. A., Garner, G., Hannaford, J., Hannah, D. M., Hess, T., Jackson, C. R., Kay, A. L., Kernan, M., Knox, J., Mackay, J., Monteith, D. T., Ormerod, S. J., Rance, J., Stuart, M. E., Wade, A. J., Wade, S. D., Weatherhead, K., Whitehead, P. G., Wilby, R. L. (2015) Climate change and water in the UK – past changes and future prospects. *Progress in Physical Geography* **39**, 6–28.

Weatherhead, E. K., Howden, N. J. K. (2009) The relationship between land use and surface water resources in the UK. *Land Use Policy* **26S**, S243–S250.

Webb, B. W., Hannah, D. M., Moore, R. D., Brown, L. E., Nobilis, F. (2008) Recent advances in stream and river temperature research. *Hydrological Processes* **22**, 902–918.

Whitehead, P. G., Wilby, R. L., Battarbee, R. W., Kernan, M., Wade, A. J. (2009) A review of the potential impacts of climate change on surface water quality. *Hydrological Sciences Journal* **54**, 101–123.

Wilby, R. L. (2005) Uncertainty in water resource model parameters used for climate change impact assessment. *Hydrological Processes* **19**, 3201–3219.

Wilby, R. L. (2010) Evaluating climate model outputs for hydrological applications. *Hydrological Sciences Journal* **55**, 1090–1093.

Wilby, R. L., Davies, G. (1997) Operational hydrology. In: Wilby, R. L. (ed.) *Contemporary Hydrology: Towards Holistic Environmental Science*. John Wiley & Sons, Ltd, Chichester, pp. 195–240.

Wilby, R. L., Fowler, H. J. (2010) Regional climate downscaling. In: Fung, F., Lopez, A., New, M. (eds.) *Modelling the Impact of Climate Change*. Blackwell, Oxford, pp. 34–85.

Wilby, R. L., Greenfield, B., Glenny, C. (1994) A coupled synoptic-hydrological model for climate change impact assessment. *Journal of Hydrology* **153**, 265–290.

Wilby, R. L., Whitehead, P. G., Wade, A. J., Butterfield, D., Davis, R. J., Watts, G. (2006) Integrated modelling of climate change impacts on water resources and quality in a lowland catchment: River Kennet, UK. *Journal of Hydrology* **330**, 204–220.

Young, A. R. (2006) Stream flow simulation within UK ungauged catchments using a daily rainfall-runoff model. *Journal of Hydrology* **320**, 155–172.

Zhang, H., Hiscock, K. M. (2010) Modelling the impact of forest cover on groundwater resources: A case study of the Sherwood Sandstone aquifer in the East Midlands, UK. *Journal of Hydrology* **392**, 136–149.

CHAPTER 10

Changing discharge contributions to the Río Grande de Tárcoles

Matthew Marsik[1], Peter Waylen[2] and Marvin Quesada[3]

[1] Land Use and Environmental Change Institute, University of Florida, Gainesville, FL, USA
[2] Department of Geography, University of Florida, Gainesville, FL, USA
[3] Departamento de Ciencias Sociales, Universidad de Costa Rica Sede Occidente, San Ramón, Costa Rica

Introduction

Water quantity issues are of increasing global concern. Many countries in Central America rely on hydroelectric power generation and on surface water withdrawals for agricultural and water supply. Costa Rica generates more than 80% of its national power from hydroelectricity and constructed more than 30 dams in the 1990s (Anderson et al. 2006). Decreased flow from major tributaries would imperil reliable power generation, diminish surficial and groundwater storage accumulated during the rainy season and create water quality issues for municipal and agricultural sectors. In the Río Grande de Tárcoles catchment, groundwater comprises 6% of total water used for industrial and urban purposes and provides 60% of all water for agricultural irrigation (Blomquist et al. 2005). A reduction in water supply coupled with anthropogenic alteration of the surface hydrological cycle has critical implications for water availability.

The relative contributions of two major tributaries of the Río Grande de Tárcoles, Costa Rica, namely, the Río Grande de San Ramón and the Río Virilla, appear to have changed around 1975

(Figure 10.1). Due to the geographic complexity of Costa Rica, particularly in the central valley region, differences in land cover conditions and the intra-annual variability of precipitation input between the two sub-basins would influence their respective streamflow contributions to the larger Tárcoles. We use an applied, mixed-method approach of statistical analysis and hydrological simulation modelling to investigate these potential causes for the observed change in discharge. The change could result from the non-linear runoff responses to land cover change, a change of climate forcing through precipitation inputs, or both.

The remaining chapter is divided into various sections. First, we briefly review tropical hillslope hydrology focusing specifically on undisturbed and altered biophysical conditions that affect the hydrological response and flow pathways. We then describe the methodological approach with a description of the data used, the statistical analysis and the model construction. Next, we present and discuss the results of the statistical analysis and model simulations with respect to the three discharge-changing hypotheses. We conclude with the contribution

Monitoring and Modelling Dynamic Environments, First Edition. Edited by Alan P. Dykes, Mark Mulligan and John Wainwright.

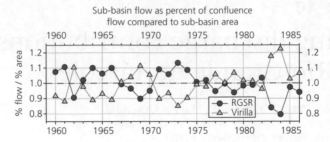

Figure 10.1 Sub-basin annual runoff as a percentage of confluence runoff compared to sub-basin area.

of a mixed-method approach used here to investigate land cover and climate changes in tropical, mesoscale catchments.

Undisturbed and altered tropical hillslope hydrology

The hydrological response of a catchment to rainfall depends on the interactions between biophysical (e.g. geologic, topographic and climatic) factors and land cover characteristics. Other important characteristics include the hydraulic conductivity of the soil at different depths, rainfall intensity and duration and slope morphology (Dunne 1978). Generally speaking, infiltration capacities of undisturbed forest soils are sufficient to accommodate most rainfall intensities (Bruijnzeel 1990). Overland flow in tropical catchments is spatially restricted to areas of less permeable soils, on steeper slopes and to where the soils are near saturation (Bonell 2004, 2010). The three flow pathways of runoff generation – infiltration excess Hortonian overland flow (HOF), saturation excess overland flow (SOF) and subsurface flow (SSF) – have all been observed in undisturbed forests (Bruijnzeel 1990; Bonell and Balek 1993; Bonell 2004, 2010; Bruijnzeel 2004).

HOF (Horton 1933) is critical in humid areas with disturbed soils and vegetation but is restricted temporally and spatially within a catchment. Overland flow is unlikely on mid- to upper slopes unless a shallow impeding layer is present (Bruijnzeel 1990) and occurs as SOF only on lower floodplain areas (e.g. Nortcliff and Thornes 1981). Regardless of position and land

cover, during high-intensity precipitation, rapid runoff (e.g. HOF) may dominate the hydrological response of a catchment. SOF generally prevails in flat bottomed valleys with gentle slopes and thin soils (Dunne 1978) and occurs with no obvious topographic or landscape control (Elsenbeer and Cassel 1991). However, SSF can contribute significantly to stormflow through unsaturated ground on deeply incised convex hillslopes with well-drained, deep and highly permeable soils (Dunne 1978). SSF is propagated in undisturbed forests by macropores and soil pipes on slopes (Bruijnzeel 2004) with high-field saturated hydraulic conductivities. SSF may also prevail mainly at depth (Nortcliff and Thornes 1981) and through topographic hollows below water tables 'perched' within the soil layer (Dykes and Thornes 2000).

Conversion of tropical forest may produce permanent changes in the hydrological response of a catchment (Bruijnzeel 2004) resulting from soil compaction, soil crusting and removal of organic forest litter and, when coupled with high rainfall intensities, results in greater HOF over the long term (Bonell and Balek 1993). Bosch and Hewlett (1982) examined altered flow path responses in 94 catchments ranging from 1 to 2500 ha and concluded that increased water yield resulted from reduced vegetation. Bonell (1998) reviewed paired catchment studies and found that they specifically indicated that >20% loss of forest cover was required to appreciably increase the total annual water yield. Furthermore, increases in water yield were conditional upon the spatial and temporal variability of rainfall and degree of surface disturbance (Bonell 1998).

Low saturated hydraulic conductivities associated with consolidated surfaces contribute disproportionately to stormflow responses via HOF during events of small rainfall totals and low intensities. Through alteration of land cover conditions, reduced surficial saturated hydraulic conductivity produces a greater likelihood of HOF, while reduced subsurface saturated hydraulic conductivity may increase the frequency of non-HOF mechanisms (Ziegler et al. 2004). Bonell et al. (2010) and Zimmermann et al. (2010) posited that human disturbance diminishes subsurface saturated hydraulic conductivity, thereby disrupting subsurface hydrological pathways and enhancing the generation of non-HOF (i.e. return flow) on fragmented hillslopes (Ziegler et al. 2004). Land cover disturbances can further alter runoff in a non-linear manner through processes such as infiltration and evaporation.

Catchment responses to altered hillslope hydrology are readily observable and easily investigated in experimental and small scale catchments (<10 km) (Dunne 1978; Bosch and Hewlett 1982) but become complicated as we scale up to larger mesoscale (>1000 km) catchments (Bruijnzeel 1990, 2004; Bonell 2004, 2010). At this scale, multiple factors impede a clear signal of altered hydrological flow paths. The mosaic of land cover types may cancel out hydrological responses of altered flow paths; at this scale, climate variability influences precipitation regimes and thus runoff; human modifications of groundwater withdrawals and impoundment compound and obscure the response; and fieldwork to catalogue the hydrological responses from different land cover and other biophysical conditions becomes impractical. A mixed-method approach of inferential statistical analysis and dynamic hydrological simulation modelling offers one way to offset the challenge of analysing coupled human–climate effects on catchment runoff responses (Refsgaard et al. 1989).

Tropical mesoscale catchment studies

Few land cover change and hydrological modelling studies have been conducted in tropical mesoscale (i.e. >1000 km²) catchments (Bonell 2004; Bruijnzeel 2004). To disentangle the effects of changing land cover and climate forcing, statistical and hydrological modelling methods are required (Refsgaard et al. 1989). Lorup et al. (1998) employed this combination and detected no change in annual runoff in six semi-arid catchments (200–1000 km²) in Zimbabwe as a result of increasing population density and urbanised areas. In addition, Wilk et al. (2001) observed no statistically significant differences in precipitation, discharge and evapotranspiration in 1957 and 1995 despite a decrease in forest cover of 80–30% in the 12,100 km² Nam Pong catchment in northeast Thailand during this period. In contrast, a 78% increase in runoff observed in the Comet River catchment (16,440 km²) of Central Queensland, Australia (Siriwardena et al. 2006), following forest clearing and conversion to grasses and cropland was attributed partially to an 8.4% increase in rainfall. Applications of an annual water balance model and a simple conceptual daily rainfall–runoff model suggested that forest clearing increased runoff by 58 and 40%, respectively (Siriwardena et al. 2006). Furthermore, despite no significant changes in precipitation, Costa et al. (2003) linked significant changes in mean annual and high-flow season discharge on the Tocantins River, southeastern Amazonia, Brazil, to a 20% increase in agricultural lands from 1960 to 1995. Large-scale hydrological modelling studies in Central America and Costa Rica are fewer in numbers.

To our knowledge, only two studies in Costa Rica have used dynamic simulation hydrological models (Colby 2001; van Loon and Troch 2002); however, both models were conceptual in formulation and spatially averaged. Krishnaswamy et al. (2001) used dynamic linear regression modelling to investigate land cover and hydroclimatic effects on discharge and sediment production in the Terraba catchment (4767 km²).

These studies provided some knowledge about hydrological responses to land cover change and climate variability but are limited by the type of model development. The use of physically based and spatially explicit hydrological models can better assess the impacts of changing land cover (Refsgaard et al. 1989).

The main objectives of this study are as follows: (i) to identify changes in discharge and precipitation at annual and monthly timescales in the Río Grande de Tárcoles, (ii) to construct a hydrological model of the two contributing sub-basins to identify non-linear runoff responses conditioned upon land cover and precipitation, and (iii) to analyse the potential role of climate variability as a cause of discharge changes. Our overarching hypothesis is that altered vegetated and urban land cover conditions modify hillslope hydrology and produce non-linear catchment responses. We developed three separate but potentially interrelated working hypotheses to detect the causes of changing discharge contributions. They are the following:

- Hypothesis 1: Precipitation input to the Río Virilla sub-basin increased after 1975, forcing a greater contribution to discharge in the Río Grande de Tárcoles.
- Hypothesis 2: Differences in the proportions of vegetated versus urban land covers caused the sub-basins to act as non-linear amplifiers of precipitation through differences in soil moisture and evapotranspiration fluxes.
- Hypothesis 3: Differential spatial responses to precipitation influenced by El Niño Southern Oscillation (ENSO) and Atlantic sea surface temperatures (SSTs) result in distinct discharge responses even though the basins are adjacent.

Study area

The Río Grande de Tárcoles catchment (1745 km²) (Figure 10.2) lies in Costa Rica's central tectonic depression, flanked to the north by the northwest–southeast trending Cordillera

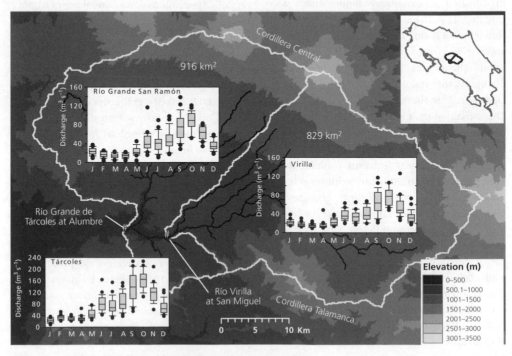

Figure 10.2 Study area and monthly runoff regimes.

Central mountains and to the south by the Cordillera Talamanca mountains. The catchment encompasses much of the valley and contains most of the metropolitan population of Costa Rica. Two major sub-basins, the Río Grande de San Ramón (916 km²) and the Río Virilla (829 km²), comprise the Tárcoles catchment. The area is topographically varied and rugged, with steep slopes geomorphologically dominated by ancient volcanic pyroclastic flows and lacustrine depositions. Alfisols, entisols, inceptisols, ultisols and vertisols are the primary soil orders in the catchment (CCT 1989). Land cover differences between the Río Grande de San Ramón and Río Virilla are marked. The former is dominantly forested, pastoral and agricultural lands (~98%) with <2% urban areas (MAG 1992). By contrast, the Gran Metropolitan urban area of Costa Rica occupies about 15% of the Río Virilla sub-basin in which agriculture is the dominant vegetated land cover class (~33%).

Due to Costa Rica's location between the Caribbean and the Pacific, precipitation-generating mechanisms are complex (Waylen et al. 1998). The northeast trade winds bring ample precipitation to the Caribbean coast, leaving the Pacific side in rain shadow during the dry season (November–April) except for the southeastern corner of the Río Virilla sub-basin, which is exposed to a wind gap in the Cordillera Central. The Río Virilla sub-basin has complex rainfall climatology caused by the interaction of the Pacific influences and an incursion of Caribbean air to the southeast. As the inter-tropical convergence zone (ITCZ) of the Eastern Pacific (Hastenrath 2002) migrates northwards through the boreal summer, convective precipitation falls over much of the Pacific coast and montane areas from May to November. Low flows during the dry season are sustained by depletion of groundwater and soil water storage from the previous rainy season. Precipitation at the onset of the wet season begins to replenish the stores. July and August witness the *veranillo*, a temporary dry period caused by intensification of northeast trade winds. This seasonal reduction

is a prominent feature of the precipitation climatology throughout Central America (Magana et al. 1999). During the post-*veranillo* months of September and October, soil moisture stores are saturated or near saturated, producing high runoff regardless of precipitation totals and intensity. The southward migration of the ITCZ heralds the dry season in November, initiating the release of groundwater stores and flow recession.

Methods

We used a mixed-method approach to disentangle the cause of observed changes in discharge into the Río Grande de Tárcoles. First, we describe the data used throughout the analysis. Statistical analysis comprised the first portion of the analysis. We created sub-basin monthly precipitation data and used double-mass curves to determine the best interpolation method. We used the hypergeometric distribution to detect changes in historical annual runoff and precipitation data using 1975 as the base year from which deviations were calculated. Ordinary least squares linear regression determined changes in runoff and precipitation at the monthly timescale. We created spatial maps of standard precipitation deviates conditioned upon SSTs to analyse spatial and temporal climate variability patterns. We used the GIS-based hydrologic model, SWAT (Soil and Water Assessment Tool), to investigate the effects of changing land cover conditions and precipitation regimes on the combined flow into the Río Grande de Tárcoles.

Data

The Costa Rican Institute of Electricity (ICE) supplied daily discharge data for five gauging locations within the catchment. The length of record spanned 23 years, from 1964 to 1986, for the gauges on the Río Poas at Tacares, the Río Virilla at San Miguel and Linda Vista, the Río Tiribi at Electriona and the Río Tárcoles at Balsa. We aggregated the daily discharge data for the

Tárcoles and Virilla to monthly and annual timescales. We estimated monthly discharge for the Río Grande de San Ramón, currently ungauged, by subtracting the discharge of the Virilla (immediately above the confluence) from that of the Tárcoles (below confluence). We then aggregated the remaining gauge data similarly as input for hydrological model sensitivity analysis and autocalibration routines.

The MIRENEM (1988) and the ICE provided daily and monthly precipitation data. Sixty-seven stations had periods of varying lengths of monthly totals from 1960 to 1986 with 4–79% missing records. Of these, only three reported complete records. Nine stations had daily records of varying lengths from 1964 to 1986 with 1–79% missing records. To produce a temporally continuous record, we calculated missing daily and monthly precipitation data using inverse distance-weighted spatial interpolation. Six to twenty-five monthly precipitation stations and up to eight daily stations required interpolation. From the interpolated monthly records, we created seasonal and annual time series for each station. We created two random sampling designs for the input precipitation to SWAT and for the stochastic weather generation. Based on each sampling point location, we extracted precipitation data from the daily interpolations.

The NCEP/NCAR reanalysis programme (Kalnay et al. 1996) provided surface atmospheric data at both $2 \times 2°$ and $2.5 \times 2.5°$ resolution and daily time steps from 1964 to 1986. We extracted relative humidity, minimum and maximum temperatures and insolation from global data sets and calculated potential evapotranspiration estimates conditioned upon mean surface temperatures and average wind speeds calculated from the U and V vector wind data.

We assigned soil properties of percentages of silt and clay, profile depth, texture and organic content (Vasquez 1980) to a 1:200,000 GIS soil layer based on the great group classification (CCT 1989). Median bulk soil densities were based on soil order (Alvarado and Forsythe 2005).

We calculated hydraulic soil parameters of the available water content, field capacity and saturated hydraulic conductivity for each great group using the Rosetta pedotransfer programme (Schaap et al. 2001).

Using Landsat images from 1975 (MSS 3) and 1986 (TM 4), we created spatial land cover data. Image preprocessing included georeferencing the MSS image to the TM image and radiometrically calibrating both images. We used the RuleGen (Loh and Shih 1997) decision tree classifier in ENVI to classify each Landsat image with five land cover classes: forest, developed grassland, cropland and shrubland. Derived products included in the decision tree classification included tasselled cap transformation and local Moran's I at a spatial lag of three pixels. To improve classification accuracies, we included elevation, slope and cloud and shadow masks in the RuleGen classifier. We resampled the classified images to 57×57 m resolution and assigned Manning's n overland flow values following Bedient and Huber (1988).

HydroSHEDS (Lehner et al. 2006) furnished elevation data at 15 arc s (~92×92 m) and yielded slope data for use in ArcSWAT. We used the ArcSWAT catchment delineation module to further process the elevation data and to enforce observed stream drainage network, digitised from 1:50,000 scale maps. In this way, the delineated streams closely followed actual stream courses, resulting in better accuracy of flow length and slope. We calculated stream channel dimensions using ArcSWAT and assigned Manning's n channel roughness values following Bedient and Huber (1988).

Statistical analysis: Change detection

We created monthly precipitation inputs for each sub-basin using four interpolation methods – arithmetic mean, Thiessen polygons, inverse distance weighted and spline interpolation (Dingman 2002) – using a custom application in ArcGIS©. To decide the best interpolation method for monthly precipitation, we compared the double-mass curves of sub-basin precipitation (derived from each interpolation method) and runoff for each sub-basin using the Kolmogorov–Smirnov goodness-of-fit statistic. We found no statistically significant differences ($\alpha = 0.20$) between the interpolation methods at the monthly timescale; therefore, we averaged the precipitation estimates from each interpolation method to calculate a representative monthly precipitation input for each sub-basin.

We calculated the statistical significance ($\alpha = 0.05$) of changes in annual runoff contributions from each sub-basin before and after 1975 using a hypergeometric distribution (Eq. 10.1), which defines the probability of drawing a sample, n (without replacement), containing x successes, from a finite population, N, consisting of R successes; thus,

$$f(x) = \binom{R}{x}\binom{N-R}{n-x} \bigg/ \binom{N}{n} \qquad (10.1)$$

In this case, x is the number of above (below) long-run median annual runoff (or precipitation) values before and after 1975. We analysed deviations of annual and monthly precipitation inputs (sub-basin volumes) from their respective long-run medians for the Virilla and Grande de San Ramón in a similar fashion to runoff.

To determine if one sub-basin contributed more precipitation to runoff than expected, we expressed monthly runoff volumes and precipitation from the Río Virilla as percentages of the combined volume observed in the Río Grande de Tárcoles for each year. We fit simple linear regression to each monthly time series to detect the presence of statistically significant trends ($\alpha = 0.05$) in percentage contribution of the Río Virilla.

Statistical analysis: Climate variability

Standardisation of precipitation conditioned upon SSTs provides direct comparison of seasonal precipitation and relative influences of climatic variability (Waylen and Quesada 2002).

We aggregated monthly precipitation totals from 1959 to 1986 into five seasons (JFMA, MJ, JA, SO and ND) at all stations to calculate the standard normal deviates (i.e. Z-scores). We then categorised precipitation deviates according to the joint states (above/below median) of SSTs in the equatorial Pacific (ITCZ) and tropical Atlantic (northeast trades) to generate four possible combinations of oceanic states: a warm Pacific and a warm Atlantic, a warm Pacific and a cold Atlantic, a cold Pacific and a warm Atlantic and a cold Pacific and cold Atlantic. We used inverse distance-weighted interpolation to create composite maps of seasonal standard deviates to indicate the seasonally varying importance of the Pacific and Atlantic SSTs and their interaction on precipitation variability. Spatial interpolation of seasonal deviates across the sub-basins should reveal any geographic patterns of climatic influences and any seasonal changes within each sub-basin (Waylen and Quesada 2002).

Soil and Water Assessment Tool model

The Soil and Water Assessment Tool (SWAT) (Arnold et al. 1998; Arnold and Fohrer 2005) is a semi-distributed, process-based, computer hydrological model that simulates the land-based hydrological cycle using a water balance. Complex mesoscale catchments are partitioned into hydrological response units (HRUs) comprised of unique land cover and soil combinations. Hydro-climatic inputs drive the relative importance of each hydrological component in SWAT. After SWAT simulates canopy storage and interception, the model handles excess precipitation in one of three ways: (i) infiltrates the water via the land surface with potential for redistribution within the soil, (ii) moves water out of a HRU by subsurface lateral flow, or (iii) moves water over the land surface as surface runoff. The model calculates flood water accumulations for main channels and routed, preserving channel mass flow, through the stream network and reservoirs, ponds or lakes to the catchment outlet. Previous researchers (Arnold and Fohrer 2005) have applied SWAT primarily in the United States and Europe with a few applications in tropical countries (e.g. Kenya and India).

SWAT model construction

Following data preprocessing, SWAT model construction proceeded using the ArcSWAT (Winchell et al. 2007) interface for ArcGIS. The interface provided tools for catchment delineation with outlet and stream definition, the creation of HRUs, the specification of climatic variables and the creation of input files and operation of the model. HRUs define unique combinations of land cover, soils and slope, which act as the spatial control in the conversion of effective precipitation into runoff. Using ArcSWAT interface, we delineated 61 HRUs for each land cover configuration observed in 1975 and 1986. In addition to model construction and simulation, ArcSWAT provided tools for calibration, sensitivity and uncertainty analyses.

We divided the daily precipitation and discharge records into three periods for calibration, validation and simulation dependent upon the land cover conditions under which we ran the model. For the 1975 land cover, the time periods 1964–1970 represented the calibration; 1971–1975, the validation; and 1976–1986, the simulation. Under the 1986 land cover, we calibrated SWAT using 1976–1981 data, validated based upon 1982–1986 and simulated from 1964–1975. For each calibration and validation period, we compared observed and simulated monthly runoffs using linear regression used to assess the initial model fit with an R^2 threshold of 0.60–0.80 (Neitsch et al. 2002). To create more robust indicators of model fit, we then calculated the OLS bisector slope (Isobe et al. 1990) and the Nash–Sutcliffe (NS) efficiency metric (Nash and Sutcliffe 1970).

SWAT has a built-in sensitivity analysis routine (Green and van Griensven 2008) using Latin hypercube combined with a one-factor-at-a-time sampling. Given the large number of model parameters potentially adjusted during calibration, we used the sensitivity analysis to identify those parameters most likely to improve model performance during automatic calibration. We then selected parameters following Neitsch et al. (2002). Table 10.1 lists the range of parameter values and the method by which the parameter adjusted during the sensitivity analysis. We ran 1000 combinations of 11 parameters of the calibration periods under each land cover condition. We then ranked the model parameters according to the sensitivity of influence on the simulated runoff. We omitted the three least sensitive from the subsequent autocalibration as recommended by Green and van Griensven (2008).

SWAT employs the parameter solution (PARASOL) autocalibration method (van Griensven and Meixner 2004; Green and van Griensven 2008) coupled with uncertainty analysis of model parameters to improve the overall model fit. PARASOL is based on the shuffled complex evolution (SCE) algorithm, a global search algorithm that minimises the sum of squares residuals between observed and simulated runoff, based on multiple model parameters, and PARASOL is run and evaluated following the manual calibration and sensitivity analyses. Initial autocalibration produced unreasonable results as parameters controlling runoff were preferentially selected for adjustment and unlikely ranges; consequently, only parameters controlling subsurface and groundwater flow were selected in the autocalibration step. The Virilla sub-basin seemed particularly sensitive during manual calibration; therefore, autocalibration was dominantly focused on improving the modelled runoffs with a more feasible set of model parameters.

SWAT uses Sources of Uncertainty Global Assessment Using Split Samples (SUNGLASSES) (van Griensven and Meixner 2004) to assess uncertainty in model output attributable to model formulation rather than parameter uncertainty. To better evaluate model predictive power and prediction errors, SUNGLASSES

Table 10.1 Model parameter and their lower and upper bounds adjusted during calibration and sensitivity and uncertainty analyses.

Parameter	Variation method*	Lower bound	Upper bound	Sensitivity rankings			
				RGSR		RV	
				1975	1986	1975	1986
Baseflow alpha coefficient (days)	1	0	1	8	6	6	8
Manning's n for channel	1	0	1	7	8	7	7
Initial SCS CN II value	3	−25	25	3	3	3	3
Soil evaporation compensation factor	1	0	1	1	2	1	2
Groundwater delay time (days)	2	−10	10	6	7	8	6
Groundwater 'revap' coefficient	2	−0.036	0.036	4	4	4	4
Threshold water depth in shallow aquifer for flow	2	−1000	1000	2	1	2	1
Threshold water depth in shallow aquifer for 'revap'	2	−100	100	5	5	5	5
Average slope steepness	3	−25	25	10	10	10	11
Average slope length (m)	3	−25	25	11	11	11	10
Surface runoff lag	1	0	10	9	9	9	9

* Model parameter variation methods are defined by 1 = replacement, 2 = addition and 3 = multiplication (Winchell et al. 2007).
RGSR, Río Grande de San Ramón; RV, Río Virilla.
Low and high rankings provided in the last four columns indicate the greatest and lowest parameter sensitivities.

examines the validation parameter set separate from the calibration set (van Griensven and Meixner 2004). Global optimisation criterion (chi-squared, $\alpha = 0.05$) assesses the fit between the observed and simulated runoffs. We replicated the procedure 2000 times for the simulation periods for both sub-basins and land cover conditions.

Once we achieved acceptable calibration and validation, we reran SWAT under both land cover conditions for the simulation periods in the sub-basins, thereby creating four scenarios based on combinations of pre- and post-1975 precipitation, coupled with the two different land cover conditions (Figure 10.3). Scenarios 1 and 4 actually occurred, while 2 and 3 assist in identifying the relative dominance of precipitation regime and land cover over runoff. Differences in runoff from each scenario provide the basis for the following hypotheses:

- H0: Precipitation and land cover do not influence runoff (scenarios 1 and 4 not different).
- HA1: Land cover changed, so runoff changed (scenarios 1 and 3, and 2 and 4, different).

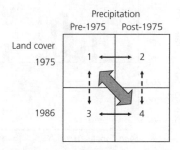

Figure 10.3 Matrix of SWAT model simulation scenarios based on land cover year (1975 and 1986) and precipitation periods (1964–1975 and 1976–1986) as applied to each sub-basin. Grey arrow connects scenarios 1 and 4 denoting observed historical land cover and precipitation conditions.

- HA2: Precipitation changed, so runoff changed (scenarios 1 and 2, and 3 and 4, different).

The greatest differences in simulated runoffs might be expected between scenarios 1 and 4 due to combined changes in precipitation regime and hydrological processes affected by land cover. Large differences under the other hypotheses would implicate one particular variable as the cause of the change in runoff.

A Mann–Whitney U test of medians applied to the mean monthly runoff values under each scenario quantified the relative importance of each control on the simulated runoff.

Results

Statistical analysis

Time series (Figure 10.1) of sub-basin annual runoff as a percentage of total runoff from the Tárcoles compared to sub-basin area shows that, prior to 1975, neither sub-basin consistently contributes greater proportional runoff than the other (Figure 10.4). However, after 1975, the Virilla contributes a greater proportion of runoff in all but 2 years. The analysis reveals statistically significant changes in the counts above/below median annual runoff pre- and post-1975 in both sub-basins (Figure 10.4a), even though the patterns are less marked in the Río Virilla. Regression parameters (Figure 10.5a) from the monthly time series indicate a statistically significant increase in percentage contribution from the Río Virilla from January to August.

Annual precipitation (Figure 10.4b) displays similar temporal patterns; however, no significant change in annual precipitation to the Grande de San Ramón was detected at all, opposite to the significances observed in flows. Regression applied to the monthly precipitation time series (Figure 10.5b) only yields significant linear trends in March (positive) and June (negative). The former is one of the driest months, and the negative trend in the latter is opposite to the positive trend observed in flows.

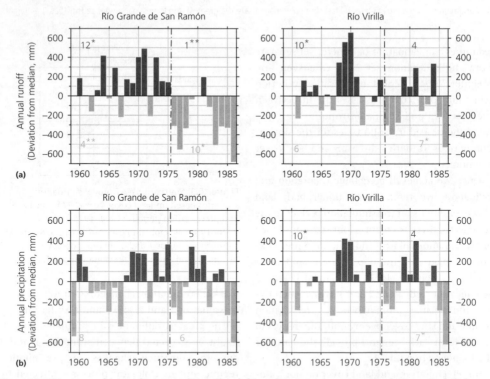

Figure 10.4 Measures of (a) median annual runoff and (b) median annual precipitation deviates. Dark/light grey bars represent total counts (given by numbers in each graph quadrant) above/below median annual values before and after 1975. We tested statistical significance at the alpha = 0.05 level using the hypergeometric distribution; * and ** represent a greater and smaller, respectively, number of counts than expected.

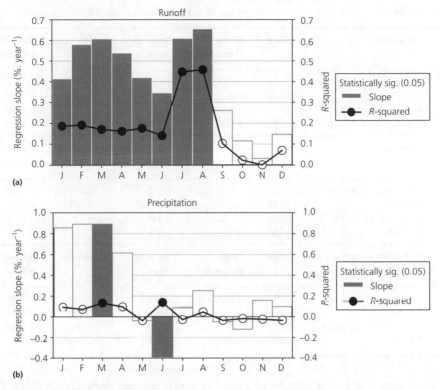

Figure 10.5 Graphical summaries of linear regression slope and R-squared metrics for (a) monthly runoff and (b) precipitation. Statistically significant values ($\alpha = 0.05$) are represented by grey bars and black dots. Values represented by white bars and dots are not statistically significant.

Small changes in sub-basin inputs appear to have produce disproportionate changes in flows.

Model calibration and validation

The principal purpose of the SWAT in this application is to determine the causes for observed changes in runoff rather than to focus on model construction performance; however, its credibility in performing the former is predicated on the latter. After manual calibration, NS efficiency measures (Table 10.2) suggest acceptable model fit – Nash and Sutcliffe (1970) report NS values >0.5 as satisfactory. Acceptable model runs have equivalent calculated and theoretical bisector slope values that are close to one and an intercept close to zero (Isobe et al. 1990). Measures of fit from validation trials should be equal to, or slightly less than, those from manual calibration

as witnessed for both sub-basins under 1986 land cover. However, validation of the Río Virilla under 1975 land cover yields an extremely low NS, and all validation measures for the Río Grande de San Ramón under earlier land cover are higher than the respective values for the manual calibration.

The sensitivity analyses identified model parameters that affected the simulation response, particularly those that could be adjusted further to improve model performance. Runoff appears most sensitive (Table 10.1) to groundwater parameters, with the runoff curve number (RCN) and Manning's n for channel flow also rated highly.

Initial experimentation with autocalibration produced unsatisfactory NS values and bisector slopes (Table 10.2). Although the variance about

Table 10.2 Measures of model fit for manual calibration, model validation and automatic calibration.

Year	Calibration	Slope	Slope theoretical	Variance of slope	Intercept	Nash–Sutcliffe efficiency
1975	Río Grande de San Ramón	1.171	1.321	0.270	−27.041	0.5234
	Río Virilla	1.121	1.229	0.092	11.492	0.4018
1986	Río Grande de San Ramón	1.017	1.034	0.037	14.125	0.7055
	Río Virilla	1.211	1.397	0.434	−29.584	0.5486
Year	**Validation**	**Slope**	**Slope theoretical**	**Variance of slope**	**Intercept**	**Nash–Sutcliffe efficiency**
1975	Río Grande de San Ramón	1.035	1.068	0.084	−28.690	0.6477
	Río Virilla	1.107	1.200	0.304	32.169	0.0768
1986	Río Grande de San Ramón	0.984	0.968	0.131	35.335	0.6032
	Río Virilla	1.259	1.482	0.815	−17.856	0.4357
Year	**Automatic calibration**	**Slope**	**Slope theoretical**	**Variance of slope**	**Intercept**	**Nash–Sutcliffe efficiency**
1975	Río Grande de San Ramón	0.861	0.729	0.047	−28.495	0.0696
	Río Virilla	1.120	1.227	0.091	11.648	0.4023
1986	Río Grande de San Ramón	0.913	0.816	0.012	20.381	0.7207
	Río Virilla	1.098	1.482	0.175	−22.010	0.6866

the bisector slope decreased considerably for the Grande de San Ramón under the 1975 land cover, the NS value decreased substantially compared to the manual calibration. Specifically, the autocalibration for the Grande de San Ramón under 1975 land cover performed worse than the manual calibration, and there was a general lack of consistency in the response of Grande de San Ramón to autocalibration. Further support for rejection of the autocalibrated models included increase in intercept (all scenarios) and the bisector slope (Río Virilla 1975 land cover), although the variance about the bisector slope was reduced consistently for each simulation.

Computational demands of the PARASOL and SUNGLASSES processes permitted 2000 trial simulations and led to an incomplete analysis of parameter and model uncertainties, as evidenced by the lack of sampling over the parameters range, especially for the baseflow alpha coefficient. However, the results do provide insight into the ranges of uncertainty for more sensitive groundwater flow parameters. Other challenges of premature termination of

optimisations and convergence of parameter populations may be corrected by increasing the number of optimisation simulations tenfold. This was not a feasible option during this study.

Model simulations

Absolute differences between simulated runoffs under the four different precipitation and land cover scenarios can best be viewed as paired scenarios (Figure 10.6). Regardless of sub-basin, land cover or precipitation regime, the null hypothesis of no change in runoff is rejected. Sensitivities of runoff within the Río Grande de San Ramón (Figure 10.6a) follow the hydroclimatology regime of the region. Land cover dominates runoff in the drier months and precipitation in the wetter ones. Comparing the pre- and post-1975 influences of precipitation and land cover on runoff differences, equal influence is implied during January–April; however, land cover prevails at the beginning of the rainy season (May and June), when soil moisture storage is at a minimum. Reduced precipitation in July and August, the *veranillo*,

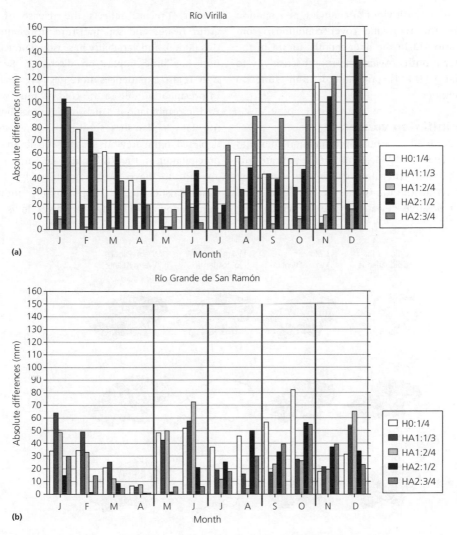

Figure 10.6 Mean monthly runoff differences for (a) the Río Virilla and (b) the Río Grande de San Ramón under the land cover and precipitation scenarios (Figure 10.2). Black vertical lines divide the months into seasons: season 1, JMFA; season 2, MJ; season 3, JA; season 4, SO; and season 5, ND. Runoff differences (dark and light grey bars) resulting from the land cover scenarios can be paired for comparison, and those from precipitation (black and medium grey) can be paired and viewed separately.

renders little difference between simulations; however, sensitivities switch to precipitation during the height of the rainy season (September–November). These observations are consistent with those of Bruijnzeel (2004) and Bonell (2004).

The Río Virilla (Figure 10.6b) displays a greater sensitivity to changes in precipitation (scenarios 1/2 and 3/4), as expected given the more extensive urban cover which proportionally converts more rainfall to runoff than a vegetated surface. The trend of runoff sensitivities to changing precipitation persists throughout the year, except during June and July, when land cover exerts control over runoff. The change in runoff sensitivities during the dry season and the *veranillo* is not apparent. However, similar to the Grande de San Ramón, a marked reversal is seen of runoff sensitivities to changing precipitation under the two land cover conditions, from scenario 1/2 in the drier months to 3/4 in the wetter months. Again, the runoff sensitivity to scenario 3/4 is pronounced from June to September.

Precipitation variability

Standardised seasonal precipitation deviates from 1959 to 1986 (Figure 10.7) indicate low precipitation (negative deviate) in season 1 (JFMA) for the Río Grande de San Ramón and slight positive deviate for the Río Virilla. The discharge record reflects the release of the groundwater and soil moisture stores during this period, and variability in seasonal precipitation has little impact on discharge. Season 2 (MJ) also experiences more negative deviates in precipitation due to the cold Atlantic–warm Pacific combination. Much of the precipitation normally replenishes subsurface water stores depleted in the previous season. The diminished precipitation in the Río Grande de San Ramón appears to have been amplified by basin characteristics (e.g. topographically induced rain shadow), producing even lower flows.

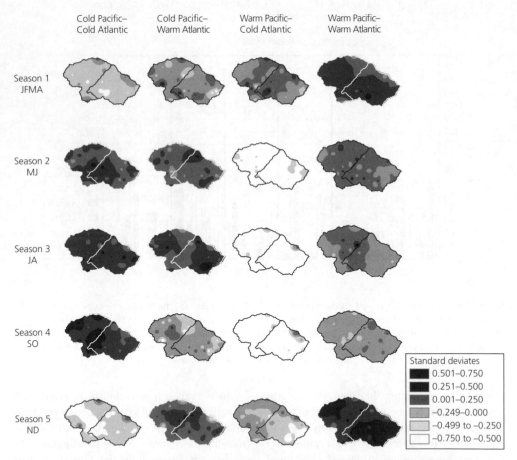

Figure 10.7 Seasonal standardised precipitation deviates for the Ríos Grande de San Ramón and Virilla. Seasonal delineations follow Waylen et al. (1996) and abbreviations are the following: JFMA–January, February–March and April; MJ–May and June; JA–July and August; SO–September and October; and ND–November and December.

Season 3 (JA) encompassing the *veranillo* has been shown to be particularly sensitive to the El Niño signal. Droughts associated with warm phases and cooler Atlantic SSTs are particularly apparent in the Río Grande de San Ramón. However, regardless of the diminished precipitation, subsurface water stores are still filling in both seasons 2 and 3. Drought-like conditions persist during season 4 (SO) with an even greater deviations. Although the Río Grande de San Ramón shows a greater response than the Río Virilla, discharges do not reflect this as the subsurface water stores were full, and both sub-basins respond similarly (as evidenced by the non-linearity of basins responses). In season 5 (ND), the signal flips from the previous season, and both sub-basins show a positive response to the ENSO–Atlantic influence. In the Río Grande de San Ramón, this was relatively unimportant since the rainy season was ending and discharge enters the recession stage. The Río Virilla has a higher deviate explainable by the gap over the Cerro Carpinteral near Cartago, which allows increased precipitation influenced by *nortes* (Waylen et al. 1998), cold fronts that extend from the North American continent and interact with the northeast trades.

Discussion

Changes in runoff contributions into the Río Grande de Tárcoles stem from both land cover change and climate variability and are particularly pronounced at the monthly and seasonal scales. Annual discharges in both rivers have declined with the flows in the Río Grande de San Ramón even more so with a decrease in its proportional input to Río Grande de Tárcoles since 1975. Flow in the Río Virilla has increased as a percentage of the overall flow in January through August but not in the remaining months. Annual basin precipitation inputs show no statistically significant change (increase or decrease) in Río Grande de San Ramón and a

weak decline in the Río Virilla. We found no statistically significant trends for monthly precipitation inputs to the Río Virilla as proportion of total monthly precipitation except in March, when precipitation is lowest, and a slight decline in June precipitation.

Model simulation results indicating changes in sub-basin land cover can act as non-linear amplifiers in the conversion of precipitation to runoff, but this effect is seasonal. The non-linearity of runoff response from the Río Grande de San Ramón became evident from analysis of the simulation model results (Figure 10.6). In the Río Grande de San Ramón, dry season flows are more affected by changes in land cover, while precipitation exerts a greater control in the wet season. The Río Virilla shows a strong sensitivity to precipitation regardless the season and land cover effects that are minimal. Soil moisture may be, in particular, an important control as its effect is inversely related to the amount of urban area in each sub-basin. This is particularly pronounced at the end of September when the soil moisture stores are full from the rainy season.

The combined effects of the Pacific and the Atlantic SSTs anomalies (Figure 10.7) may explain precipitation trends highlighted by the statistical analyses and simulation modelling. Coupling of SSTs in the two ocean basin has been shown to heavily influence precipitation inputs to the area (Poveda et al. 2006). Although the sub-basins are spatially contiguous, historical data indicate slightly different responses to El Niño forcing, while land cover changes may further enhance this response. The post-1975 decline in the Río Grande de San Ramón may result from the occurrence of warm-phase El Niño (drought) conditions, whose effects are amplified by a persistently cooler Atlantic. The Río Grande de San Ramón shows a response in precipitation variability more typical of basins on the Pacific slope, while the Río Virilla has influences from the Pacific and the Caribbean (Waylen et al. 1996). Chavez et al. (2003) identified a global climate shift in the mid-1970s with a colder Atlantic coupled with a warmer Pacific (El Niño). In the time series used here, the Southern Oscillation Index (SOI) indicates warmer-phase El Niño (i.e. droughts in the basins) than cold-phase La Niña (i.e. excess rainfall) after 1975. The Atlantic also shows a change around 1975, with predominately positive SST anomalies earlier and negative SST anomalies post-1975. In this region of Costa Rica, warm-phase El Niño enhanced by a cold Atlantic produces drought-reducing precipitation inputs to discharge (Waylen et al. 1996).

Non-linear rainfall–runoff responses of sub-basins

The reversals of runoff sensitivities in both sub-basins to changes between precipitation periods for the changing land cover and vice versa are perplexing (Figure 10.6). For example, the scenario combination 1/3 of pre-1975 precipitation under changing land cover conditions from 1975 to 1986 produces sensitivity in runoff during the dry season (January to March), the

veranillo (July and August) and October and November and then reverses to scenario combination 2/4 for April to June, September, and December. This reversal may be explained by the non-linear conversion of precipitation into runoff. With respect to this non-linear amplification of rainfall signal in runoff, there are two other notable aspects of land cover and the 1975 switch in runoff responses.

As an illustration of non-linear rainfall–runoff responses of drainage basins in converting precipitation to runoff, observed annual precipitation and runoff in both sub-basins were plotted against one another (Figure 10.8). If all precipitation went to runoff (i.e. evapotranspiration losses equal to 0), then dots would fall along the 1:1 line. As precipitation increases, the absolute value of evapotranspiration decreases, and evapotranspiration as a percentage of precipitation declines in both sub-basins though more markedly in the Río Grande de San Ramón. In years of diminished precipitation, a lower proportion goes to runoff, while in

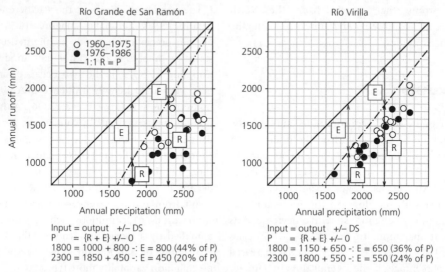

Figure 10.8 Non-linear responses of runoff, precipitation and evapotranspiration for sub-basins. Actual runoff for a given precipitation is defined by dots. The 1:1 line (solid grey diagonal line) represents the special case of no evapotranspiration losses. An upper enveloping line (dashed line) represents the maximum expected runoff (black vertical line) for a given annual precipitation. The difference between the enveloping line and the 1:1 line represents the minimum possible evapotranspiration.

rainy years, the opposite is true. Observations of precipitation in the Río Grande de San Ramón post-1975 clearly plot below those of pre-1975. So although rainfall has declined little since 1975, evapotranspiration appears to have increased, an effect that is not as marked in the Río Virilla where the groundwater stores are bypassed due to more impervious surface. Being a more urban sub-basin, more runoff goes directly to the rivers, bypassing potential sources for evapotranspiration in soil and groundwater stores.

The more rural and vegetated conditions seen in the Río Grande de San Ramón (Table 10.3) modify the incoming precipitation differently than the Río Virilla, thus producing differences seen in discharge contributions. The vegetated conditions in the Río Grande de San Ramón permit greater infiltration capacities to the soil water store, reducing the amount of effective precipitation going to runoff. This maintenance of evapotranspiration contributes to reduced runoff under vegetated conditions. The Río Virilla is less non-linear in transforming precipitation to runoff due to impervious areas, which omit evaporation and infiltration processes from partitioning precipitation into runoff (Figure 10.8). The Río Virilla sub-basin will produce greater runoff due to increased impervious urban areas that decrease evapotranspiration contributions and decreased infiltration.

The observed changes in the Río Virilla (Figure 10.6b) from rainfall-dominated responses under 1975 land cover in seasons one and two to sensitivities under the 1986 land cover are physically plausible. The 'more' naturally vegetated conditions of 1975 could produce a stronger response during drier conditions (Figure 10.7). Under wetter conditions, more disturbed land cover (e.g. urban and pasture) produces a greater proportional response in runoff. This sensitivity is most apparent after the soil and groundwater stores have been depleted during the dry season. In June, sensitivities are almost equal, particularly in the simulation under pre-1975 rainfall and 1975 land cover. Despite no significant increase, rainfall does appear to dominate the differences in runoff in the simulation results.

Mixed results as to the effects of land cover change on climatic and hydrological responses are common in the literature. Wilk et al. (2001) reported that despite a >50% reduction in forest from 1957 to 1995, no changes in seasonal or annual rainfall totals were detected. Lorup et al. (1998) found no indication of increased runoff or changes in land use, although population increased significantly. In contrast to previous research (Bruijnzeel 1990; Bonell and Balek 1993), Costa et al. (2003) proposed that alterations in hydrological regime arising from land use changes would be more evident in the rainy season and that higher discharges would be

Table 10.3 Percentages of land cover and land cover change per sub-basin.

Sub-basin	Land cover class	1975 area (%)	1986 area (%)	Change from 1975 to 1986
Río Grande de San Ramón	Agricultural crops	4.6	1.1	−3.5
	Forest – deciduous	10.1	7.0	−3.1
	Pasture	14.2	27.3	13.1
	Range – brush	19.6	13.2	−6.4
	Residential	0.9	0.9	0.0
Río Virilla	Agricultural crops	4.2	1.4	−2.9
	Forest – deciduous	10.5	8.1	−2.4
	Pasture	15.4	24.1	8.7
	Range – brush	14.3	10.1	−4.1
	Residential	6.2	6.9	0.8
	Totals	100.0	100.0	

expected from more intense land use. Reduced infiltration, although insufficient to affect dry season flow, increased surface runoff during the rainy season and increased discharge throughout the year through an associated reduction in evapotranspiration (Costa et al. 2003). Similar to these studies, we found mixed results as to which control (i.e. land cover or precipitation) dominates at the seasonal and annual timescales.

Future challenges for mesoscale catchment modelling

Although this study provides one approach using mixed methods for mesoscale catchment studies, improvements could be made within this methodological framework. Future improvements could include better spatial distribution of rain gauges within each sub-basin; a limitation of SWAT is that only one gauge station was permitted per sub-basin. The atmospheric parameters of wind speeds, temperatures and humidity could be downscaled better using regional climate models of Central America to capture the spatial variability of evapotranspiration across the catchment as controlled by the complex topography (Figure 10.2). HRUs delineated from SWAT would be targeted for future fieldwork to collect more detailed data about the nature of the soils and land cover characteristics as opposed to parameter values from the literature. In particular, important soil hydraulic parameters such as depth to shallow aquifer and groundwater delay, important during sensitivity and uncertainty analysis, could be empirically measured.

These modelling challenges and limitations are not unique to our study. Other researchers (Bruijnzeel 1990, 2004; Bonell 2004) have echoed the challenges of conducting modelling efforts in tropical mesoscale catchment due to the lack of field data and the use of the literature or derived parameters to parameterise models. As catchment size increases, traditional paired catchment studies become unfeasible due to excessive cost of instrumentation and difficulty in controlling land cover treatments (Bosch and Hewlett 1982). Practical and operational data collection methods are needed to parameterise models. Delineation and mapping of HRUs used by the SWAT model can direct field sampling to measure key hydrological properties needed for model parameterisation. A stratified sampling scheme applied to the HRUs can target unique combinations of soils, land cover and topography that control hillslope hydrological processes. Preliminary sensitivity analyses can further guide field sampling by focusing data collection on those input parameters that most impact the simulated runoff.

Hydrological impacts may be muted by the heterogeneous catchment characteristics and mosaic of multiple land covers, underlain by variability in soils, geology and topography (Bruijnzeel 2004). Land cover changes are spatially and temporally heterogeneous and can impede the detection of changes in discharge and other variables in the water balance (Wilk et al. 2001). The use of spatially explicit hydrological models and HRUs, both used by the SWAT model, can help to overcome catchment and land cover change heterogeneity by simplifying the complexity that results from combinations of soils, geology and spatial and temporal distribution of rainfall. The use of HRUs may detect increased surface runoff from sub-basins that generate more discharge than expected due to unique biophysical characteristics and landscapes altered by humans. The additional use of remotely sensed data for precipitation (Andersen et al. 2002), surficial soil parameters (Schmugge et al. 2002) and land cover (this study) can provide empirical data necessary to account for complex heterogeneity of mesoscale catchments.

Conclusions

The application of the process-based simulation model SWAT in Costa Rica, a first to our knowledge, characterises catchment heterogeneity by identifying non-linear hydro-climatic responses using the integrative approach of hydrological modelling and statistical analyses. This research augments existing mesoscale catchment studies combining simulation modelling and statistical analysis to distinguish the hydrological impacts from land cover from those due to climate variability on discharge contributions within a tropical catchment. This is currently an under-studied area with potential impacts on water withdrawals for municipal and agricultural use not only in Costa Rica's central valley but also in other global metropolitan areas.

This work expands our understanding of the coupled effects of land cover change and climate (i.e. precipitation) variability in tropical mesoscale catchments. Applied to the Río Grande de Tárcoles catchment, a mixed-method approach of statistical analysis and simulation modelling identified seasonal and spatial controls of land cover change and precipitation variability. This is a preferred approach to analyse coupled with land cover and climate influences in mesoscale catchment studies (Refsgaard et al. 1989; Lorup et al. 1998). Statistical analysis identified changes in annual and monthly runoff and precipitation after which runoff simulations identified distinctive seasonal combinations of land cover and precipitation that generate a unique runoff response for each sub-basin. The spatially explicit representation allowed us to investigate the hydrological effects of changing land cover conditions under different historical precipitation regimes. To deal with complex heterogeneity of mosaic of land cover and topography and to adequately represent catchment runoff generating processes, the SWAT model grouped soils, land cover and topography into distinct HRUs. The process-based nature of the SWAT model helps to identify the catchments processes that hold seasonal influence over the partitioning of excess precipitation into surface and subsurface runoff. The unique climate variability signal in the seasonal precipitation deviates matches that are seen in the simulated runoffs.

The Río Grande de Tárcoles is a complex tropical catchment that experiences climatic interplay between the Atlantic, the Caribbean and the Pacific at various timescales (5–7 years ENSO, several decades for Atlantic). Changes in the resultant climate variability need to be accounted for when seeking hydrological consequences to changing land cover conditions. Further exploration of climate variability might include the division of the simulation periods conditioned upon ENSO phase and investigation of differences in the mean monthly runoff. In addition to the standard precipitation deviates, the influences of ENSO and Atlantic SSTs can be controlled providing a description of the runoff responses most expected under different ENSO phases.

Acknowledgements

Peter Waylen was fortunate to have been taught and mentored by John Thornes, in the lecture room, the lab and the field – not to mention various hostelries – for 3 years as an undergraduate. John's open and supportive attitude towards students, his constant inquisitiveness, his positive challenging of an individual's abilities and his prescient words of advice about the direction our discipline have provided lofty goals to strive for. Undoubtedly, John is one of the greatest influences shaping Peter's professional life and one for which he is ever grateful.

References

Alvarado, A., Forsythe, W. (2005) Densidad aparente en suelos de Costa Rica. *Agronomía Costarricense* **29**, 85–94.

Anderson, E. P., Pringle, C. M., Rojas, M. (2006) Transforming tropical rivers: an environmental perspective on hydropower development in Costa Rica. *Aquatic Conservation* **16**, 679–693.

Andersen, J., Dybkjaer, G., Jensen, K. H., Refsgaard, J. C., Rasmussen, K. (2002) Use of remotely sensed precipitation and leaf area index in a distributed hydrological model. *Journal of Hydrology* **264**, 34–50.

Arnold, J. G., Fohrer, N. (2005) SWAT2000: current capabilities and research opportunities in applied watershed modelling. *Hydrological Processes* **19**, 563–572.

Arnold, J. G., Srinivasan, R., Muttiah, R. S., Williams, J. R. (1998) Large area hydrologic modeling and assessment – Part 1: Model development. *Journal of the American Water Resources Association* **34**, 73–89.

Bedient, P. B., Huber, W. C. (1988) *Hydrology and Floodplain Analysis*. Addison-Wesley, Reading, MA.

Blomquist, W. A., Ballestero, M., Bhat, A., Kemper, K. (2005) *Institutional and Policy Analysis of River Basin Management: The Tárcoles River Basin, Costa Rica*. World Bank, Washington, DC.

Bonell, M. (1998) Possible impacts of climate variability and change on tropical forest hydrology. *Climatic Change* **39**, 215–272.

Bonell, M. (2004) Runoff generation in tropical forests. In: Bonell, M., Bruijnzeel L. A. (eds.) *Forests, Water and People in the Humid Tropics: Past, Present and Future Hydrological Research for Integrated Land and Water Management*. International Hydrology Series. Cambridge University Press, Cambridge, UK, pp. 314–406.

Bonell, M. (2010) The impacts of global change in the humid tropics: selected rainfall-runoff issues linked with tropical forest-land management. *Irrigation and Drainage Systems* **24**, 279–325.

Bonell, M., Balek, J. (1993) Recent scientific developments and research needs in hydrological processes of the humid tropics. In: Bonell, M., Hufschmidt, M. M., Gladwell J.S. (eds.) *Hydrology and Water Management in the Humid Tropics* (1st Ed.). Cambridge University Press, Cambridge, UK, pp. 167–260.

Bonell, M., Purandara, B., Venkatesh, B., Krishnaswamy, J., Acharya, H. A. K., Singh, U. V., Jayakumar, R., Chappell, N. (2010) The impact of forest use and reforestation on soil hydraulic conductivity in the Western Ghats of India: implications for surface and sub-surface hydrology. *Journal of Hydrology* **391**, 47–62.

Bosch, J. M., Hewlett, J. D. (1982) A review of catchment experiments to determine the effect of vegetation changes on water yield and evapo-transpiration. *Journal of Hydrology* **55**, 3–23.

Bruijnzeel, L. A. (1990) *Hydrology of Moist Tropical Forests and Effects of Conversion: A State of Knowledge Review*. UNESCO, Paris.

Bruijnzeel, L. A. (2004) Hydrological functions of tropical forests: not seeing the soil for the trees? *Agriculture Ecosystems and Environment* **104**, 185–228.

Centro Científico Tropical (1989) *Tipo de suelos en Costa Rica*. Centro Científico Tropical, San José, Costa Rica.

Chavez, F. P., Ryan, J., Lluch-Cota, S. E., Niquen, M. (2003) From anchovies to sardines and back: multidecadal change in the Pacific Ocean, *Science* **299**(5604), 217–221.

Colby, J. D. (2001) Simulation of a Costa Rican watershed: resolution effects and fractals. *Journal of Water Resources Planning and Management* **127**, 261–270.

Costa, M. H., Botta, A., Cardille, J. A. (2003) Effects of large-scale changes in land cover on the discharge of the Tocantins River, Southeastern Amazonia. *Journal of Hydrology* **283**, 206–217.

Dingman, S. L. (2002) *Physical Hydrology*. Prentice Hall, Upper Saddle River, NJ.

Dunne, T. (1978) Field studies of hillslope processes. In: Kirkby, M. J. (ed.) *Hillslope Hydrology* (1st Ed.). John Wiley & Sons, Ltd, Chichester, pp. 227–294.

Dykes, A. P., Thornes, J. B. (2000) Hillslope hydrology in tropical rainforest steeplands in Brunei. *Hydrological Processes* **14**, 215–235.

Elsenbeer, H., Cassel, D. K. (1991) The mechanisms of overland flow generation in a small catchment in western Amazonia. In: Braga, B. P. F., Fernandez-Jauregui, C. A. (eds.) *Water Management of the Amazon Basin Symposium: Proceedings of Manaus Symposium August 1990* (1st Ed.). UNESCO, Montevideo, pp. 275–288.

Green, C. H., van Griensven, A. (2008) Autocalibration in hydrologic modeling: using SWAT2005 in small-scale watersheds. *Environmental Modelling & Software* **23**, 422–434.

Hastenrath, S. (2002) The intertropical convergence zone of the eastern pacific revisited. *International Journal of Climatology* **22**, 347–356.

Horton, R. E. (1933) The role of infiltration in the hydrologic cycle. *Transactions of the American Geophysical Union* **14**, 446–460.

Isobe, T., Feigelson, E. D., Akritas, M. G., Babu, G. J. (1990) Linear-regression in astronomy. *Astrophysical Journal* **364**, 104–113.

Kalnay, E., Kanamitsu, M., Kistler, R., Collins, W., Deaven, D., Gandin, L., Iredell, M., Saha, S., White, G., Woollen, J., Zhu, Y., Chelliah, M., Ebisuzaki, W., Higgins, W., Janowiak, J., Mo, K. C., Ropelewski, C., Wang, J., Leetmaa, A., Reynolds, R., Jenne, R., Joseph, D. (1996) The NCEP/NCAR 40-year reanalysis project. *Bulletin of the American Meteorological Society* **77**, 437–471.

Krishnaswamy, J., Halpin, P. N., Richter, D. D. (2001) Dynamics of sediment discharge in relation to land-use and hydro-climatology in a humid tropical watershed in Costa Rica. *Journal of Hydrology* **253**, 91–109.

Lehner, B., Verdin, K., Jarvis, A. (2006) *Hydrological Data and Maps Based on Shuttle Elevation Derivatives at Multiple Scales (HydroSHEDS) – Technical Documentation*. World Wildlife Fund US, Washington, DC.

Loh, W. Y., Shih, Y. S. (1997) Split selection methods for classification trees. *Statistica Sinica* **7**, 815–840.

Lorup, J. K., Refsgaard, J. C., Mazvimavi, D. (1998) Assessing the effect of land use change on catchment runoff by combined use of statistical tests and hydrological modelling: case studies from Zimbabwe. *Journal of Hydrology* **205**, 147–163.

Magana, V., Amador, J. A., Medina, S. (1999) The midsummer drought over Mexico and Central America. *Journal of Climate* **12**, 1577–1588.

Ministerio de Agricultura y Ganadería (MAG) (1992) *Uso de la tierra en Costa Rica en 1992.* MAG, San José.

Ministerio de Recursos Naturales, Energía y Minas (MIRENEM) (1988) *Instituto Meteorológico Nacional Año del Centenario 1888–1988: Catastro las Series de Precipitaciones Medidas en Costa Rica.* MIRENEM, San José, Costa Rica.

Nash, J. E., Sutcliffe, J. V. (1970) River flow forecasting through conceptual models part I: a discussion of principles. *Journal of Hydrology* **10**, 282–290.

Neitsch, S. L., Arnold, J. G., Kiniry, J. R., Williams, J. R. (2002) *Soil and Water Assessment Tool, Version 2000, Theoretical Documentation.* Texas Water Resources Institute, College Station, TX.

Nortcliff, S., Thornes, J. B. (1981) Seasonal variations in the hydrology of a small forested catchment near Manaus, Amazonas, and the implications for its management. In: Lal, R., Russell, E. W. (eds.) *Tropical Agricultural Hydrology* (1st Ed.). John Wiley & Sons, Ltd, Chichester, pp. 37–57.

Poveda, G., Waylen, P., Pulwarty, R. (2006) Annual and interannual variability of present climate in northern South America and southern Mesoamerica. *Paleogeography, Paleoclimatology, Paleoecology* **234**, 3–27.

Refsgaard, J. C., Alley, W. M., Vuglinsky, V. S (1989) *Methodology for Distinguishing Between Man's Influence and Climatic Effects on the Hydrological Cycle.* UNESCO, Paris.

Schaap, M. G., Leij, F. J., van Genuchten, M. T. (2001) ROSETTA: a computer program for estimating soil hydraulic parameters with hierarchical pedotransfer functions. *Journal of Hydrology* **251**, 163–176.

Schmugge, T. J., Kustas, W. P., Ritchie, J. C., Jackson, T. J., Rango, A. (2002) Remote sensing in hydrology. *Advances in Water Resources* **25**, 1367–1385.

Siriwardena, L., Finlayson, B. L., McMahon, T. A. (2006) The impact of land use change on catchment hydrology in large catchments: the Comet River, Central Queensland, Australia. *Journal of Hydrology* **326**, 199–214.

van Griensven, A., Meixner, T. (2004) Dealing with unidentifiable sources of uncertainty within environmental models. In: Pahl-Wostl, C., Schmidt, S., Rizzoli, A. E., Jakeman, A. J. (eds.) *Complexity and Integrated Resources Management: Transactions of the 2nd Biennial Meeting of the International Environmental Modelling and Software Society* (1st Ed.). University of Osnabrück, Osnabrück, pp. 1045–1052.

van Loon, E. E., Troch, P. A. (2002) Tikhonov regularization as a tool for assimilating soil moisture data in distributed hydrological models. *Hydrological Processes* **16**, 531–556.

Vasquez, A. (1980) *Metodología para la Determinación de la Capacidad de Uso de la Tierra.* Ministerio de Agricultura y Ganadería, San José.

Waylen, P., Quesada, M. E. (2002) The effect of Atlantic and Pacific sea surface temperatures on the mid-summer drought of Costa Rica. In: Garcia-Ruiz, J. M., Jones, A. A., Arnáez, J. (eds.) *Environmental Change and Water Sustainability.* Instituto Pirenaico de Ecología, Consejo Superior de Investigaciones Científicas, Zaragoza, pp. 197–209.

Waylen, P. R., Quesada, M. E., Caviedes, C. N. (1996) Temporal and spatial variability of annual precipitation in Costa Rica and the Southern Oscillation. *International Journal of Climatology* **16**, 173–193.

Waylen, P., Quesada, M., Caviedes, C., Poveda, G., Mesa, O. (1998) Rainfall distribution and regime in Costa Rica and its response to the El Nino-Southern Oscillation. *Conference of Latin Americanist Geographers Yearbook* **24**, 75–84.

Wilk, J. Andersson, L., Plermkamon, V. (2001) Hydrological impacts of forest conversion to agriculture in a large river basin in northeast Thailand. *Hydrological Processes* **15**, 2729–2748.

Winchell, M., Srinivasan, R., Di Luzio, M., Arnold, J. G. (2007) *ArcSWAT interface for SWAT2005 – User's Guide.* Blackland Research Center, Texas Agricultural Experiment Station and Grassland, Soil and Water Research Laboratory, and USDA Agricultural Research Service, Temple, TX.

Ziegler, A. D., Giambelluca, T. W., Tran, L. T., Vana, T. T., Nullet, M. A., Fox, J., Vien, T. D., Pinthong, J., Maxwell, J. F., Evett, S. (2004) Hydrological consequences of landscape fragmentation in mountainous northern Vietnam: evidence of accelerated overland flow generation. *Journal of Hydrology* **287**, 124–146.

Zimmermann, B., Papritz, A., Elsenbeer, H. (2010) Asymmetric response to disturbance and recovery: changes of soil permeability under forest-pasture-forest transitions. *Geoderma* **159**, 209–215.

CHAPTER 11

Insights on channel networks delineated from digital elevation models: The adaptive model

Ashraf Afana[1] and Gabriel del Barrio[2]

[1] Department of Geography, Science Labs, Durham University, Durham, UK

[2] Estación Experimental de Zonas Áridas (EEZA), CSIC, Cañada de San Urbano, Almeria, Spain

Introduction

In landscape studies, delineation of channel networks is a major problem. Its effect goes further than the limit of one discipline and restricts not only the results expected but also the methodologies used in the desired studies. Recent advents in digital cartographic data, mainly digital elevation models (DEMs), have provided new capacities and rational objective estimation of geomorphological parameters for hydrological modelling at several scales (Gyasi-Agyei and De Troch 1995). For channel networks, the analysis of their components obtained from DEMs provided deeper insights into their structure properties to be incorporated in the quantitative analysis of land surface models (Grayson and Blöschl 2001).

Early procedures to describe channel networks from DEMs were based on the notion of Gilbert (1909) that divergent-convex landforms are linked to hillslope processes and convergent-concave surfaces are associated with fluvial-dominated erosion and hence valleys and channel networks. Critically, this approach defines channel-network position in the landscape, and to define their limits, the threshold concept was incorporated, which quantifies the drainage accumulation at each cell in the DEM (Tarboton et al. 1991). Consequently, all cells which have a user-specified threshold (i.e. critical support area that defines the minimum drainage area) are considered to be on a channel network. As a rule of thumb, channel-network limits are defined by a user-specific threshold. This point determines where channels begin in the landscape, and it is widely known as the 'critical or optimum threshold' and symbolised as A_s. The choice of the appropriate A_s used to define the optimum channel network is highly related to DEM resolution, local environmental factors and landscape complexity (Hancock 2005; Thompson et al. 2001). This observation implies that natural channel networks are scale invariant, whereas streams derived from DEMs are scale dependent (Tarolli and Dalla Fontana 2009). Theoretically, a single value of A_s is only applicable under homogeneous landscape conditions (Bischetti et al. 1998; Vogt et al. 2003), and a multi-fractal approach is needed to cope consistently with the properties of measured field data (De Bartolo et al. 2000).

Monitoring and Modelling Dynamic Environments, First Edition. Edited by Alan P. Dykes, Mark Mulligan and John Wainwright.

Hence, applying a unique A_S value under varying landscape conditions could be of low suitability to reflect natural variability of drainage density.

The debate over the optimality of A_S and whether it is sufficient to determine channel initiation is of great importance, since several morphometric and topographic indices depend on it (Lin et al. 2006; Wilson et al. 2000). Numerous models have been proposed to handle A_S optimally, such as constant threshold area (Tribe 1992), slope-dependent threshold (Montgomery and Dietrich 1989, 1992), constant drop analysis (CDA) (Tarboton et al. 1991), grid order threshold (Peckham 1995), probability distribution of cumulative drainage area (Perera and Willgoose 1998), the logistic regression (Heine et al. 2004), probability distribution of energy index (McNamara et al. 2006) and probability density function of curvature (Lashermes et al. 2007). Other approaches proposed combinations of these methods in order to enhance stream network delineation (Passalacqua et al. 2010). These approaches demonstrated varying capabilities in relation to landscape complexity, scale variability and data spacing, where these factors exert a limiting effect on the applied approach (Poggio and Soille 2011). Under these conditions, we believe that defining the optimum channel network using DEM data under limited conditions of data availability and scale variability is still a basic requirement for hydrologic and geomorphologic studies. Such problems should be handled by efficient models capable to describe the dimension of scale dependency in order to describe landscape dissection, irrespective of terrain heterogeneity. Thus, an adequate solution, according to our judgement, could be achieved by using algorithms that best convey spatial heterogeneity, represent dominant processes and make use of available data.

Herein, we propose a compound model that delineates channel networks in relation to the intrinsic landscape information. Such an approach aims to depict landscape dissection in relation to data availability (i.e. DEM resolution) and prevailing landscape heterogeneity. Thus, the general objective of the present work is to define the optimal channel network that best describes landscape dissection at a determined scale and resolution. Such an objective highlights the need for a recursive examination of scale properties of the landscape, mainly hillslope–channel relationships. Another associated objective is the application of a generalised analysis of network complexity to other areas of distinct scale and resolution in order to obtain the best approach for the depiction of channel networks. The methodological approach of the present work consists of two main parts. The first is the extraction and the delineation of different stream networks of different origins (e.g. natural vs. automated). The second consists of a robust validation approach that permits a justified quantitative comparison between natural and automatic delineated stream networks.

Methods for channel-head definition

Major approaches to define an optimal A_S for channel-network delineation can be divided into those models that incorporate local factors as correction parameters and models that support the use of DEMs as the sole source of information. Examples related to the first approach include the slope-dependent threshold (Montgomery and Dietrich 1989, 1992) and the compound model (Vogt et al. 2003). These approaches predict variable A_S values between regions associated with different local factors, for example, climate, relief, lithology, vegetation cover and land use and stage of landscape evolution (Istanbulluoglu et al. 2002; Tucker and Bras 1998). Examples related to the second approach are diverse, but it is worth highlighting the constant slope–area threshold approach (Tarboton et al. 1991), which predicts channel heads to be associated with the transition from slope-dependent hillslope transport to discharge fluvial domain (Montgomery and Dietrich 1994).

The slope–area relationship is the most widely applied common algorithm to delineate channel-network limits in the landscape. Constant threshold area or constant critical support area (A_s) predicts that channel heads are associated with a change in the relation between local slope (S) and upslope contributing area (A). Mathematically, this relationship can be expressed by

$$S = cA^{-\theta} \qquad (11.1)$$

where c is a constant and θ is a scaling coefficient.

In a log–log plot of S against A, the transition from convex hillslopes to concave valleys is expressed by a characteristic change from a positive to negative trend. Tarboton et al. (1991) proposed using the value of the A at this break as the critical contributing area (A_s). However, difficulties extracting the appropriate value of θ are hampered by several factors, for example, DEM resolution, variability in methods used to assess θ and landscape complexity (Giannoni et al. 2005; Tarolli and Dalla Fontana 2009), giving rise to misleading interpretations on stream limits in the landscape.

Modified methods to assess and extract values of θ have been proposed. For example, Istanbulluoglu et al. (2002) introduced a probabilistic model to evaluate variations in channel initiation dependent on c in Equation 11.1 and concluded that a uniform probability distribution of channel initiations is needed to match results observed at the field for channel-head locations. Tarboton et al. (1991) had earlier proposed the use of the constant drop property (CDP) for the definition of A_s. The CDP (i.e. fall in elevation) analysis is based on the use of the smallest weighted support area threshold that produces a channel network where the mean stream drop in first-order streams is not statistically different from the mean stream drop in higher-order streams, evaluated by the t statistic

of difference between means and using it as the optimal value of A_s.

This chapter presents an ordered sequence of the main steps involved in channel delineation from a DEM that may serve as a practical exercise on geomorphometric modelling applied to environmental studies. The CDP approach is used as an accepted benchmark based on automated channel networks, and a new approach, named the adaptive model, is used to deal with some improvements required to account for the scale variability of channel networks as found in natural landscapes. Throughout the present work, A_s is estimated as the number of cells and is used as a proxy to define the area needed to verify where channels begin in the landscape.

Study area and data type

The study was conducted for the Tabernas basin, located in the south-eastern part of the Iberian Peninsula (Figure 11.1), which is widely known as 'The Desert of Tabernas' due to extensive badlands dominating the landscape. The Tabernas basin occupies an area of $567\,\mathrm{km}^2$, which consists of several landscape units associated with various lithologic and tectonic formations, as well as different hydrological and geomorphological processes marked with varying climatic conditions. In general, the geomorphology of the region is influenced by two factors: active tectonics and semi-arid to arid climates, which characterise the actual mesoform context of Tabernas basin in particular and the southeast of Spain in general. In Tabernas, model testing and validation were carried out at two major scales. First, the Tabernas basin as a whole, representing a highly heterogeneous landscape, is verified using a DEM at $30\,\mathrm{m}$ grid spacing. Second, within the Tabernas basin, a limited unit of homogeneous erosional landscape, known as the El Cautivo site, is represented by a DEM at $1\,\mathrm{m}$ grid spacing (Figure 11.1).

Figure 11.1 Locations of the Tabernas basin (30 m resolution DEM – centre) and the corresponding sub-catchment of El Cautivo (1 m resolution DEM – right).

The adaptive model approach

Model origin and derivation

The present work is based on the Horton–Strahler ordering system (Horton 1945; Strahler 1952a), which assumes (i) channels that originate at a source and have no tributaries are defined as first-order stream (ω); (ii) when two streams of order ω join, a stream of order $\omega + 1$ is created; and (ii) when two streams of different orders join, the channel segment immediately downstream has the higher of the orders of the two combining streams. Exterior links are stream segments that connect source areas, whereas interior links connect the different stream junctions. Description of properties of the drainage network structure was first proposed by Horton (1945) with two main ratios that describe the organisation of stream bifurcation and length, widely known as 'Horton's laws' or 'the structure regularity framework'. Mathematically, the ratios are

$$\frac{N_{\omega-1}}{N_\omega} \approx R_B \rightarrow \omega = 2,3,\ldots,\Omega \qquad (11.2)$$

$$\frac{\overline{L}_\omega}{\overline{L}_{\omega-1}} \approx R_L \rightarrow \omega = 2,3,\ldots,\Omega \qquad (11.3)$$

where Ω is the total network order, N_ω is the number of streams of order ω, $\overline{L}_\omega$ is the mean length of streams of order ω and R_B and R_L are the bifurcation and length ratios, respectively.

Opposed to the regularity approach, Shreve (1966) proposed the random topology model based upon the concept that networks of given magnitude, under the absence of geologic control, are comparable in topological complexity. Both approaches of structure regularity and randomness have been widely confirmed by observation on natural channel networks (Jarvis 1977; Smart 1972a).

Starting with the constraint that a DEM is the only available information to delineate a drainage network, we propose a new technique to select the optimal value of A_S for a specific location based on the intrinsic properties of the channel network. In this approach, we assume that DEMs are self-contained structures, capable of conveying the landscape dynamics, and that channel complexity is best explained by its corresponding intrinsic properties. This complexity is reflected by the combination of structure regularity framework with topological random approach. The sharing point between the two approaches is reflected in the ratio between interior and exterior link lengths, widely known as R_A (Schumm 1956):

$$R_A = \frac{\overline{l_i}}{\overline{l_e}} \qquad (11.4)$$

where $\overline{l}$ and $\overline{l_e}$ are the average length of interior and exterior links, respectively.

Herein, we postulate that R_A bears direct and indirect information on channel-network characteristics and age, which could be used to reveal basin dissection and maturity. This technique consists of examining the curve relationship between the R_A ratios and their corresponding thresholds (Figure 11.2). The resulting ratios change throughout the axis of threshold values generating a varying-tendency curve. The value of A_S is the threshold that reflects drainage density and hence landscape dissection, for which it represents, on the one hand, drainage evolution and hence basin age and, on the other hand, landscape complexity, since

different R_A values reflect distinct geometric and topologic properties. Accordingly, we propose the following starting hypothesis: the R_A tendency curve is regular and steady in young or homogeneous landscapes and unsteady irregular in mature or heterogeneous landscapes.

The R_A index was widely applied to verify channel-network structure (Abrahams 1984): several studies confirmed R_A as an independent geomorphic parameter (Smart 1981) for which each landscape structure has its own specific R_A value. The first attempt to determine channel limits with R_A was realised by Shreve (1974), who used exterior and interior link lengths as separated variables and associated them with area to define the source of channel heads. Montgomery and Foufoula-Georgiou (1993) used R_A of different thresholds to examine whether statistical properties of channel networks are useful for estimating parameters of a slope–area relationship. They concluded that these properties do not change systematically with the imposed A_S. In the interpretation of the results, it seems that Montgomery and Foufoula-Georgiou anticipated a significant statistical relationship for the varying ratios of R_A, which seems to be improbable. Evidently, they disregard the

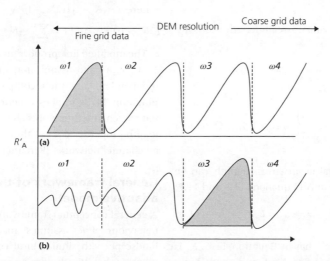

Figure 11.2 A conceptual framework for R_A' behaviour in (a) a hypothetical homogeneous landscape and (b) a hypothetical heterogeneous landscape. Shaded area indicates the stability zone (SZ) in the curve relationship.

effect of local factors over channel-network properties, for which the resulting curve should be interpreted in terms of changing phases rather than as a single trend.

The current synthesis on model derivation is an essential step in the formulation of the working hypothesis. The R_A ratio represents a simple case in a dynamic complex landscape, in which the applicability of this dimensionless index is limited to homogeneous landscapes. In addition, R_A confirms more to the Horton's law of lengths ratio rather than to Shreve's random model. In natural landscapes, however, channel networks are the final result of a complex evolutionary process throughout time that integrates different local and environmental factors (Tucker and Slingerland 1997). Importantly, these conditions imply the need for a more comprehensive model capable of simulating natural relief formations and at the same time integrating random and regularity concepts in stream delineation.

The structure regularity framework of Horton consists of R_B and R_L, defined in Equations 11.3 and 11.4, which have been expressed by Smart (1968, 1972b) in a topological form by

$$\bar{L}_\omega = \bar{l}_1 \prod_{\omega=2}^{\Omega} \frac{(N_{\omega-1} - 1)}{(2N_\omega - 1)} \rightarrow \omega = 2, 3, \ldots, \Omega \quad (11.5)$$

whereas the individual stream length ratios are given by

$$\lambda_2 = \frac{\bar{L}_\omega}{\bar{L}_1} = R_A \frac{(N_1 - 1)}{(2N_2 - 1)} \quad (11.6)$$

$$\lambda_{w'} = \frac{\bar{L}_{\omega'}}{\bar{L}_{\omega'-1}} = \frac{(N_{\omega'-1} - 1)}{(2N_{\omega'} - 1)} \rightarrow \omega' = 3, 4, \ldots, \Omega \quad (11.7)$$

Hence, the total mean stream length can be expressed as the sum of Equations 11.6 and 11.7:

$$\lambda_w = \lambda_2 + \lambda_{w'} = \bar{L}_w \quad (11.8)$$

Smart explained that in Equation 11.6, R_A is required for λ_2 because $\bar{l}_e \neq \bar{l}_i$ and there is no theoretical model for relating $\bar{l}_e$ and $\bar{l}_i$. Getting

back to the Horton's laws of stream number and stream lengths, it is accepted that the number of streams N_ω of order ω decreases as a geometric series with R_B (Eq. 11.2), and mean length of streams $\bar{L}_w$ of order ω increases as a geometric series with R_L (Eq. 11.3). Accordingly, the individual stream length ratio of Equation 11.3 could be rearranged as

$$\lambda_w = \frac{\bar{L}_w}{\bar{L}_{w-1}} = \bar{L}_w \sim R_L \quad (11.9)$$

If we assume that channel networks are space filling with a fractal dimension of 2 in the plane (Tarboton et al. 1989), where Horton laws hold exactly at all scales in the network, one may accept the assumption of Smart, in the case of moderately large N_ω, that is,

$$\lambda_\omega \sim \frac{R_B}{2} \approx R_B = 2\lambda_\omega \quad (11.10)$$

Rearranging Equations 11.6, 11.7 and 11.8 into 11.3 and 11.9 and substituting into 11.10, we can get a modified value of R_A:

$$R_A' = \frac{\left[2 * \left(\Delta + (\Lambda * R_A)\right)\right]}{\Gamma} \quad (11.11)$$

where $\Delta = (N_1 - 1) / (2N_2 - 1)$, $\Lambda = \sum_{\omega=3}^{\Omega} \left((N_{\omega-1} - 1) / (2N_\omega - 1)\right) = \lambda_3$ and $\Gamma = \sum_{\omega=2}^{\Omega} (N_{\omega-1} / N_\omega)$.

The modified link proportion (R_A') in Equation 11.11 is a valid assumption in all landscapes, independently of their complexity. Moreover, Equation 11.11 implies a better description of natural channel networks than bifurcation and length properties does as they virtually describe all channel networks (Kirchner 1993).

General framework of the adaptive model

A general conceptual framework that covers the behaviour of R_A' assumes, in a homogeneous landscape with similar local conditions everywhere, that R_A' holds a constant-tendency change within the same order in the channel network and varying tendencies between orders (Figure 11.2a). Conversely, in a heterogeneous, complex landscape, R'_A shows variable changes as channel order changes (Figure 11.2b). Such variability is maintained until a stabilisation stage is reached, where the model is capable of recognising all of the existing relief forms. In a wide sense, the relationship between R'_A and A_S contains several pieces of geomorphic information that can be used in drainage network interpretation, among which is the point of scale breaking that describes the change in dominant processes (Tarboton et al. 1989, 1991) or limits between channelled and unchannelled areas (Montgomery and Dietrich 1992).

Contrary to the constant threshold approach, the behaviour of R'_A assumes that a range of A_S values would serve as optimal thresholds for delineation of channel limits. In practice, this delineation is achieved by plotting growing numbers of A_S values against their corresponding R'_A ratios, and a curve of tendencies is formed indicating changes in each order of the channel network (Figure 11.2). These changes are measured by a rate of change, which estimates the area under curve between two successive minima. The total curve will be represented by different rate of change areas (RCA) that should reflect the ability of the model to depict landscape complexity in relation to DEM resolution. RCA is formally defined as the area under the curve formed by plotting R'_A against A_S and thus reflects the change in link ratio as a function of DEM resolution as a dimensionless index. First, under spatially equivalent tectonic and environmental conditions, all RCAs are similar and each will reflect a change in the order of the channel network, that is, $\omega = 2, 3, ..., \Omega$ (Figure 11.2a). In this case, the first RCA will be accepted to reflect present landscape dissection. This area will be designated as a stability zone (SZ). Alternatively, under heterogeneous landscapes, the generated curve is irregular (Figure 11.2b) and the SZ is the highest RCA. In this case, changes between the distinct RCA are not related to the order change in the channel

network, rather to the extent to which the smallest landforms in relation to DEM resolution can be resolved in the model. In both cases, the SZ holds a range of thresholds with the optimum A_S value(s) for the available complexity and resolution. In each SZ, a local minimum and maximum are detected. The local minimum represents the maximum complexity of the generated drainage network with minimum feathering (i.e. false streams represented in unchannelled areas) in a heterogeneous, complex landscape. Likewise, the local maximum represents the least complexity with the minimum feathering in a homogeneous simple landscape. Thus, and to ensure minimum feathering in the analysed landscape, both local minimum and maximum of the SZ are evaluated for heterogeneous and homogeneous landscapes, respectively.

The SZ theory seems to work best for homogeneous relief settings than heterogeneous ones. Therefore, if it is possible to determine the heterogeneity of the landscape, the SZ area should approach the ideal range of A_S values. In relation to local relief conditions, R'_A is related to the stage of geomorphic development that reflects landscape heterogeneity. In this direction, scientists have linked the geomorphic forms and erosion processes of catchments to the evolution stage of their channel networks (Pérez-Peña et al. 2009; Sternai et al. 2011). A link between channel networks and their geomorphic development has been quantified by the hypsometric integral (HI). HI is a nondimensional index that describes the area under the hypsometric curve, thus measuring the frequency distribution of elevation; this integral has been employed to resolve erosional stage and uplift tectonics (Strahler 1952b; Willgoose and Hancock 1998). An important attribute of the HI is that basins of different sizes and erosional stages can be compared (Pérez-Peña et al. 2010). As a rule, HI usually takes values between 0 and 1, where if HI ≥ 0.60, then uniformity of the erodible materials is the dominant aspect in the landscape (Hurtrez et al. 1999). In the present work, we will assume a direct link between basin sizes and uniformity of eroded materials, the smaller the basin size the higher the uniformity, and *vice versa*.

In order to detect homogeneous landscape units in relation to values of R'_A representing similar geomorphic processes, a recursive stratification procedure (*RSP*) has been integrated into the aforementioned approach. In this process, the intrinsic properties of the channel network will control the stratification process of catchment units, which allows for a simple reclassification of the generated sub-catchments of decreasing orders (i.e. $\Omega-1, \Omega-2,...,\Omega=2$). Such classification provides as many A_S values in relation to the classified sub-catchments, which usually approximates to homogeneous landscape conditions of similar geomorphic processes. As soon as the model recognises the smallest unit in the terrain, based on the prevailing geomorphic domain, the SZ is formed and the optimal A_S can be estimated. Concisely, the

procedure to use in order to define the optimal drainage network in the studied landscape can be summarised as follows (Figure 11.3):

1 Definition of the SZ for the whole catchment from the relationship between A_S and R'_A.

2 Definition of the optimal A_S: in general, a local minimum will be selected as the optimal A_S, except under the following primary conditions when the local maximum will be used: (i) the studied basin forms a unique SZ area, which indicates a homogeneous landscape; (ii) the SZ starts from a ridge or a catchment divide (i.e. $A_S \le 2$); and (iii) HI value ≥ 0.60. Under these conditions, the majority of the constructed curve relationships are formed by either only one SZ or several SZ values but with equal RCAs, which confirms previous conclusions of homogeneity in relation to HI values.

3 The application of RSP, which consists of the following steps: (i) applying the selected A_S from the defined SZ; (ii) in the generated

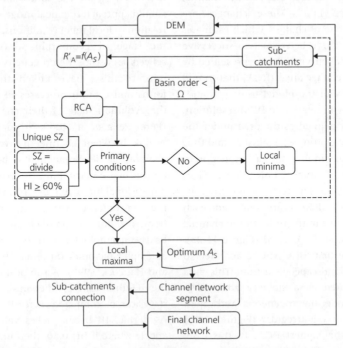

Figure 11.3 A schematic flowchart representing the R'^t_A approach and related processes. The shaded area represents the recursive stratification procedure (RSP).

drainage network, all sub-catchments of order $\Omega - 1$ are selected and the SZ of each one is defined; (iii) the process of RSP is repeated until the classified sub-catchments reached one of the primary conditions mentioned earlier in step 2; (iv) again, the RSP is repeated with sub-catchments of order $\Omega - 2$ and the SZ is defined until the sub-catchments reach the primary conditions; and (v) the process is repeated as necessary for all sub-catchments of successive descending order until all the sub-catchments have been classified and corresponding channel networks are defined.

4 Finally, the sub-classified catchments are reconnected in order to build the final channel network.

It is important to underline that SZ and RSP processes are successive and complementary steps in the present methodology approach, and, in our opinion, the two processes are complementary to find optimum drainage networks in both complex and uniform landscapes. Hereafter, the combination of R'_A and RSP will be assigned as the R'^t_A approach, whereas the R'_A algorithm will be denoted as the adaptive model.

Experimental validation

A robust validation procedure is essential (Güntner et al. 2004) not only for a reliable quantitative comparison between channels of different sources but also to stress the efficiency of one approach over the others. Mode and type of comparison are between the several factors that may limit validation procedures between DEM-generated channel networks (i.e. automated) and the natural streams represented by the digitised blue lines (BLs) from topographic maps. Thus, a set of morphometric indices were selected (Table 11.1) and applied in the comparison process that cover a wide range of river system properties (e.g. geometric, topologic, fractal and optimality).

A pair of channel networks (i.e. automated vs. its blue line equivalent) should be compared in terms of their overall resemblance as described by the 16 indices of Table 11.1, rather than using such indices individually. For that purpose, a multidimensional dissimilarity index such as the Gower metric (GM) can be used. The GM measure of association (Gower 1971) has been used to determine the degree of dissimilarity between pairs of vectors such as those formed by the morphometric indices computed for the different channel networks. The GM value is given by

$$D = \frac{1}{m} * \sum_1^m \frac{\left| D_{ik} - D_{jk} \right|}{k_n} \qquad (11.12)$$

where D_{ik} and D_{jk} are the values of morphometric indices k in channel networks i and j, k_n is the range of k across all the examined channel networks, and m is the number of morphometric indices used in the comparison.

In this study, two categories of channel network will be processed. The first is the BLs originated from topographic maps at 1:50,000 and 1:500 scales, approximately equivalent to 30 and 1 m grid resolutions, respectively. The second is the automated channel networks defined from DEMs, of 30 m and 1 m resolutions, with two different methods for A_s definition: the R'^t_A and the CDA. Data extracted from BLs are referred to as observed values, whereas those extracted from DEMs are assigned as expected values. The final result is presented in two forms, either as a percentage (%) of overall level of agreement (OLA) if comparison is carried out between various catchments of different sizes or directly if different channel networks within the same catchment unit are compared.

Analysis of Tabernas basin

Channels from the CDA and R'^t_A approaches were extracted and directly compared to the BLs of the area (Figure 11.4). The R'_A algorithm in the Tabernas basin provided a trend of

Table 11.1 Morphometrical indices proposed for the comparison and validation procedure between different drainage networks.

No.	Morphometrical indices	References	Symbol	Expression
1	Order of the channel network	Horton (1945) and Strahler (1957)	Ω	
2	Longest stream in the channel network	Hack (1957)	La	
3	Drainage network density	Horton (1945)	Dd	$Dd = \dfrac{L_1}{A}$
4	Magnitude of the channel network	Shreve (1966)	μ	
5	Ratio of average stream length	Schumm (1956)	inR_A	$inR_A = \dfrac{\overline{l_e}}{\overline{l_i}}$
6	Macroscopic interior link density	Abraham (1980)	K_i	$\dfrac{\overline{l_i}}{\overline{a_i}} = k_i$
7	Horton bifurcation ratio	Horton (1945)	R_B	$\dfrac{N_{\omega-1}}{N_\omega} \approx R_B$
8	Channel frequency	Horton (1945)	F_s	$F_s = \dfrac{N}{A}$
9	Exceedence probability slope of stream length	Tarboton et al. (1988)	P_s	$P_s = \dfrac{m}{n+1} *$
10	Optimal channel network or catchment energy	Rodriguez-Iturbe et al. (1992)	OCN	$OCN = \sum_{i}^{N} liAi^{0.5}$
11	Theoretical stream network diameter	Werner and Smart (1973)	D_{cal}	$D_{cal} = 2 * \sqrt{\pi * \mu}$
12	Probability of drawing a link of magnitude μ	Shreve (1966)	$p(\mu)$	$p(\mu) = \dfrac{2^{-(2\mu-1)}}{2\mu-1}\left(\dfrac{2\mu-1}{\mu}\right)$
13	Jarvis index	Jarvis (1972)	E	$E = \dfrac{\sum \mu \overline{l_i}}{\sum \mu \overline{l_e}}$
14	Stream network development index	Strahler (1957)	Isd	$Isd = \dfrac{L}{P}$
15	Fractal dimension of the channel network	Tarboton et al. (1988)	ε	
16	Melton ratio	Shreve (1967)	K	$K = \dfrac{(2\mu-1)}{LDd}$

*Where m is the ranking from longest to shortest stream length and n is the number of streams in the sample dataset (Tarboton et al. 1988).

varying RCAs with a clear SZ at the final part of the curve (Figure 11.5) with $R'_A = 6.703$. At the working resolution, the SZ of the Tabernas basin extends over a high range of A_s values that oscillate between 4650 and 11850 cells representing the local minimum and maximum, respectively. According to the R'^t_A methodology, none of the primary conditions was fulfilled in the first iteration, and the RSP continued until the primary conditions were met and the final network was assembled (Figure 11.4c). The final result of applying the R'^t_A procedure is a highly branched channel network that divides the landscape and hence the total basin to different levels of details. Broadly, the RSP produced 330 sub-catchments of different sizes

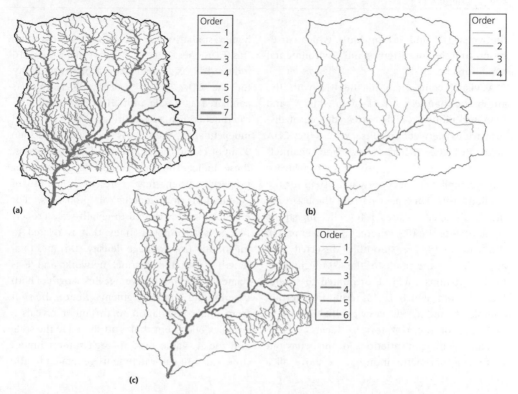

(a)

(b)

(c)

Figure 11.4 Channel network in Tabernas basin: (a) the digitised blue lines, shown here in grey (BLs), (b) channel networks delineated with CDA technique with $A_s = 500$ cells and (c) channel network defined by the $R_A'^t$ procedure.

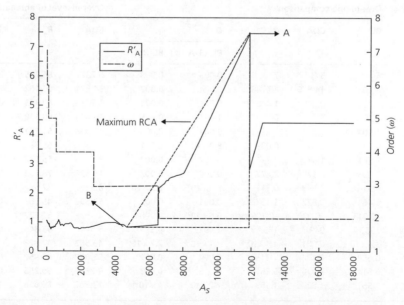

Figure 11.5 The curve relationship between R_A' and its corresponding A_s values for Tabernas basin, showing a local maximum at A and a local minimum at B. The shaded area explains the concept of maximum rate of change (MRC) or the stability zone (SZ) concept in the curve relationship.

(between 0.21 and 567.265 km^2), which correspond to different tectonics and environmental conditions.

A visual comparison of the BLs with the automatic channel networks defined by $R_A'^t$ and CDA techniques reveals a clear degree of dissimilarities between results. While the CDA delineates the major valleys and channels (Figure 11.4b), the $R_A'^t$ procedure provides a more complex channel network system for the studied basin. The upper parts of the basin are highly dissected (Figure 11.4c) reflecting strong similarities to the BLs, whereas the lower part of the catchment are smoothly dissected that approximates the results of the CDA technique. Such variations could be attributed to several factors, from which DEM resolution, terrain complexity and $R_A'^t$ efficiency are highlighted.

Results of the GM test in Tabernas basin revealed significant variations for the behaviour of the morphometric indices, but above all a

higher similarity for the BLs–$R_A'^t$ set over the BLs–CDA ones. Table 11.2 shows that 4 out of 16 indices (inR_A, K_l, k and P_S) reveal higher similarity for the CDA–$R_A'^t$ set if Tabernas basin was compared as a sole unit. However, if all sub-catchments resulted from the RSP are used independently, the OLA is decreased to include 3 out of 16 indices (k, $p(\mu)$ and E). In both cases, these indices do not pertain to a particular property; rather, they represent a mixture of geometrical and topological properties. For example, inR_A is related to length-scale properties; k is a complex index that is related to magnitude (μ), drainage density (Dd) and total length (L) of the channel network; and E is related to (μ) and the average link length of both exterior and interior segments. Hence, the shift from one combination to the other reveals a complex behaviour that is attributed to the scale and the A_S value used in each sub-catchment unit. The 330 sub-catchments generated by the

Table 11.2 Comparison between morphometric indices of the digitised BLs and automated channel networks delineated by the constant drop analysis (CDA) and the $R_A'^t$ approaches.

No.	Index	One-to-one comparison					Overall level of agreement (OLA)		
		BL	CDA	$R_A'^t$	D		CDA	$R_A'^t$	Similar
					BL–CDA	BL–$R_A'^t$			
1	Ω	6	5	7	0.003	0.003	6.722	**64.705**	29.411
2	La	46.041	44.570	46.748	0.004	**0.002**	34.871	**64.102**	1.025
3	Dd	1.762	0.535	1.448	0.004	**0.001**	9.743	**89.743**	0.512
4	μ	407	51	689	1.041	**0.825**	13.559	**80.508**	5.932
5	inR_A	1.504	1.235	0.577	**0.001**	0.003	45.378	**53.781**	0.840
6	K_l	0.049	0.019	0.089	**8.8 × 10^{-5}**	0.001	29.896	**69.587**	0.515
7	R_B	2.581	1.930	2.771	0.002	**0.001**	38.983	**55.932**	5.084
8	Fs	1.433	0.178	2.427	0.004	**0.003**	15.833	**78.333**	5.833
9	H_μ	0.514	0.543	0.511	8.3 × 10^{-5}	**9.9 × 10^{-6}**	8.333	**82.5**	9.166
10	k	0.461	0.622	1.157	**0.001**	0.002	**53.636**	45.563	0.801
11	P_S	−0.943	−0.920	−0.799	**6.6 × 10^{-5}**	0.0004	41.525	**57.627**	0.847
12	OCN	1358.7	949.3	1557.3	1.197	**0.581**	5.641	**93.846**	0.512
13	$p(\mu)$	3.4 × 10^{-5}	0.001	2.4 × 10^{-5}	2.1 × 10^{-6}	**2.7 × 10^{-8}**	**49.562**	47.878	2.560
14	E	0.653	2.174	1.3157	0.004	**0.002**	**51.020**	47.959	1.020
15	lsd	6.514	1.976	5.351	0.013	**0.003**	9.230	**90.256**	0.512
16	ε	1.551	1.527	1.553	7.2 × 10^{-5}	**4 × 10^{-6}**	12.307	**86.666**	1.025

The degree of similarity (D) test implies one-to-one comparison for Tabernas basin as a whole and the overall level of agreement (OLA), measured in %, for all the analysed sub-basins. Bold values indicate the best values.

RSP and used in the comparison process imply high variations in tectonic and local geomorphic conditions, which directly link channel-network properties to different morphometric attributes. Therefore, each attribute is directly linked to the degree of landscape dissection described by the applied A_S value. Moreover, the sensitivity and variations of the morphometric attributes are also altered by DEM-data spacing (Orlandini et al. 2011) and underlying geology (Hancock 2005). Hence, the types of morphometric indices used to differentiate quantitatively between different channel networks, properties have to be evaluated in relation to the earlier factors.

The adaptive model in heterogeneous landscapes

The adaptive model basically consists of achieving an equilibrium state between geometric and topologic properties of the channel network, which is reflected in the SZ in the relationship between A_S and R'_A (Figure 11.2). This relationship tries to achieve the optimal approximation between bifurcation and length-scale properties as well as the R_A ratio that best describes channel-network structure. A general conceptual framework that explains R'_A behaviour in a hypothetical landscape (i.e. homogeneous vs. heterogeneous) is the fluctuating pattern described in Figure 11.2. The oscillation in the curve relationship represents the trenching process of Schumm's (1977) experiment that characterises stream formation and evolution, and hence the process of stream initiation. Schumm attributed such oscillations to the exceeding of an intrinsic geomorphic threshold, which corresponds to homogeneous landscape conditions (Figure 11.2a). In this case, each RCA will reflect a change in the channel-network order (i.e. from ω_1 to ω_2, ω_2 to ω_3,..., ω_{n-1} to ω_n), which is directly related to a change in terrain properties and hence changes in erosion mechanisms (e.g. stream incision, debris accumulation and sediment transport) and time-varying uplift (Willgoose 1994). However, in nature, this relationship is not the case (Bull and Kirkby 2002), and channel

networks are the result of integrating different extrinsic and intrinsic factors that form the different stages of the landscape evolution (Howard 1994). In this case, the adaptive model seeks for an equilibrium point between terrain complexity and DEM-data spacing to depict the smallest stream link interpreted in presence of major prevailed processes (i.e. main valleys) in catchment area. Thus, the R'_A curve is of steady RCA in homogeneous landforms and unsteady in heterogeneous relief leading to variable RCAs depending on DEM capacity to resolve the finest terrain forms at the working resolution (Orlandini et al. 2011). Such changes in the curve direction will reflect variations in terrain properties, but the most significant is the SZ that reveals the optimal point at which resolution and channel complexity are most fully explained by the model. These observations underline the capacity of the $R'^{\prime}_A$ approach to depict landscape dissection under varying landscape complexities.

A complex drainage system is usually characterised by a diversity of channel-initiation areas controlled by different erosion mechanisms (Perron et al. 2008) that limit stream extents into the hillslopes. In this case, the channel-network system is composed of varying sub-catchments, each of which have particular geometric and topological properties. Hence, and in relation to DEM resolution, the adaptive model verifies each sub-catchment in relation to its complexity and interprets it in unsteady–irregular form in the curve relationship (Figure 11.2b); that is, the higher the unexplained heterogeneity, the lesser the stability in the curve relationship and vice versa. Critically, the geomorphic fluctuating, as indicated in Figure 11.2, will not provide a single A_S value, but a range of values that are directly related to variations of local factors in the channel-initiation area (Gandolfi and Bischetti 1997). This result is broadly in agreement with earlier findings on the multi-scale approach over the simple one (De Bartolo et al. 2000) and the non-realistic assumption of one constant A_S value in channel networks of varying incision domains (Henkle et al. 2011).

Analysis of El Cautivo basin

In the El Cautivo basin, the $R_A'^t$ technique provides an improved representation of the drainage network over the CDA method and even
the BLs. While field visits have confirmed location and limits for all the defined streams by the
$R_A'^t$ approach, presence or absence of some
streams was quite obvious between the BLs and
the CDA approaches (Figure 11.6). In general,
the BLs show high depiction for the stream network in the mid part of the basin and absence of
clear streams in the upper and lower-mid parts
of the catchment (Figure 11.6a), whereas the
CDA resulted with feathering in parts of the
basin (Figure 11.6b). Finally, the $R_A'^t$ highlighted
a well-dissected channel network mainly for
first-order streams both position and extent to
approximate natural dissection of the landscape
(Figure 11.6c).

The morphometrical indices for the different
methods were calculated, and the GM test
confirmed the high similarity for BLs–$R_A'^t$ over
BLs–CDA (Table 11.3). In the El Cautivo basin,

three indices – inR_A, R_B and E – revealed a better
approximation for the BLs–CDA, whereas the
remainder underlined greater similarities for
the BLs–$R_A'^t$. Such results confirm again the
potential and the capabilities of the morphometric attributes as a mode of comparison
between stream networks.

The adaptive model in homogeneous landscapes

In the El Cautivo basin, the SZ occupied a range
of A_S values that extend between 26 and 50 cells
(Figure 11.7), down which a continuous fluctuation in the A_S and R_A' curve relationship was
observed and above which the channel network
is limited to three segments. This range breaks
down our expectations on the presence of a
complete homogeneous catchment of one prevailing domain, because the primary conditions
were not achieved and the RSP should be
applied. The oscillations in each segment are not
steady ones; rather, they are characterised by a
continuous fluctuation throughout the curve.
Indeed, the earlier fluctuation reflects local

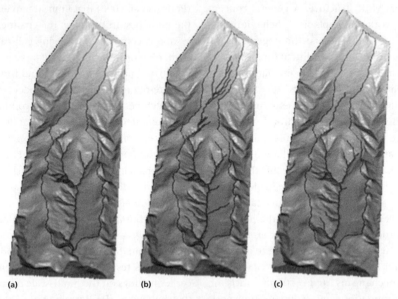

(a)   (b)   (c)

Figure 11.6 Comparison of the different channel networks in the El Cautivo basin: (a) digitised channel
networks (BLs), (b) channel network delineated by *CDA* technique (A_S=8 cells) and (c) channel network
delineated by $R_A'^t$ technique.

Table 11.3 One-to-one comparison values using the GM test in the El Cautivo.

No.	Index	BLs	CDA	$R_A'^t$	BLs–CDA	BLs–$R_A'^t$
1	Ω	3	3	3	0	0
2	La	235.4	377.9	312.9	0.4165	**0.2263**
3	Dd	0.0262	0.0578	0.0291	8.9×10^{-5}	**8.6×10^{-6}**
4	μ	13	35	13	0.0643	**0.0000**
5	inR_A	1.0764	1.1363	0.8858	**0.0002**	0.0005
6	K_i	0.0018	0.0023	0.0022	1.41×10^{-6}	**1.35×10^{-6}**
7	R_B	2.3325	1.9362	2.7738	**0.0012**	0.0013
8	Fs	0.0013	0.0034	0.0013	6.1×10^{-6}	**0.0000**
9	H_μ	0.5927	0.5527	0.5927	0.00011	**0.0000**
10	k	1.9209	1.1228	1.5513	0.0023	**0.0012**
11	P_s	−0.9685	−0.7763	−0.8797	0.0006	**0.0003**
12	OCN	6392.9	10559.3	6410.9	12.1824	**0.0526**
13	$p(\mu)$	0.0061	0.0013	0.0061	1.4×10^{-5}	**0.0000**
14	E	0.1589	0.1043	0.0709	**0.0002**	0.0003
15	lsd	0.5953	1.2953	0.6624	0.0020	**0.0002**
16	ε	1.8475	1.8922	1.8707	0.0001	**6.9×10^{-5}**

Bold values indicate best results.

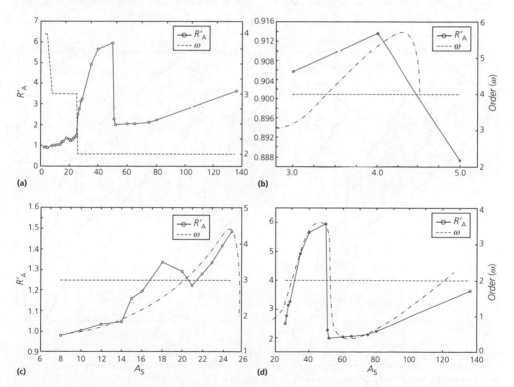

(a) (b) (c) (d)

Figure 11.7 (a) Curve relationship for the R_A' model in the El Cautivo basin. Parts (b), (c) and (d) are the segmented R_A' in relation to ω. Dot-dash lines indicate a theoretical rate of change area (RCA) under a completely homogeneous landscape.

factor effects and the presence of varying domains in the catchment analysed.

In order to understand the behaviour of the adaptive model, a closer inspection was performed of the relationship between A_s and R'_A, in which the curve was segmented according to order change in the constructed drainage network providing three order sections. The first section is associated to A_s between 3 and 5 cells (Figure 11.7b) and corresponds with a channel network of $\Omega=4$ with high feathering and diffused aspect. The second section corresponds with a channel network of $\Omega=3$ (Figure 11.7c), which is related to A_s that extends from 6 to 25 cells. In this case, the curve is fairly fluctuated with approximately 3 clear RCA indicating more recognition to local factors. Finally, the last stage corresponds with $\Omega=2$, and A_s ranges from 26 to 135 cells (Figure 11.7d). Herein, two clear RCAs were identified. The first one extends from 26 to 51 cells with RCA=124.96. The second extends from 52 to 135 cells and RCA=46.15. According to the R'^t_A approach, the SZ corresponds to maximum RCA with optimal A_s of 26, since neither of the primary conditions has been observed. Under these conditions, the resulted channel network is characterised by a moderately dissected aspect that adapts well to relief landforms of the El Cautivo basin, as well as a completely vanishing of feathering features (Figure 11.6c). The order (Ω) and magnitude (μ) of the resulted channel network approximate fairly to those of the BLs, with Ω value of 3 for both and μ value of 16 and 13, respectively.

The second step in the R'^t_A procedure involves the application of the RSP. Such process verifies three and five sub-catchments of $\Omega=2$ and 1, respectively (Figure 11.8a). Again, the SZs were defined and the optimal A_s values for each sub-catchment has been selected. The final

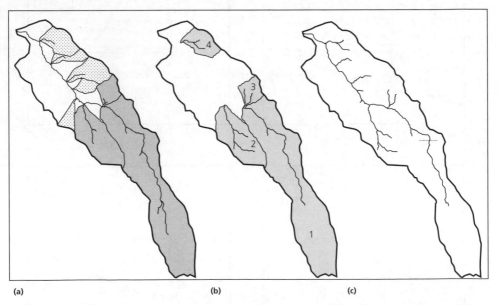

(a)   (b)   (c)

Figure 11.8 Generated sub-catchments in the El Cautivo basin using the RSP method: (a) channel network delineated by the R'_A model, in which solid grey highlighted sub-catchments of first-order streams and dotted are second order streams ($\Omega=2$), (b) amount of improvement in the reclassified sub-catchments in the El Cautivo basin after applying the R'_A model for each sub-catchment and (c) the final channel networks defined by the R'^t_A procedure. The dotted line in (c) indicates the initiation of the extreme headcut resulted from piping erosion.

result of the $R_A^{'t}$ procedure is an evident read-justment in the generated drainage network, which is reflected in an apparent enhancement in depicting landscape dissection for the deline-ated sub-catchments of the El Cautivo basin (Figure 11.8c). Such enhancement is expressed in two distinct forms: the first as an apparent increase in the length of the channel networks and hence landscape dissection, such as the cases in sub-catchments 2, 3 and 4 (Figure 11.8b); and the second form shows a clear decrease in the channel-network length attributed to a smooth relief structure, such as the case of sub-basin 1.

In order to explain this type of result, Montgomery and Dietrich (1994) indicated that processes controlling each channel-head-initiation mechanism are related to a specific threshold, according to which the landscape is divided into process regimes where each erosional zone needs a particular threshold value to define channel-head location. A detailed inspection of the studied area reveals that the dominant erosional pro-cesses are rill, gully, splash and piping erosion (Solé-Benet et al. 2009). Thus, at least three A_s values are needed to represent prevailing ero-sional processes required to depict convergent and divergent processes and hence channelled area. In view of that observation, a field visit to the channel heads in the studied site allowed for the definition of three basic types of prevailing soil erosion (piping, rill and inter-rill erosion) that are associated with the sub-catchments of Figure 11.8b (Table 11.4). The dominant erosion

processes in the sub-catchments reflect different types of geomorphic channel-head formations. In the first sub-catchment, a mixture of piping and inter-rill erosion prevails that explains the behav-iour of the channel network. Inter-rill erosion is dominated in the upper part of this unit resulting in a smooth, concave hillslope structure that extends to an extreme headcut; this headcut emerged in the middle part of the unit (Figure 11.8c) and is attributed to the piping ero-sion. In the second sub-catchment (Figure 11.8b), piping is dominant and channels are incised directly by such effects. Finally, in sub-catchments 3 and 4, rill erosion dominates and the gradual incision at the channel head is attributed directly to this process.

Summary points

Identification of stream limits in terms of critical contributing area to initiate a stream (A_s) is essential for a reliable parameterisation of hydro-logic and geomorphologic models. Although automated methods have provided quantitative approaches based on mathematical algorithms, their efficiency under varying local and environmental conditions is still a matter of debate. In this work, a new approach has been proposed to define an optimal threshold or thresholds based on the combination of the intrinsic properties of drainage network struc-ture (R_A') and an RSP, denoted as the 'adaptive model'. Basically, R_A' combines the exterior and

Table 11.4 The optimum A_s values defined according to the $R_A^{'t}$ approach in El Cautivo basin and the corresponding information of the reclassified sub-catchments of Figure 11.8.

No.	SZ	Primary conditions	A_s	Incision process	Ω	μ	D_d (m/m²)
1	3–35	The minimum is a divide	35	Inter-rill and piping erosion	2	2	0.3064
2	8–18	None	8	Piping erosion	2	2	0.3584
3	5–22	HI = 62.1	22	Rill erosion	2	3	0.3559
4	12–20	None	12	Rill erosion	2	3	0.3706

interior link-length ratio with length and bifurcation properties described in terms of structure regularity framework and topological random approach. The RSP implies a hierarchical stratification of the main catchment to small subunits based on the intrinsic model. Such technique provides various critical A_s in relation to DEM-data resolution and landscape complexity. Hence, the capacity of $R_A^{'}$ could be extended as a potential morphometric attribute to describe landscape behaviour based on prevailed geomorphic domain.

In order to determine the capabilities of the model under varying terrain and environmental conditions, a robust validation approach using one-to-one comparison was suggested. The procedure consists of a direct comparison of the morphometric attributes using the Gower index of dissimilarity. The comparison was performed between two automatic techniques: the CDA and the $R_A^{'t}$ approach, in which the digitised BLs were served as referenced values. Results revealed a significant enhancement in the delineation of channel networks with the $R_A^{'t}$ approach over the CDA technique, since its function depends on intrinsic properties of the drainage network, being at the same time objective and easy to implement. Likewise, the present work underlines the importance not only of the algorithm used to delineate stream limits but also the need to a complementary broad process of comparison and validation.

The morphometrical indices are highly specialised attributes that should form part of any quantitative description of the channel-network morphology. However, importance and significance of each attribute is to be evaluated in relation to the mode of validation and type of the test used in these processes. The morphometrical indices are sensitive attributes that may comprise one or several parameters to describe specific properties of the channel network. Such description involves the geometry, topology, optimality and fractal properties of natural landscapes. While in some cases few parameters may achieve significant conclusions, a wide range of descriptors is desirable, because each attribute may describe one or several structure properties. However, the morphometric properties vary considerably with A_s values, and thus parameters reported without their associated A_s value could lead to erroneous conclusions.

Finally, a self-critique of the present work highlighted two types of interrelated problem. The first is inherent to the stratification procedure leading to an exhaustive examination of first-order streams near the divides and one restricted to channels that drain near to the outlets. Such a problem will lead to varying degrees of landscape depiction promoted by the available resolution to decode the present heterogeneity of the basin unit studied. This problem is widely related to the second, which highlights the capacity of the $R_A^{'}$ model under such conditions (resolution vs. complexity). The complex structure of natural streams is well described by geometric, topologic, fractality, self-organisation and optimality parameters, which integrate to form the actual shape and structure of any drainage network system. Since the $R_A^{'t}$ approach incorporates just the first two properties, future work should develop the integration of other properties in the adaptive model in order to enhance landscape depiction.

Acknowledgements

A special dedication in honour of Dr. John Thornes, who has introduced me to basic knowledge in hydrologic and geomorphologic modelling. The authors would like to thank Dr. Albert Sole-Benet and Dr. Nick Rosser for their contributions to this paper. Part of this work has been supported by the CANOA project (characterisation and modelling of hydrological processes and regimes in gauged catchments for the prediction in ungauged catchments) financed by the Spanish Ministry of Science and Education (Ref. CGL2004-04919-C02-01).

References

Abrahams, A.D. (1980) Channel link density and ground slope. *Annals of the Association of American Geographers* **70**, 80–93.

Abrahams, A. D. (1984) Channel networks: A geomorphological perspective. *Water Resources Research* **20**, 161–168.

Bull, L. J., Kirkby, M. J. (2002) Dryland river characteristics and concepts. In: Bull, L. J., Kirkby, M. J. (ed.) *Dryland Rivers: Hydrology and Geomorphology of Semi-arid Channels.* John Wiley & Sons, West Sussex, pp. 3–14.

Bischetti, G., Gandolfi, C., Whelan, M. J. (1998) The definition of stream channel head location using digital elevation data. In: Proceedings of the HeadWater '98: Hydrology, Water Resources and Ecology in Headwaters. IASH, Meran, Italy, 20–23 April 1998, pp. 545–552.

De Bartolo, S. G., Gabriele, S., Gaudio, R. (2000) Multifractal behaviour of river networks. *Hydrology and Earth System Sciences* **4**, 105–112.

Gandolfi, C., Bischetti, G. B. (1997) Influence of the drainage network identification method on geomorphological properties and hydrological response. *Hydrological Processes* **11**, 353–375.

Giannoni, F., Roth, G., Rudari, R. (2005) A procedure for drainage network identification from geomorphology and its application to the prediction of the hydrologic response. *Advances in Water Resources* **28**, 567–581.

Gilbert, G. K. (1909) The convexity of hilltops. *Journal of Geology* **17**, 344–350.

Gower, J. C. (1971) A general coefficient of similarity and some its properties. *Biometrics* **27**, 857–871.

Grayson, R., Blöschl, G. (eds.) (2001) *Spatial Patterns in Catchment Hydrology: Observations and Modelling.* Cambridge University Press, Cambridge, UK.

Güntner, A., Seibert, J., Uhlenbrook, S. (2004) Modelling spatial patterns of saturated areas: An evaluation of different terrain indices. *Water Resources Research* **40**, W05114.

Gyasi-Agyei, Y. G., De Troch, F. P. (1995) Effects of vertical resolution and map scale of digital elevation models on geomorphological parameters used in hydrology. *Hydrological Processes* **3**, 363–382.

Hack, J. T. (1957) Studies of longitudinal stream profiles in Virginia and Maryland. *U.S. Geological Survey Professional Paper* **294-B**, 97p.

Hancock, G. R. (2005) The use of DEMs in the identification and characterization of catchments over different grid scales. *Hydrological Processes* **19**, 1727–1749.

Heine, R. A., Lant Raja, C. L., Sengupta, R. (2004) Development and comparison of approaches for automated mapping of stream channel networks. *Annals of the Association of American Geographers* **94**, 477–490.

Henkle, J. E., Wohl, E., Beckman, N. (2011) Locations of channel heads in the semiarid Colorado front range, USA. *Geomorphology* **129**, 309–319.

Horton, R. E. (1945) Erosional development of streams and their drainage basins: hydrophysical approach to quantitative morphology. *Geological Society of America Bulletin* **56**, 275.

Howard, A. D. (1994) A detachment-limited model of drainage basin evolution. *Water Resources Research* **30**, 2261–2285.

Hurtrez, J. E., Sol, C., Lucazeau, F. (1999) Effects of drainage area on the hypsometry from an analysis of small scale drainage basins. *Earth Surface Processes and Landforms* **24**, 799–808.

Istanbulluoglu, E., Tarboton, D. G., Pack, R. T. (2002) A probabilistic approach for channel initiation. *Water Resources Research* **38**, 1325.

Jarvis, R. S. (1972) New measure of the topologic structure of dendritic drainage networks. *Water Resources Research* **8**, 1265–1271.

Jarvis, R. S. (1977) Drainage network analysis. *Progress in Physical Geography* **1**, 271–295.

Kirchner, J.W. (1993) Statistical inevitability of Horton's laws and the apparent randomness of stream channel networks. *Geology* **21**, 591–594.

Lashermes, B., Foufoula-Georgiou, E., Dietrich, W. E. (2007) Channel network extraction from high resolution topography using wavelets. *Geophysical Research Letters* **34**, L23S04.

Lin, W. T., Chou, W. C., Lin, C. Y., Huang, P. H., Tsai, J. S. (2006) Automated suitable drainage network extraction from digital elevation models in Taiwan's upstream watersheds. *Hydrological Processes* **20**, 289–306.

McNamara, J. P., Ziegler, A. D., Wood, S. H., Vogler, J. B. (2006) Channel head locations with respect to geomorphologic thresholds derived from a digital elevation model: A case study in northern Thailand. *Forest Ecology and Management* **224**, 147–156.

Montgomery, D. R., Dietrich, W. E. (1989) Source areas, drainage density and channel initiation. *Water Resources Research* **25**, 1907–1918.

Montgomery, D. R., Dietrich, W. E. (1992) Channel initiation and the problem of landscape scale. *Science* **255**, 826–830.

Montgomery, D. R., Dietrich, W. E. (1994) Landscape dissection and drainage area-slope thresholds.

In: Kirkby, M. J. (ed.) *Process Models and Theoretical Geomorphology.* John Wiley & Sons, Inc, New York, pp. 221–246.

Montgomery, D. R., Foufoula-Georgiou, E. (1993) Channel network source representation using digital elevation model. *Water Resources Research* **29**, 3925–3934.

Orlandini, S., Tarolli, P., Moretti, G., Dalla Fontana, G. (2011) On the prediction of channel heads in a complex alpine terrain using gridded elevation data. *Water Resources Research* **47**, W02538.

Passalacqua, P., Trung, T. D., Foufoula-Georgiou, E., Sapiro, G., Dietrich, W. E. (2010) A geometric framework for channel network extraction from lidar: nonlinear diffusion and geodesic paths. *Journal of Geophysical Research* **115**, F01002.

Peckham, S. D. (1995) New results for self-similar trees with applications to river networks. *Water Resources Research* **31**, 1023–1029.

Perera, H., Willgoose, G. (1998) A physical explanation of the cumulative area distribution curve. *Water Resources Research* **34**, 1335–1343.

Pérez-Peña, J. V., Azor, A., Azañón, J. M., Keller, E. A. (2010) Active tectonics in the Sierra Nevada (Betic Cordillera, SE Spain): Insights from geomorphic indexes and drainage pattern analysis. *Geomorphology* **119**, 74–87.

Pérez-Peña, J. V., Azañón, J. M., Booth Rea, G., Azor, A., Delgado, J. (2009) Differentiating geology and tectonics using a spatial autocorrelation technique for the hypsometric integral. *Journal of Geophysical Research: Earth Surface* **114**, F02018.

Perron, J. T., Dietrich, W. E., Kirchner, W. J. (2008) Controls on the spacing of first-order valleys. *Journal of Geophysical Research* **113**, F04016.

Poggio, L., Soille, P. (2011) A probabilistic approach to river network detection in digital elevation models. *Catena* **87**, 341–350.

Rodríguez-Iturbe, I., Rinaldo, A., Rigon, R., Bras, R. L., Marani, A., Ijjász-Vásquez, E. (1992) Energy dissipation, runoff production, and the three-dimensional structure of river basins. *Water Resources Research* **28**, 1095–1103.

Schumm, S. A. (1956) Evolution of drainage systems and slopes in badlands at Perth Amboy, New Jersey. *Bulletin of the Geological Society of America* **67**, 597–646.

Schumm, S. A. (1977) *The Fluvial System.* John Wiley & Sons, Ltd, Chichester.

Shreve, R. L. (1966) Statistical law of stream numbers. *Journal of Geology* **74**, 17–37.

Shreve, R. L. (1967) Infinite topologically random channel networks. *Journal of Geology* **77**, 397–414.

Shreve, R. L. (1974) Variation of mainstream length with basin area in river networks. *Water Resources Research* **10**, 1167–1177.

Smart, J. S. (1968) Statistical properties of stream lengths. *Water Resources Research* **4**, 1001–1014.

Smart, J. S. (1972a) Channel networks. *Advances in Hydroscience* **8**, 305–345.

Smart, J. S. (1972b) Quantitative characterization of channel network structure. *Water Resources Research* **8**, 1487–1496.

Smart, J. S. (1981) Short communications link lengths and channel network topology. *Earth Surface Processes and Landforms* **6**, 77–79.

Solé-Benet, A., Cantón, Y., Lázaro, R., Puigdefábregas, J. (2009) Meteorización y erosión en el Sub-Desierto de Tabernas, Almería. *Cuadernos de Investigación Geográfica* **35**, 141–163.

Sternai, P., Herman, F., Fox, M., Castelltort, S. (2011) Hypsometric analysis to identify spatially variable glacial erosion. *Journal of Geophysical Research* **116**, F03001.

Strahler, A. N. (1952a) Dynamic basis of geomorphology. *Geological Society of America Bulletin* **63**, 923–938.

Strahler, A. N. (1952b) Hypsometric (area-altitude) analysis of erosional topography. *Geological Society of America Bulletin* **63**, 1117–1142.

Strahler, A. N. (1957) Objective field sampling of physical terrain properties. *Annals of the Association of American Geographers*, **47**(2), 179–180.

Tarboton, D. G., Bras, R. L. Rodriguez-Iturbe, I. (1988) The fractal nature of river networks. *Water Resources Research* **24**, 1317–1322.

Tarboton, D. G., Bras, R. L., Rodriguez-Iturbe, I. (1989) Scaling and elevation in river networks. *Water Resources Research* **25**, 2037–2051.

Tarboton, D. G., Bras, R. L., Rodriguez-Iturbe, I. (1991) On the extraction of channel networks from digital elevation data. *Hydrological Processes* **5**, 81–100.

Tarolli, P., Dalla Fontana, G. (2009) Hillslope-to-valley transition morphology: New opportunities from high resolution DTMs. *Geomorphology* **113**, 47–56.

Thompson, J. A., Bell, J. C., Butler, C. A. (2001) Digital elevation model resolution: Effects on terrain attribute calculation and quantitative soil-landscape modeling. *Geoderma* **100**, 67–89.

Tribe, A. (1992) Automated recognition of valley lines and drainage networks from grid digital elevation models: A review and new method. *Journal of Hydrology* **139**, 263–293.

Tucker, G. E., Bras, R. L. (1998) Hillslope processes, drainage density, and landscape morphology. *Water Resources Research* **34**, 2751–2764.

Tucker, G. E., Slingerland, R. (1997) Drainage basin response to climate change. *Water Resources Research* **33**, 2031–2047.

Vogt, J. V., Colombo, R., Bertolo, F. (2003) Deriving drainage network and catchment boundaries: A new methodology combining DEM data and environmental characteristics. *Geomorphology* **53**, 281–298.

Werner, C., Smart, J. S. (1973) Some new methods of topologic classification of channel networks. *Geographical Analysis* **5**, 271–295.

Willgoose, G. R. (1994) A statistic for testing the elevation characteristics of landscape simulation models. *Journal of Geophysical Research: Solid Earth* **99**, 13987–13996.

Willgoose, G. R., Hancock, G. (1998) Revisiting the hypsometric curve as an indicator of form and process in transport-limited catchment. *Earth Surface Processes and Landforms* **23**, 611–623.

Wilson, J. P., Repetto, P. L., Snyder, R. D. (2000) Effect of data source, grid resolution, and flow-routing method on computed topographic attributes. In: Wilson, J. P., Gallant, J. C. (eds.) *Terrain Analysis Principles and Applications*. John Wiley & Sons, Inc, New York, pp. 133–161.

CHAPTER 12

From digital elevation models to 3-D deformation fields: A semi-automated analysis of uplifted coastal terraces on the Kamena Vourla fault, central Greece

Thomas J.B. Dewez[1] and Iain S. Stewart[2]

[1] Direction of Risks and Prevention – BRGM – French Geological Survey, Orléans-la-Source, France
[2] Centre for Research in Earth Sciences, School of Geography, Earth, & Environmental Sciences, University of Plymouth, Plymouth, UK

Introduction

Coastal terraces are widely used as geomorphic markers of tectonic deformation (Lajoie 1986; Burbank and Pinter 1999; Burbank and Anderson 2001; Keller and Pinter 2002). Their widespread utility for tracking tectonism arises from three facets. Firstly, because they originally formed at sea level, marine terraces (or shore platforms) provide reference datums against which the vertical uplift or subsidence of former shorelines can be measured (e.g. Merritts 1996). Secondly, the relatively simple geometry of these platforms – broad and flat surfaces dipping gently seawards – allows the spatial pattern of deformation (faulting, warping, tilting) on active tectonic structures to be resolved (e.g. Valensise and Ward 1991; Ward and Valensise 1994; Perg et al. 2001). Thirdly, by supplying dateable material and palaeoenvironmental diagnostics, the evolution of those tectonic structures can be reconstructed over time (e.g. Collier et al. 1992; Armijo et al.

1996; Morewood and Roberts 1999; Houghton et al. 2003).

In recent years, numerical simulations of terrace development (e.g. Anderson et al. 1999; Trenhaile 2000, 2002) and better direct dating methods and glacio-eustatic correlations (e.g. Perg et al. 2001; Dumas et al. 2005; Ferranti et al. 2006) have strengthened the value of coastal terraces as tectonic geomorphic markers. Despite these methodological advances, there has been little or no change in the way coastal terraces are identified and defined. This is largely because their distinctive appearance makes former marine platforms relatively easy to recognise in the landscape. Furthermore, the position of their sea-level index point – the intersection between the shore platform and the landward cliff (termed 'shore angle') – can be measured directly in the field or estimated from analysis of air photos or topographic maps. Even where this critical inner margin is obscured by cliff erosion or sediment accumulation, 2-D topographic profiles can give a reasonable

Monitoring and Modelling Dynamic Environments, First Edition. Edited by Alan P. Dykes, Mark Mulligan and John Wainwright.

geometric approximation of its position. Alongshore variations in the elevation of these index points are widely used to detect changes in the style or amount of deformation along a tectonically mobile coast (Cucci et al. 1996; Baldi et al. 2005). Yet even a precise discrimination of a 2-D landscape element provides only an incomplete descriptor of the 3-D deformation field that evolves over time as active faults and folds develop and grow.

To capture this 3-D deformation picture, the full geometric representation of the coastal terrace needs to be resolved. Digital elevation models (DEMs) offer the potential to do this. The use of DEMs in tectonic studies has been largely limited to producing shaded-relief maps or perspective views highlighting distinctive tectonic landforms (Burbank and Anderson 2001; Goldsworthy and Jackson 2001; Formento-Trigilio et al. 2002). However, the numerical information contained in the DEM grid permits more powerful analyses, especially in classifying continuous topography into discrete landforms (Irvin et al. 1997; Ventura and Irvin 2000; Wood 1996) and in quantifying landscape deformation (Carozza and Delcaillau 1999; Lavé and Avouac 2000; Formento-Trigilio et al. 2002).

Coastal terraces, with their simple geometry, are ideal objects for discrete classification. For this reason, we explore a simple semi-automated computational method to delimit coastal terraces in a consistent and readily reproducible manner in DEMs. Morphometric attributes of terraces (dip and dip direction, elevation, surface area) are then extracted from the DEM and subjected to quantitative analysis. To determine the degree to which a terrace's 3-D disposition differs from its original geometry, we use classical stereographic restoration approaches to retrieve deformation components in a way that can shed light on the causative tectonic activity. We demonstrate the approach by way of an investigation of an enigmatic suite of terraces developed on an active normal fault in central Greece.

Coastal terraces of the Arkitsa headland, northern Gulf of Evia

While some tectonically emerging seaboards preserve 'classic' well-resolved terrace sequences, most famously in the Gulf of Corinth (central Greece) or in the Gulf of Taranto (Southern Italy), most upwardly mobile coasts only retain a patchy assemblage of incomplete terrace remnants. In these settings, lateral correlation of the remnants based on elevation is uncertain, dating is often made difficult by a dearth of appropriate material, and relationships with underlying faults and other tectonic structures are elusive. Exactly this situation prevails on the Arkitsa headland in the northern Gulf of Evia, central mainland Greece (Figure 12.1), where a well-expressed suite of coastal terraces is developed at the end of a major active normal fault – the Kamena Vourla fault system.

Even though the Kamena Vourla fault system has been studied (Philip 1974; Roberts and Jackson 1991; Jackson and McKenzie 1999) (Figure 12.1), its slip history remains poorly resolved. In the long term, the Kamena Vourla fault zone began its activity sometime in the last 1 Ma when tectonic activity jumped northwards from the adjacent Kalidromon fault (Goldsworthy and Jackson 2001). Its more recent tectonic history is equally uncertain. The fault lacks a documented historical surface rupture along it, though the fresh topographic expression of the range-front escarpment provides compelling evidence for recent reactivation (e.g. Jackson and Mc Kenzie 1999). In particular, the easternmost segment, known as the Arkitsa fault segment (Figure 12.1), preserves a spectacular fault plane exhumed by quarrying of the overlying colluvium (Figure 12.2). This is undoubtedly the cumulative effect of many tens of repeated earthquakes (Jackson and McKenzie 1999). Along the base of the escarpment slope is a dark-grey weathered limestone fault scarp with a distinct unweathered basal band (Figure 12.2), which

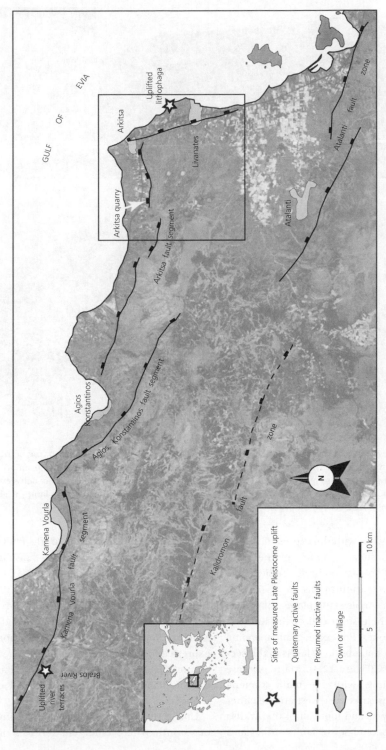

Figure 12.1 Location map of the study area. The Kamena Vourla fault zone comprises a series of 10–15 km long left-stepping normal fault segments. The easternmost segment of the fault zone occurs at Arkitsa, where recent coastal uplift is attested to by raised Holocene uplifted *Lithophaga* borings (Pirazzoli et al. 1999) as well as higher elevated coastal terraces.

(a)

(b)

Figure 12.2 (a) View of the Arkitsa fault plane (see Figure 12.1 for location), exposed by quarrying of the hanging wall colluvial deposits. The inset photo shows the detail of the slickensided principal slip surface. (b) The same fault plane prior to quarrying exhibited a smooth band of limestone at the contact between the scree and the fault plane, which is likely to represent the last coseismic surface rupture of the fault (height ~1.5 m). Gerald Roberts for scale.

probably corresponds with the expression of the last (Holocene?) surface rupture of the fault, though to date no palaeoseismic studies have been undertaken to confirm this.

The prominent Arkitsa fault plane is the main active trace in the eastern end of the Kamena Vourla fault zone, but geological mapping reveals a network of subsidiary faults in this sector (Philip 1974) (Figure 12.3). Many of these faults have affinity with the main Kamena Vourla fault: they strike east-west and lie either in the hanging wall and footwall of the Arkitsa escarpment. A second set of faults strikes at a high angle to the Kamena Vourla fault zone. The most notable structure in this subsidiary set is the Livanates fault (Figure 12.3), which trends NNW–SSE and downthrows to the east (Philip 1974). The sharp switch from E–W striking to NNW–SSE striking faults at the end of the Kamena Vourla fault zone is topographically expressed in the dramatic bend of the coastline around the Arkitsa headland. Wrapped around this headland is a prominent flight of coastal terraces.

Unlike the shores of the neighbouring Gulf of Corinth where emergent terraces provide

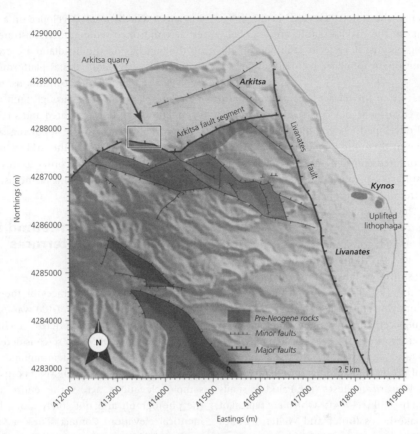

Figure 12.3 Structural map of the Arkitsa headland (modified from Philip 1974). The headland is cut by faults with a broadly E–W trend (WNW–ESE to WSW–ENE). The Livanates fault, which separates the coastal plain of Livanates from the hills to the west, was mapped by Philip (1974) but is otherwise unstudied. Grey polygons delimit areas of pre-Neogene sediments, and the remainder of the headland comprises Neogene or Quaternary sediments.

excellent indicators of associated fault histories (e.g. Armijo et al. 1996; Morewood and Roberts 1999; Houghton et al. 2003; McNeill and Collier 2004; Di Martini et al. 2004), the coastal landforms along the Kamena Vourla shores are poorly resolved. In the west, a pair of former river terraces stranded 20 m and 8 m above the present riverbed have been dated by optically stimulated luminescence (OSL) dating at approximately 19.5 ka and approximately 7.6 ka, respectively, and interpreted as having been abandoned by uplift on the coastal-bounding fault at a rate of around 1 mm year^{-1} (Walker et al. 2010). (Radiocarbon ages from both terraces yielded much younger ages – and consequently, higher slip rates – but are considered suspect due to carbonate contamination and so have not been considered further.) In the east, near Livanates (Figure 12.1), raised Holocene marine notches cut into the limestone rocky shore indicate a net uplift of 1.4 m in the recent millenia (Pirazzoli et al. 1999) but appear to reflect a complex interplay of coseismic uplift and subsidence (Goldsworthy and Jackson 2001; Walker et al. 2010). Higher terraces on the Arkitsa headland lie at different elevations and on both the footwall and hanging wall of this Arkitsa segment of the main Kamena

Vourla fault; the terraces elevated in the hanging wall of the Arkitsa fault are especially intriguing because long-term fault slip on the main fault ought to induce tectonic subsidence here.

The nature of the terraces themselves is also elusive. Occasional outcrop sections through the well-developed topographic benches reveal that some of them are capped by a flat, 1–2 m thick, indurated conglomerate and all exhibit a terrace surface separated by an angular unconformity from tilted siltstone and sandstone beds below. The cemented conglomerate caprock covering certain terrace remnants comprises a mixture of sand, gravel and pebbles set in a hard calcite matrix. In thin section, the matrix occasionally shows a drusy fabric, indicative of underwater crystallisation, but in places that is absent, implying subaerial crystallisation. These contrasting cement styles appear typical of coastal environments. A beach-like environment is also suggested by the well-rounded nature of the pebbles and by abundant conical shells of *Viviparus* molluscs (the Paludina snails described by Philip (1974)), which are typical of Balkan lakeshores (Butot and Welter-Schultes 1994). Taken together, this evidence suggests that the caprock unit is typical of a supra-tidal beach deposit, or 'cayrock' (Gischler and Lomando 1997).

While coastal in origin, the *Viviparus* deposit implies either a marine or a lacustrine setting. The possibility of a non-marine origin means that the associated terraces cannot simply be tied to global glacio-isostatic sea-level curves. For that reason, they are not correlatable with the Pleistocene eustatic chronology that has been widely applied to date equivalent features along the southern shores of the Gulf of Corinth (e.g. Armijo et al. 1996). Although not useful chronometric indicators, the Arkitsa coastal terraces remain potentially valuable geometric markers of fault activity. The *Viviparus* caprock constitutes an especially distinctive marker horizon because it caps only one low-lying terrace level. Where it is absent, the terraces are

capped instead by a soil developed on a sharply eroded and tilted sedimentary substrate. Thus, all the benches on the headland are erosional, presumably wave-cut, coastal platforms. What evolved as a flight of once continuous, multiple shoreline platforms has, through fault activity, been left displaced, fragmented and stranded at different elevations. In the following sections, we explore how a DEM of the Arkitsa headland can be used to bring some order to this assemblage of disparate geomorphic elements.

A semi-automated method for delimiting coastal terraces

The DEM

To analyse the coastal terraces of the Arkitsa headland, a 10 m resolution DEM was extracted using digital photogrammetric techniques. Photogrammetric DEMs were considered more appropriate for gradient determinations than simple contour-derived DEMs because they sample elevations across the entire field of the photograph and not solely along lines of identical elevation. Contact slides of 1:40,000 black and white air photos (July 1986 flights of the Hellenic Military Geographical Survey) were scanned on a professional Leica–Helava photogrammetric scanner. Elevation was then extracted automatically with Erdas Imagine Orthomax (8.3) using image correlation techniques and post-processed with an improved version of the failure warning model (Gooch and Chandler 2001). The failure warning model filtering technique is designed to flag and remove unreliable elevation pixels and quantify the reliability of the remaining ones. After post-processing, elevation measurements appeared repeatable to within ±1.2 m (1 sigma) under varying extraction algorithm parameters. In addition to this intrinsic precision, DEM accuracy was checked in the field with more than 200 points collected with differential global positioning system (DGPS). DEM elevations represented true ground elevations to within

±1.9 m (1 sigma). To confirm its fitness for purpose, the attitude of best-fit planes fitted through DGPS and identical DEM sample locations yielded virtually identical attitude values (aspect ±3° and slope ±0.05°).

The terrace delimitation algorithm

The derived DEM generated clearly shows the distinct coastal terraces of the Arkitsa headland (Figure 12.4), but in order to use these features as effective geomorphic markers, their topographic attributes have to be defined. A terrace can be defined as an extensive and gently dipping flat surface (Sunamura 1992), but this description equally applies to many different landforms. For this reason, a fully automated terrace extraction algorithm will likely lead to classification mistakes and should be avoided. Nevertheless, a semi-automated procedure can be of value in delimiting terraces because while different geologists are likely to debate the precise edges of a specific terrace, they are likely to agree on a particular point on the ground being located securely within a terrace. Consequently, our terrace algorithm uses expert knowledge (i.e. the field scientist) to roughly locate terrace surfaces and refines the result with automated procedures.

Our algorithm works in three steps. Firstly, an automated set of rules crudely classifies the entire DEM into two categories: one category for pixels likely to qualify as terraces and another category where the geometry is demonstrably not that of a terrace. Secondly, an operator selects locations ('seed points') where he/she recognises a terrace in the landscape. Thirdly, a matching algorithm analyses each seed point in turn and retrieves the surrounding patches of land sharing identical characteristics. The final product is a raster map of 'terrace' pixels.

The rules to classify pixels as terraces in step 1 use simple map algebra to discern (i) shallow slope, (ii) relatively planar geometry and (iii) position elevated above sea level (Figure 12.5). Accordingly, input data layers are elevation, slope, aspect and aspect divergence. Elevation, slope and aspect are usual morphometric functions found in virtually all GIS packages, but aspect divergence is a concept developed here to characterise the 'flatness' of land. Aspect divergence measures the maximum range of slope directions found inside a restricted neighbourhood of pixels, in our case a 3 × 3 kernel of eight pixels surrounding any given pixel. This is essentially the same as computing a planform curvature but is easier to implement in raster-based

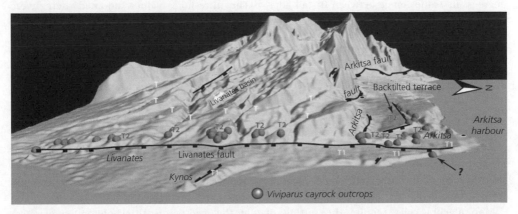

Figure 12.4 Perspective view of the Arkitsa headland highlighting the main terraces (T) visible in the landscape. T1/T2: tentative terrace levels. The spheres represent locations of *Viviparus* cayrock outcrops which, with the exception of one site, are located in the uplifted side (footwall) of the Livanates fault.

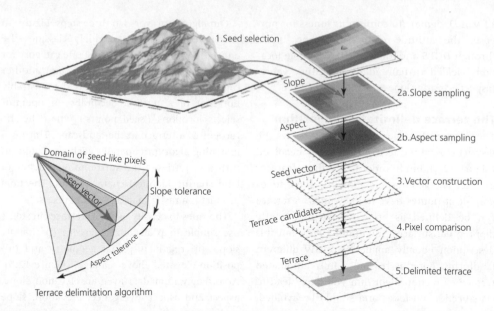

Figure 12.5 Conceptual diagram of the terrace delimitation algorithm. The operator identifies a seed point in the field or on-screen. Slope and aspect values are sampled at the location of the seed and retained. Terrace candidate pixels correspond to all the pixels having the same slope and aspect value (within pre-specified tolerance limits) as the seed point and that are contiguous to it. For the algorithm to work, one needs a set of seed points, a DEM, slope and aspect maps and defined threshold criteria for similarity between a seed and its neighbours.

geographic information systems that have a limited range of morphometric filters implemented. Aspect divergence is also more intuitive than a curvature parameter or a distance vector because it expresses changes of azimuth in degrees. Its values range between 0° (all adjacent pixels dip in the same direction, i.e. planar terrace) and 180° (at least two pixels in the neighbourhood have opposite dip directions, e.g. convergent slopes at valley bottom or divergent slopes along sharp ridges). Of course, measuring aspect divergence on very shallow-dipping surfaces is meaningless because elevations are influenced by the random noise of automated photogrammetric extraction. For this reason, for slopes shallower than 0.5°, the computed slope direction and aspect divergence values were disregarded.

Map algebra rules require the setting of thresholds to decide what constitutes a *gently dipping* piece of land, a relatively *planar* surface

and an *elevated* position. As is always the case with thresholds, setting tight values incurs the risk of excluding meaningful pixels, while setting loose thresholds may incorporate rogue pixels and risk qualifying irrelevant landforms. The threshold values adopted here were calibrated with slope and aspect divergence values observed on a training set of four well-developed terraces (Figure 12.6). In step 1 (automatic thresholding of slope and aspect divergence), where the purpose is to guide the interpreter to the most relevant places, slope and aspect divergence thresholds were defined rather loosely in order to include all relevant landforms. Thus, slopes shallower than 8° and aspect divergence smaller than 30° qualified as 'terraces'. In step 3, where thresholding constrains the limits of the terraces designated by the seed points, threshold values were set more strictly (2.5° of slope and 20° of aspect divergence).

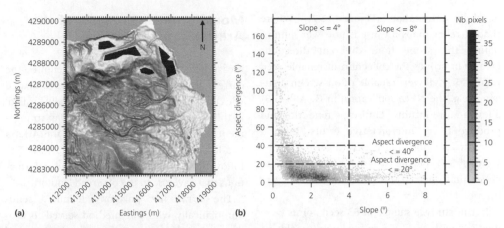

(a) Eastings (m)   (b) Slope (°)

Figure 12.6 (a) Set of user-defined training terraces (black patches) used to calibrate slope and aspect thresholds. (b) Density plot of observed slope and aspect divergence of pixels composing the training sets. Slope and aspect divergences were computed with a 3×3 kernels. The majority of pixels fall between 0 and 4 degrees of slope and between 0 and 20 degrees of aspect divergence.

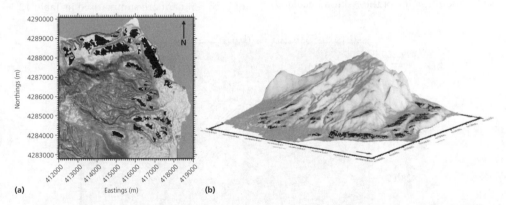

(a) Eastings (m)   (b)

Figure 12.7 Example of 2-D rendering techniques. (a) Shaded relief representing the topography as if a light source was illuminating the landscape. In this instance, the scene is illuminated from the SW (215°) with an elevation of 45°. (b) The relief is visualised through a slope map coded in light greys for shallow slopes and dark grey for steep slopes. In map B, the edges of landforms stand out in any case because they do not depend on a lighting source.

The seed points

Seed points are sets of locations provided by the expert user (e.g. the geomorphologist) and positioned anywhere on a terrace. The seed point location is left to the user's discretion but ideally should be informed by field surveys and validated by a handheld GPS. When field verification is not possible (or at an exploratory stage of a study), it is instead necessary to digitise seed

points on the screen using a 2-D or 3-D DEM rendering technique.

For the 2-D rendering option, shaded-relief images can be used (Figure 12.7a), but our experience with 2-D rendering techniques led us to prefer slope maps (displayed with an appropriate white-to-grey/black colour ramp) because they reveal sharp topographic features such as ridges and breaks of slope independently of their orientation (Figure 12.7). Where shaded-relief maps are used, however, illuminating the scene from different directions greatly improves the collection of reliable seed points. 3-D viewers capable of on-screen digitising (e.g. the NVIZ application in GRASS GIS) allow a more intuitive landscape rendering and produce fewer interpretative errors. Even if the software does not permit digitising in the viewer, 3-D perspective views are useful to check that selected seeds fall where they were intended.

In this study, a sample of 63 seed points were digitised with NVIZ. These seed points yielded a set of 47 'terrace-like' landforms (Figure 12.8). The product of this semi-automated terrace delimitation algorithm is a *terrace map*, and it is that map which is the basis for discriminating tectonic deformation of the Arkitsa headland.

Morphometric analysis of the Arkitsa coastal terraces

Measuring landscape deformation requires the current disposition of a landform to be compared with that expected when it was originally formed. In this section, we detail the present-day arrangement of the coastal terraces around the Arkitsa headland by means of a DEM analysis. The following section discusses the tectonic deformations that gave rise to the observed landforms.

The terrace-like polygons delimited semi-automatically with our method served to produce a data set of elevation, slope and aspect of every pixel belonging to each terrace remnant. We chose to describe each terrace remnant with the following attributes: X, Y, Z, slope, aspect and surface area (Figure 12.8). The parameters of each remnant are summarised in Table 12.1.

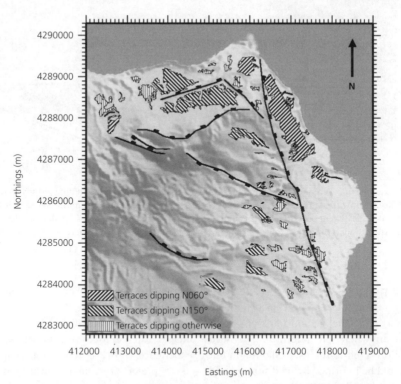

Figure 12.8 Terrace map produced with the terrace delimitation algorithm. The limits of the terraces (striped black areas) appear somewhat indented due to short wavelength variations of the DEM (possible errors) and due to surface processes that eroded the terrace surface after the abandonment of the shoreline.

Table 12.1 Morphometric parameters of the 47 terraces detected with the interactive delimitation method.

Terrace ID	Mean X (m)	Mean Y (m)	Mean Z (m a.m.s.l.)	Mean slope (°)	Mean azimuth (N°E)	Terrace area (m²)
1	415,901	4,289,231	15.8	1	49	116,000
2	415,074	4,289,273	26.2	3	139	5,900
3	415,456	4,289,271	19.2	2	30	31,000
4	414,927	4,289,100	38.7	3	20	20,400
5	415,671	4,288,852	33.4	2	93	28,200
6	413,208	4,288,924	53.9	1	201	16,800
7	414,107	4,288,844	52.9	1	151	230,900
8	416,118	4,288,692	36.6	3	197	11,900
9	414,842	4,288,427	53.3	2	156	488,900
10	412,595	4,288,432	50.5	1	95	7,500
11	412,859	4,288,453	50.0	2	344	11,700
12	416,192	4,288,381	37.6	3	215	9,100
13	413,506	4,288,589	52.4	1	181	96,200
14	416,153	4,288,244	39.2	3	353	7,400
15	412,434	4,288,295	53.1	3	10	90,000
16	412,644	4,287,849	72.0	2	48	60,600
17	415,904	4,287,546	140.8	3	153	141,300
18	416,203	4,287,146	112.8	3	152	4,000
19	416,886	4,287,871	17.3	2	62	624,200
20	416,539	4,287,022	91.8	4	128	700
21	416,571	4,286,771	80.0	4	148	3,800
22	416,525	4,286,584	86.3	5	82	10,700
23	417,662	4,286,650	19.2	1	43	65,900
24	416,258	4,286,374	124.2	4	68	20,500
25	416,500	4,286,392	100.2	5	67	7,400
26	416,785	4,286,183	94.9	3	73	11,400
27	417,002	4,286,120	77.5	5	87	9,500
28	416,704	4,285,893	84.2	4	130	49,000
29	416,289	4,285,952	116.3	2	71	1,200
30	415,904	4,285,862	132.1	3	114	8,600
31	416,252	4,285,699	108.3	4	160	50,000
32	416,806	4,285,432	73.1	3	97	20,400
33	416,382	4,285,516	96.0	2	86	9,700
34	417,513	4,285,134	31.7	3	114	16,400
35	416,575	4,284,875	73.4	2	127	25,000
36	416,128	4,284,811	90.3	3	138	75,600
37	417,060	4,284,827	56.1	5	183	19,200
38	417,683	4,284,709	17.4	3	91	48,100
39	417,327	4,284,786	47.9	4	82	33,600
40	416,698	4,284,466	68.7	2	128	42,100
41	416,080	4,284,360	108.4	4	139	73,200
42	417,487	4,284,343	37.0	2	85	18,600
43	415,063	4,284,033	166.8	5	167	103,800
44	417,498	4,284,131	34.9	3	98	21,500
45	416,316	4,283,855	86.3	2	77	3,400
46	417,180	4,283,986	46.0	3	103	37,700
47	416,961	4,283,777	59.4	3	118	17,500

It is difficult to analyse these multiple variables simultaneously, not least because of two particular preliminary issues.

First, terrace elevation and positions were computed as the average of X, Y and Z of the contributing pixels. In coastal tectonic studies, most workers use the elevation of the shore angle rather than the mean elevation of a terrace (e.g. Lajoie 1986; Merritts 1996). The shore angle is really the closest representative of sea level at the time of platform formation, but detecting the elevation of the shore angle is problematic. In the field, the landward edge of terraces becomes progressively concealed by degradation of the palaeo-cliff. Likewise, the edge of a terrace delimited by the algorithm can be blurred by its sensitivity to the thresholds set on the slope criterion. In a domain of smoothly varying slopes, it is difficult to establish a robust and absolute cut-off value. For this reason, averaging the elevation of each terrace remnant was deemed to be a more robust way of resolving the issue numerically. Bearing in mind the noise of the DEM and the limited elevation range covered by terrace remnants, it seems reasonable to consider the mean elevation as representing the inner-edge elevation, particularly given that our analysis seeks to discriminate between terrace geometries rather than simply measure their vertical displacement.

Second, determining the mean terrace aspect is problematic. Aspect variables are circular in nature and vary between 0° and 360°. The consequence is that the average between 359° and 1° should be 0° (i.e. due North) and not 180° as linear algebra dictates. To average such circular variables, it is necessary to compute in sequence (i) the orthogonal components of the direction vector (e.g. Gumiaux et al. 2003), (ii) average these orthogonal components separately and (iii) recombine them into the average azimuth.

A general map of the derived Arkitsa coastal terraces is shown in Figure 12.9, in which colours (or shades of grey) represent the elevation and vectors represent both dip direction and amount of dip. In this form, all the parameters of terrace remnants – X, Y, Z, slope, aspect and surface area – are represented simultaneously. The map reveals terraces along a coastal strip that extends 2 km inland. The terraces closest to the east coast broadly dip in a general seaward direction, while those located further away from the coast generally dip oblique to, or even away from, the sea. This dip direction discrepancy is most obvious for the terrace located west of Arkitsa (terrace #9 on Figure 12.9) which is inclined towards the south-east despite the modern shoreline lying to the north, demonstrating it to be substantially back tilted.

The spatial distribution of dip patterns is arguably better viewed as a rose diagram and a stereogram to represent the frequency of dip and dip direction, respectively (Figure 12.10). Both diagrams emphasise two principal dip directions. The most frequent one is oriented ENE (N150°±20°) and is referred to here as the 'east-dipping terrace group'; the second most frequent direction is SSE (N060°±20°), at right angles to the first group, and so is referred to as the 'south-dipping terrace group'.

The difference between the major terrace-dip directions of almost 90° cannot be attributed to measurement or DEM artefacts because the most represented aspects are not oriented along the rows and columns of the DEM grid. In addition, the slope and aspect were computed with an algorithm designed specifically to avoid usual chequerboard artefacts occurring with oversimplified aspect computations (Wood 1996). In short, we regard these perpendicular terrace-dip directions as being real.

The dramatic contrast in terrace-dip direction across the Arkitsa headland appears to reveal the importance of what had previously been regarded as a subsidiary structure on the Arkitsa headland: the Livanates fault (Figure 12.11). In particular, the south-dipping terrace group is generally confined to the west of this N–S trending normal fault, while the east-dipping terrace group is concentrated east of it. Moreover, the number of terrace levels changes across this structure. East of the Livanates fault, there is one prominent terrace level defined below the elevation of the

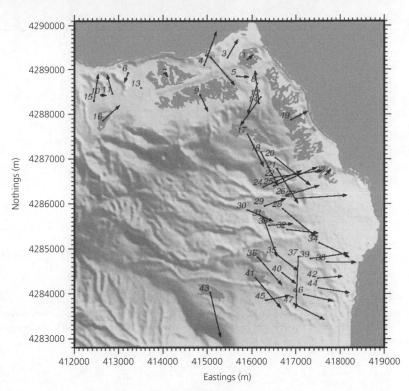

Figure 12.9 Synthetic map showing the locations (1, 2, 3, etc.), extents (shaded areas), dips and dip directions (arrow lengths and azimuths) of terraces. Although a useful overview of the data, this type of map is difficult to analyse visually because too much information is displayed at once. Numbered locations correspond to the terrace IDs in Table 12.1.

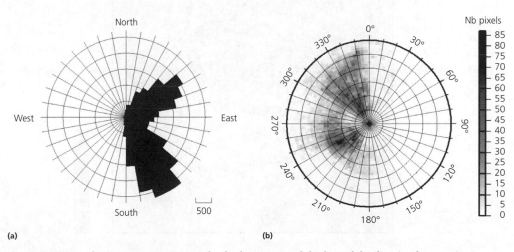

(a)   **(b)**

Figure 12.10 Rose diagrams representing (a) the dip frequency and (b) dip and dip direction frequencies observed in terrace pixels. These diagrams collate all pixels classified as terraces by the terrace delimitation algorithm. Note that (b) is a lower-hemisphere stereonet of poles to planes, which means that all azimuths must be shifted by 180° to represent true azimuths. Both diagrams imply the existence of two families of dip directions, oriented ENE and SSE, respectively.

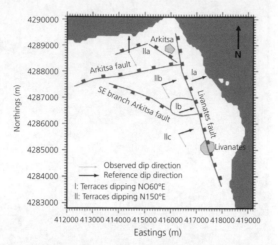

Figure 12.11 Synthetic map of terraces classified into east-dipping (N060°E) and south-dipping (N150°E) planes: (a) showing the terraces overlaid on the topography; (b) schematically showing the south-dipping terraces confined to the footwall (west block) of the Livanates fault and the east-dipping surfaces confined to the hangingwall of the Livanates fault (east block).

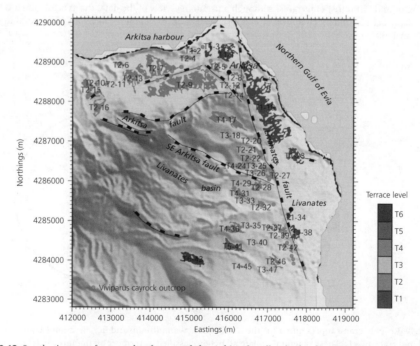

Figure 12.12 Synthetic map of terrace levels around the Arkitsa headland. The determination of terrace levels was conducted with several arguments: vertical position relative to the *Viviparus* cayrock marker horizon (outcrops marked as dots), existence of palaeo-cliffs between terrace remnants and geometric correlations based on different cross sections not shown here.

Viviparus cayrock terrace level; accordingly, we denote this lowermost platform as Terrace 1 (T1) and the *Viviparus* cayrock level as Terrace 2 (T2). West of the Livanates fault, however, it is the *Viviparus* cayrock (T2) that caps the lowest platform on both sides of the Arkitsa fault. Higher terraces above the *Viviparus* cayrock are distributed on the uplifted footwalls of both the Arkitsa and Livanates faults, and their relative relationships are difficult to gauge. In Figure 12.12, we simply group the higher benches in terms of their relative order and elevation above T2 and reconstruct their apparent minimum extent.

Tectonic interpretation of the Arkitsa terraces

Based on our derived terrace data set, some inferences can be drawn about the tectonic deformation of the Arkitsa headland

(Figure 12.13). For example, reconstructing T2 benches (Figure 12.12) to a continuous level indicates that around 75 m of displacement occurred across the Arkitsa segment of the Kamena Vourla fault since the deposition of the *Viviparus* cayrock. Perhaps more significantly, the entire terrace field west of the Livanates fault is tilted to the south (Figure 12.12), with the T2 terraces expressing a dip of around 1.1° degrees to the SSE (Figure 12.4). The timing of both the fault offset and the tilting is problematic: the *Viviparus* shells are recrystallised and therefore are unfit for radiometric dating. In the absence of absolute ages for any of the terraces, the geometry of deformation is instead considered.

Assuming that the terraces were initially carved with a seaward dip towards a coastal configuration similar to that of today and also that the axis of rotation is presumed to be horizontal, a stereographic projection can restore the tilted planes of the terraces either side of the

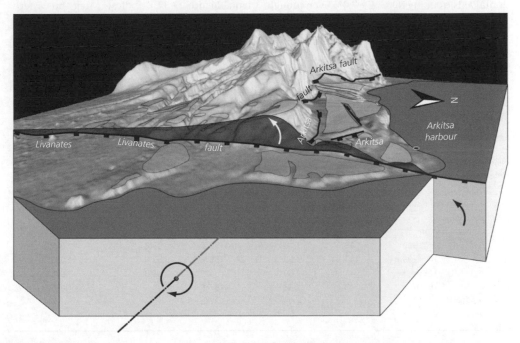

Figure 12.13 Perspective view of the Arkitsa headland with a structural sketch of the Livanates fault accommodating the rotation about a horizontal axis. In this structural scheme, the Livanates fault is envisaged as behaving as a pivot or scissor fault.

Livanates fault (Figure 12.11) (see details in Dewez 2003). The youngest terrace, T1, is apparently unaffected by rotation and currently dips seawards at 1.75°, so this was used as a reference. Thus, the east-dipping terraces (east of the Livanates fault) are restored to a dip of 1.75° on an azimuth of N060°, and the south-dipping terraces (west of the fault) are restored to a dip of 1.75° on an azimuth of N000° (Figure 12.11). In order to track if and how the tilt pattern changed over time, higher terrace levels were restored in relative sequential order (in other words, T2 was restored and the disposition of the higher terraces corrected, then T3, then T4, etc.). When a terrace level was made of several remnants, the average dip and azimuth of the terraces were computed (Table 12.2). For more details of the procedure, see Dewez (2003).

To a first approximation, a single transformation is capable of restoring both sets of terraces in a seaward direction. That transformation invokes a rotation of 4.0° ± 0.7° about a horizontal axis trending N085°E ± 7.3°. Even though none of the initial assumptions promoted this solution, this rotation axis turns out to be nearly orthogonal to the Livanates fault and parallel to the trace of the Arkitsa fault segment. Moreover, the rotation parameters are virtually identical whether one takes the terraces in sequence or whether one examines the terrace locations with respect to the Arkitsa fault segment.

The analysis prompts two key observations. First, the terraces west of the Livanates fault are equally affected by the rotation, regardless of their elevation, indicating that the rotation has not been taking places as successive terraces developed. Second, the rotation equally affected the terraces on both the Arkitsa footwall *and* hanging wall, implying that this E–W trending segment has been less dominant in recent times. We conclude that the rotation probably began after the formation of T2 but before the formation of T1, for which no rotation can be reasonably deduced. Moreover, the fact that the rotation affects the western side of the Livanates fault irrespective of whether a terrace lies on the upthrown or downthrown part of the Arkitsa fault (Figure 12.11b) indicates that the eastern and western blocks must have become decoupled across the Livanates fault.

Table 12.2 Rotation parameters needed to restore two sets of terraces to a seaward-dipping attitude.

Terrace level	Mean remnant elevation a.m.s.l.	Terrace ID	Mean slope (°)	Mean azimuth	Rotation angle (°)	Axis azimuth
Arkitsa fault hanging wall subgroup						
Reference			1.75	N000°E		
T2	53, 53 m	7–9	1.4	N153°E	3.1	N078°E
Arkitsa fault footwall subgroup						
Reference			1.75	N060°E		
T2	79 m	21	3.70	N148°E	4.0	N084°E
T3	112, 90 m	18	3.40	N152°E	3.8	N089°E
T4	140, 108 m	17-31-36	3.50	N150°E	3.9	N097°E
T5	108 m	41	3.69	N139°E	3.8	N076°E
T6	167 m	43	4.51	N167°E	5.3	N095°E
Mean rotation parameters			4.0 ± 0.7	N087°E ± 8.7°		

These were computed in relative chronological order (see Figure 12.12). For levels made of several terrace remnants, the mean slope and azimuth were computed as the average of the terrace values. The rotation parameters were obtained with the numerical restoration method (Dewez 2003).

Conclusions and implications

This paper presents the first detailed study of a complex suite of terraces developed along the prominent Kamena Vourla active normal fault zone in central Greece. Our analysis of a complex flight of terraces wrapped around Arkitsa headland reveals the existence of two families of terraces, one set that seems unaffected by tectonics (because they were still dipping in a seaward direction) and a second set dipping at high angle to present-day sea direction. The findings shed light on the mode and relative timing of activity on the Kamena Vourla fault with respect to the Livanates fault, whose activity, style and geometry were previously poorly resolved (Philip 1974). The Livanates fault trace is perpendicular to the Arkitsa segment and clearly cross-cuts this structure, implying that it formed more recently.

The present-day disposition of terraces can be most simply explained by rotational faulting on the Livanates fault rather than on the more prominent Kamena Vourla fault. In this structural scheme, the Livanates fault is envisaged as behaving as a pivot or scissor fault. In this way, the entire sequence of terraces west of the Livanates fault appears to be passively tilted, while those terraces to the east seem unaffected. Intriguingly, during this rotation, the Arkitsa segment of the Kamena Vourla fault must have continued to be active, displacing terrace level T2 vertically by ca. 75 m. Such displacement, however, appears not to have affected T1, east of the Livanates fault. The absence of significant dislocation east of the Livanates fault implies that blocks either side of this structure are decoupled. In other words, eastward propagation of the Arkitsa tip of the Kamena Vourla fault seems to have arrested at the Livanates fault once this structure began developing (between terrace episodes T2 and T1). The former's attempt to break through this structural barrier can be seen where the south-eastern branch of the Arkitsa segment and the Livanates fault intersect, expressed topographically as an unusual cluster of terraces with anomalous dips. At a broader scale, it seems possible that the previously neglected Livanates fault in fact played a pivotal role accommodating the change of fault-dip polarity across the northern Gulf of Evia, linking the Kamena Vourla fault with both the Atalanti fault to the south and the Kandili faults to the north.

Despite the wider tectonic implications that arise from these observations, the main purpose of this paper was to demonstrate the utility of high-resolution DEMs in deriving geometric information on the tectonic deformation of coastal terraces. The spatial disposition of fragmented terraces in this area can be usefully interrogated using standard and widely available GIS tools. After ensuring via DEM with DGPS field surveys that our digital data set was suitable for deriving accurate geometric characteristics of shallow-dipping planar landforms, we developed a semi-automated GIS algorithm to delimit terrace surfaces. The efficacy of this algorithm is that it combines the field geologist's ability to recognise landforms with the computer's superiority in applying systematic and reproducible criteria. The resulting discrimination of morphological quantities (slope and aspect) offers a quantitative view of the 3-D disposition of coastal terraces – though the method could just as readily be used to discriminate other potential tectonic geomorphological markers, such as fluvial terraces. The enigmatic terraces at the eastern end of the Kamena Vourla fault zone serve as an instructive testing ground because even in the absence of a detailed knowledge of fault slip history, the terraces offer insightful markers of an evolving 3-D deformation field. In this respect, we would argue that the disordered and fragmented terraces of the Arkitsa headland are likely to be far more typical of other strongly partitioned tectonic seaboards than the more continuous and well ordered 'classic' terrace staircases that track some upwardly mobile coasts.

Acknowledgements

A quantitative approach to the analysis of active faults in central Greece was encouraged by John Thornes during his supervision of IS's doctoral research in the 1980s but only realised in part a decade or so later through the doctoral work of TD, courtesy of a 3-year studentship from the Faculty of Science at Brunel University (west London, United Kingdom). That work was made possible through IGME and the Greek Ministry of Culture fieldwork permits and GPS fieldwork carried out under the guidance of Alain Demoulin (University of Liege). Dr Ir. Albert Colignon is thanked for giving access to a professional photogrammetric scanner at the Ministry of Equipment and Transport (MET) in Belgium. The paper presented here has benefitted hugely from enlightening discussions with and comments from John Moore, Ian Downman, Robert Anderson, Mike Ellis, Richard Walker, Matt Telfer and James Jackson and was further improved by reviews by two anonymous referees.

References

Anderson, R. S., Densmore, A. L., Ellis, M. A. (1999) The generation and degradation of marine terraces. *Basin Research* **11**, 7–19.

Armijo, R., Meyer, B., King, G. C. P., Rigo, A., Papanastassiou, D. (1996) Quaternary evolution of the corinth rift and its implications for the Late Cenozoic evolution of the Aegean. *Geophysical Journal International* **126**, 11–53.

Baldi, P., Fabris, M., Marsella, M., Monticelli, R. (2005) Monitoring the morphological evolution of the Sciara del Fuoco during the 2002–2003 Stromboli eruption using multi-temporal photogrammetry. *ISPRS Journal of Photogrammetry and Remote Sensing* **59**, 199–211.

Burbank, D. W., Anderson, R. S. (2001) *Tectonic Geomorphology*. Blackwell Science, Malden, MA.

Burbank, D. W., Pinter, N. (1999) Landscape evolution: the interactions of tectonics and surface processes. *Basin Research* **11**, 1–6.

Butot, L. J. M., Welter-Schultes, F. W. (1994) Bibliography of the mollusc fauna of Greece, 1758–1994. *Schriften zur Malakozoologie* **7**, 1–160.

Carozza, J.-M., Delcaillau, B. (1999) Geomorphic record of quaternary tectonic activity by alluvial terraces: example from the Tet basin (Roussillon, France). *Comptes Rendus de l'Académie des Sciences de Paris* **329**, 735–740.

Collier, R. E. L., Leeder, M. R., Rowe, P. J., Atkinson, T. C. (1992) Rates of tectonic uplift in the Corinth and Megara basins, central Greece. *Tectonics* **11**, 1159–1167.

Cucci, L., D'Addezio, G., Valensise, G., Burrato, P. (1996) Investigating seismogenic faults in Central and Southern Apennines (Italy): modeling of fault-related landscapes features. *Annali di geofisica* **39**, 603–618.

Dewez, T. (2003) Geomorphic markers and digital elevation models as tools for tectonic geomorphology in central Greece. PhD thesis. Brunel University, Uxbridge.

Di Martini, P., Pantosti, D., Palyvos, N., Lemeille, F., McNeill, L., Collier, R. (2004) Slip rates of the Aigion and Eliki Faults from uplifted marine terraces, Corinth Gulf, Greece.*Comptes Rendus Geosciences* **336**, 325–334.

Dumas, B., Gueremy, P., Raffy, J. (2005) Evidence for sea-level oscillations by the 'characteristic thickness' of marine deposits from raised terraces of Southern Calabria (Italy). *Quaternary Coastal Morphology and Sea-Level Changes* **24**, 2120–2136.

Ferranti, L., Antonioli, F., Mauz, B., Amoros, A., Dai Pra, G., Mastronuzzi, G., Monaco, C., Orru, P., Pappalardo, M., Radke, U., Renda, P., Romano, P., Sanso, P., Verrubi, V. (2006) Markers of the last interglacial sea-level high stand along the coast of Italy: tectonic implications. *Quaternary International* **145–146**, 30–54.

Formento-Trigilio, M. L., Burbank, D. W., Nicol, A., Shulmeister, J., Rieser, U. (2002) River response to an active fold-and-thrust belt in a convergent margin setting, North Island, New Zealand. *Geomorphology* **49**, 125–152.

Gischler, E., Lomando, A. J. (1997) Holocene cemented beach deposits in Belize. *Sedimentary Geology* **110**, 277–297.

Goldsworthy, M., Jackson, J. (2000) Active normal fault evolution in Greece revealed by geomorphology and drainage patterns. *Journal of the Geological Society* **157**, 967–981.

Goldsworthy, M., Jackson, J. (2001) Migration of activity within normal fault systems: examples from the Quaternary of mainland Greece. *Journal of Structural Geology* **23**, 489–506.

Gooch, M. J., Chandler, J. H. (2001) Failure prediction in automatically generated digital elevation models. *Computers & Geosciences* **27**, 913–920.

Gumiaux, C., Gapais, D., Brun, J. P. (2003) Geostatistics applied to best-fit interpolation of orientation data. *Tectonophysics* **376**, 241–259.

Houghton, S. L., Roberts, G. P., Papadonikolaou, I. D., McArthur, J. M., Gilmour, M. A. (2003) New 234U-230Th coral dates from the western Gulf of Corinth: implications for extensional tectonics. *Geophysical Research Letters* **30**, doi:10.1029/2003GL018112.

Irvin, B. J., Ventura, S. J., Slater, B. K. (1997) Fuzzy and isodata classification of landform elements from digital terrain data in Pleasant Valley, Wisconsin. *Geoderma* **77**, 137–154.

Jackson, J. A., Mc Kenzie, D. (1999) A hectare of fresh striations on the Arkitsa Fault, central Greece. *Journal of Structural Geology* **21**, 1–6.

Keller, E. A., Pinter, N. (2002) *Active Tectonics: Earthquakes, Uplift, and Landscape*. Prentice Hall, Upper Saddle River, NJ.

Lajoie, K. (1986) Coastal tectonics. In: Geophysics Study Committee (ed.) *Active Tectonics*. National Academy Press, Washington, DC, pp. 95–124.

Lavé, J., Avouac, J.-P. (2000) Active folding of fluvial terraces across the Siwaliks Hills, Himalayas of central Nepal. *Journal of Geophysical Research* **105**(B3), 5735–5770.

McNeill, L., Collier, R.E.L. (2004) Uplift and slip rates of the eastern Eliki fault segment, Gulf of Corinth, Greece, inferred from Holocene and Pleistocene terraces. *Journal of the Geological Society* **161**, 81–92.

Merritts, D. (1996) The Mendocino triple junction: active faults, episodic coastal emergence, and rapid uplift. *Journal of Geophysical Research* **101**(B3), 6051–6070.

Morewood, N. C., Roberts, G. P. (1999) Lateral propagation of the surface trace of the South Alkyonides normal fault segment, central Greece: its impact on models of fault growth and displacement-length relationship. *Journal of Structural Geology* **21**, 635–652.

Perg, L. A., Anderson, R. S., Finkel, R. C. (2001) Use of new [10]Be and [26]Al inventory method to date marine terraces, Santa Cruz, California, USA. *Geology* **29**, 879–882.

Philip, H. (1974) Etude néotectonique des rivages Egéens en Locride et Eubée nord occidentale (Grèce). PhD thesis. Université des Sciences et Techniques du Languedoc, Montpellier.

Pirazzoli, P. A., Stiros, S. C., Arnold, M., Laborel, J., Laborel-Deguen, F. (1999) Late Holocene Coseismic vertical displacements and tsunami deposits near Kinos, Gulf of Euboea, central Greece. *Physics and Chemistry of the Earth (Part A)* **24**, 361–367.

Roberts, S., Jackson, J. A. (1991) Active normal faulting in central Greece: an overview. In: Roberts, A. M., Yielding, G., Freeman, B. (eds.) *The Geometry of Normal Faults*. Geological Society Special Publication No. 56. Geological Society, London, pp. 125–142.

Sunamura, T. (1992) *Geomorphology of Rocky Coasts*. John Wiley & Sons, Inc, New York.

Trenhaile, A. S. (2000) Modelling the development of wave-cut platforms. *Marine Geology* **166**, 163–178.

Trenhaile, A. S. (2002) Modeling the development of marine terraces on tectonically mobile rocks coasts. *Marine Geology* **185**, 341–361.

Valensise, G., Ward, S. N. (1991) Long-term uplift of the Santa Cruz coastline in response to repeated earthquakes along the San Andreas Fault. *Bulletin of the Seismological Society of America* **81**, 1694–1704.

Ventura, S. J., Irvin, B. J. (2000) Automated landform classification methods for soil landscapes studies. In: Wilson, J. P., Gallant J.C. (eds.) *Terrain Analysis: Principle and Applications*. John Wiley & Sons, Inc, New York, pp. 267–294.

Walker, R.T., Claisse, S., Telfer, M., Nissen, E., England, P., Bryant, C., Bailey, R. (2010) Preliminary estimate of holocene slip rate on active normal faults bounding the southern coast of the Gulf of Evia, central Greece. *Geosphere* **6**, 583–593.

Ward, S. N., Valensise, G. (1994) The Palos Verdes terraces, California bathtub rings from a buried reverse fault. *Journal of Geophysical Research* **99**(B3), 4485–4494.

Wood, J. (1996) The geomorphological characterisation of digital elevation models. PhD thesis. University of Leicester, Leicester.

CHAPTER 13

Environmental change and landslide hazards in Mexico

Alan P. Dykes[1] and Irasema Alcántara-Ayala[2]

[1] School of Civil Engineering and Construction, Kingston University, Kingston upon Thames, UK
[2] Instituto de Geografía, Universidad Nacional Autónoma de México (UNAM), Coyoacan, Mexico

Introduction

Natural hazards and environmental change are two issues of major concern in Mexico. As in most other populated parts of the world, environmental change in Mexico is most clearly observed or experienced as changing patterns of vegetation cover (i.e. 'land cover'), the most extreme example of which is deforestation in the southeast of the country (Alcántara-Ayala and Dykes 2010a). Unlike many other parts of the world, however, the country is highly susceptible to almost the entire range of natural hazards capable of causing major disasters for vulnerable communities. Indeed, there have been suggestions that disasters have become more frequent in Mexico over the last 35 years (Saldaña-Zorrilla 2007). This may be an artefact of an increasing vulnerability of elements of the population to natural hazards such as more houses being built in potentially dangerous locations or an increasing economic reliance on local agricultural productivity. In any case, economic indicators of the impacts of adverse natural events highlight the dominance of weather-related disasters, which primarily affect rural agricultural communities and particularly subsistence farmers, most commonly in the southern states (Saldaña-Zorrilla 2007). These include storms (direct effects, i.e. rainfall impact, erosion, wind damage), floods, mass movements (mostly triggered by storm rainfall) and, particularly in the more arid northern states, drought.

John Thornes was interested in landslides as geomorphological processes, probably as a result of his earlier collaborations on theoretical geomorphology with Denys Brunsden (e.g. Thornes and Brunsden 1977; Brunsden and Thornes 1979). He was involved with landslides research in Nepal, Brunei (with APD) and Spain (with IAA) before taking an active interest in the occurrence of landslides in Mexico in his latter years. He undertook fieldwork with IAA in November 2003 (Figure 13.1), October 2004 and March 2006 for research into the role of land-use change in the occurrence of landslides through the mechanism of altered soil water budgets, especially in the case of deforestation. Typically for John, the aim was to model the soil hydrology and corresponding changes in the factor of safety (FS) (i.e. the slope stability) based on field monitoring of changing soil moisture conditions.

This chapter explores wider issues relating to the implications of land-use and/or land cover change ('LULCC') for the occurrence of natural

Monitoring and Modelling Dynamic Environments, First Edition. Edited by Alan P. Dykes, Mark Mulligan and John Wainwright.
© 2015 John Wiley & Sons, Ltd. Published 2015 by John Wiley & Sons, Ltd.

Figure 13.1 John Thornes with IAA and a reporter at Tlatlauquitepec in Puebla, Mexico, during fieldwork in 2003. Photo: Roberto C. Borja.

hazards and disasters, particularly those involving landslides, and to assess the value – and difficulties – of land cover monitoring and slope stability modelling for landslide hazard management and mitigation in Mexico.

Background

Natural hazards in Mexico

The geography of Mexico predisposes the country to significant and potentially disastrous natural hazards. It is a large country (1.92 million km²) with the Gulf of Mexico to the east, the warm waters of which feed westward-moving Atlantic hurricanes with heat energy to maintain or even intensify them, and major tectonic plate boundaries to the south and west (Figure 13.2). South of 20°N, subduction of the Cocos Plate (Pacific Ocean floor) beneath the Mexican land mass gives rise to frequent earthquakes and the Trans-Mexican Volcanic Belt (TMVB), an east–west zone of monogenetic and persistent volcanoes that extend across the centre of the country. Large areas of central Mexico are therefore covered with volcanic sediments (e.g. Bonasia et al. 2011; Covaleda et al. 2011). North of 20°N, the western coastline is defined by the southern extension of the San Andreas Fault, a transverse plate boundary that has displaced the coastal strip of North America southwards to form the Gulf of California. Northwest of 96°W, most of the country comprises a high central plateau at elevations of 1000–2000 m. This high interior, which includes the TMVB, is separated from the east and west coasts by relatively narrow coastal plains and mountainous flanks and extends north into semi-arid latitudes where mean annual rainfall is less than 125 mm and desertification is a growing concern (Orellana 2005; Huber-Sannwald et al. 2006). The southern and south-western coastal states are mountainous due to the uplift associated with the ongoing plate collision processes, with steep slopes and deep weathering driven by (sub)tropical climates with mean annual rainfall from slightly less than 1200 mm on the Yucatan Peninsula to over 4000 mm in the Chiapas Highlands. The south of Mexico is sufficiently narrow to be entirely susceptible to weather systems and tropical storms developed over the oceans on either side.

Risk assessments and responses to these natural hazards by the relevant authorities should ideally be based on the best available scientific assessment of the existence and potential

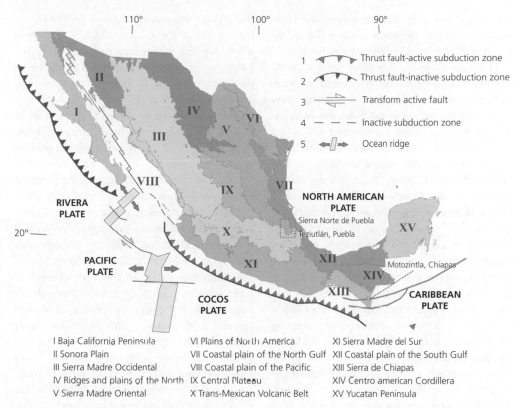

I Baja California Peninsula   VI Plains of North America   XI Sierra Madre del Sur
II Sonora Plain   VII Coastal plain of the North Gulf   XII Coastal plain of the South Gulf
III Sierra Madre Occidental   VIII Coastal plain of the Pacific   XIII Sierra de Chiapas
IV Ridges and plains of the North   IX Central Plateau   XIV Centro american Cordillera
V Sierra Madre Oriental   X Trans-Mexican Volcanic Belt   XV Yucatan Peninsula

Figure 13.2 Major physiographic and tectonic zones, also showing the locations of the study areas.

character of the possible natural hazards. Hence, there must be active and ongoing dialogue between the authorities and the scientists including the transfer of knowledge from the scientists to the authorities – and the effective communication of scientific uncertainties and their implications. This is achieved more successfully for some types of hazards for which the causal factors are well known and, at least in principle, easily explained. Tectonic hazards, that is, earthquakes and volcanic hazards, and regional-scale meteorological hazards such as hurricanes fall into this category. Local-scale meteorological hazards, such as extreme rainfall from otherwise unremarkable weather systems or localised convection cells, are more difficult to explain to non-scientists and, thus, may be afforded less attention by the authorities than is warranted by the apparent risk – although this

is changing as awareness of climate change issues improves globally (among scientists and policymakers if not the general public: Pidgeon 2012; Ratter et al. 2012). Likewise, landslides and other mass movements had often been largely ignored by officials concerned with hazard mitigation and disaster preparedness planning. This is because although individual landslide events may be common, their impacts are usually very local, and so they are recorded and managed only at a local level. Wider regional or national trends may not be apparent to national authorities. Similarly, the impacts of more widespread landsliding known as 'regional landslide events', comprising many near-simultaneous landslides triggered by unusually heavy rainfall or major earthquakes, are typically combined with those of the primary event (i.e. the earthquake or flood) in official records. Hence, the true impacts of the landslides cannot be identified, making the assessment of future landslide risk very difficult.

There are many possible strategies that can be used for landslide hazard assessment, depending on the general nature and the spatial extent of the hazard. In some locations, the landslide hazard may be a local concern such as the Super-Sauze mudslide in France (van Asch and Malet 2009) or the Valoria landslide in Italy (Ronchetti et al. 2007). At this scale, hazard assessments are normally based on detailed geomorphological and geotechnical investigations, usually with ongoing monitoring of critical variables such as ground movements and pore water pressures. Regional hazards may relate to a distinct topographic/environmental setting or geographical region (e.g. the 'flysch sector' of the Spanish Pyrenees: García-Ruiz et al. 2010). Landslide records and, ideally, a full inventory of known events would normally guide the development of appropriate geostatistical indicators using GIS techniques, incorporating factors (e.g. topography, geology, vegetation cover) for which data are available and ideally those identified as being of greater significance for landsliding in the study region. Such assessments usually take the form of spatial patterns of landslide susceptibility rather than landslide hazard, the latter implying a time-related probability of landsliding at any given location (Thiery et al. 2007; Atkinson and Massari 2011). Quantitative stability analyses may be incorporated into such GIS-based studies (e.g. Montrasio et al. 2011). In some cases, national-scale landslide susceptibility assessments may be undertaken as more extensive regional assessments, for example, Bălteanu et al.'s (2010) study of Romania's landslides. This example demonstrates the typically more qualitative approach required for such smaller map scales, relying on inventory data and expert judgements to specify the indices to be used in the spatial analysis (Glade and Crozier 2005).

A fundamental difficulty in landslide studies is that of identifying which site characteristics have the greatest influences on slope instability, not least because the critical characteristics will often vary with the prevailing causal factors. The latter may comprise the combined influences of (i) predisposing site factors such as subsurface geological structures, (ii) long-term natural preparatory factors (*sensu* Crozier 1986) such as accelerated localised weathering or river undercutting of a slope, (iii) shorter-term anthropogenic preparatory factors commonly manifest as LULCC and then (iv) the nature of the potential trigger event. These causal factors may be effective over very different spatial scales and timescales. Furthermore, it is increasingly recognised that modifications of slopes by humans may affect the long-term stability of those slopes, though to what extent and over what periods of time is often highly uncertain. The relative importance of topography, geology and soils (geotechnical controls) and vegetation cover (indirectly through different infiltration and evapotranspiration (ET) rates affecting water table positions, i.e. the focus of John Thornes' work) can be investigated by modelling the stability of individual hillslopes. This may be necessary, for representative sample hillslopes, to support regional and national landslide hazard assessments. However, if landslide susceptibility can be shown to change in response to LULCC, then monitoring of the latter must be considered a necessary, and possibly sufficient, component of a regional- or national-scale assessment.

Recent landslide disasters in context

A simple subdivision of natural hazards in Mexico is between 'weather events' and 'tectonic events'. Damaging tectonic (or 'geologic') disasters are rare but have potential to be highly destructive. There have been sixteen earthquakes since 1970 with magnitudes of 7 or more (ANSS 2010) but only one has caused nationally damaging levels of impact (Saldaña-Zorrilla 2007). The 1985 Michoacan earthquake killed 10,000 people in Mexico City (Moreno Murillo 1995), 350 km from the epicentre, and cost US$4.1 billion in total losses. However, this level of destruction arose from a previously unforeseen combination of factors, that is, the effect of a long duration of shaking on the soft unconsolidated lake sediments beneath the city (Moreno Murillo 1995), especially on buildings that were not constructed according to design codes for earthquake resistance. The 1982 eruption of the El Chichon volcano killed around 2000 people and constituted 'the largest volcanic disaster in modern Mexican history' (Klemetti 2012, webpage), a combination of unexpectedly large explosions from a volcano that had been inactive for over 600 years, the short time between the first signs of activity and the eruption and the lack of awareness of the risk among the local populations. The only other deaths from volcanic eruptions in Mexico in the 20th century were three in 1943 during Parícutin's famous inception and five climbers on Popocatépetl in 1996 (Seach 2010).

According to Saldaña-Zorrilla (2007), weather events accounted for two-thirds of the economic losses from disasters during the period 1980–2005. Ten major hurricanes during this period indicate the scale of economic losses (totalling US$6.4 billion at the 2005 value of the dollar), but a focus on hurricanes ignores many other, often even more damaging, tropical storms or undefined weather systems. A recent project that examined the direct (storms) and indirect (floods, mass movements) impacts of the 90 weather disasters recorded in

Mexico during the 20th century showed why landslides were historically overlooked as specific hazards – and demonstrates the lack of separate recording of landslide impacts (Table 13.1). Knowledge of just two events (Nos. 3 and 6 in Table 13.2), based on published accounts (Alcántara-Ayala 2004a; Caballero et al. 2006) and field visits to the affected locations by both authors of this chapter in 2003 (Sierra Norte de Puebla) and 2008 (Motozintla, Chiapas) (Figure 13.2), suggests that the data under 'mass movements' in Table 13.1, not least the monetary damage, are probably significantly under-recorded (Alcántara-Ayala 2008). Details of these locations and events are presented later in this chapter.

Alcántara-Ayala (2008) analysed Mexican landslide disasters (defined in this case as ≥10 fatalities) from 1935 to 2006. Most occurred within the mountainous flanks of the central altiplano, with a significant cluster in the Chiapas Highlands in the south, and 90% of the landslide fatalities occurred in ten events. Several significant observations can be made from these ten events: (i) two were rainfall-induced failures of engineered structures, the deaths resulting from the post-failure flood; (ii) only three of the ten were associated with hurricanes but all involved heavy rain; and (iii) two weather events (1998 and 1999) each caused two separate landslide disasters. Later sections of this chapter focus on events 3 and 6.

Table 13.1 Disasters associated with storms, floods and mass movement processes in Mexico from 1900 to 2000.

	Floods	Mass movement processes	Storms	Total
Number of disasters	35	5	50	90
People killed	3938	192	4925	9,055
People injured	569	0	1749	2,318
People affected	1,269,845	0	2,173,915	3,443,760
Homeless	162,990	120	307,450	470,560
Total affected	1,433,404	120	2,483,114	3,916,638
Damage (US$ × 1000)	1,529,800	0	5,457,110	6,986,910

Based on the OFDA/CRED database (EM-DAT undated)). Alcántara-Ayala and Dykes 2010b. Reproduced with permission of John Wiley & Sons.

Table 13.2 The ten most disastrous landslides in Mexico from 1935 to 2006 (after Alcántara-Ayala 2008).

Rank	Locality	State	Event type*	Deaths	Date	Source
1	La Paz	Baja California Sur	Dam failure (Hurricane Liza)	1000	1 October 1976	*Excelsior* and *El Universal* newspapers
2	Minatitlan	Colima	Regional landslide event (Hurricane 15)	871	27 October 1959	Lugo and Flores (pers. comm., 1997, based on *Excelsior* newspaper)
3	Sierra Norte de Puebla	Puebla	Regional landslide event	247	4–6 October 1999	SEPROCI (pers. comm., 1999)
4	Acapulco	Guerrero and Oaxaca	Regional landslides and flash floods (Hurricane Pauline)	228	8–9 October 1997	CENAPRED (1997)
5	Tlalpujahua	Michoacan	Tailings dam failure associated with heavy rains	176	27 May 1937	Lugo and Flores (pers. comm., 1997, based on *Excelsior* and *El Universal* newspapers)
6	Motozintla	Chiapas	Regional landslide event (Tropical Storm 'Earl')	171	6–12 September 1998	*La Jornada* newspaper
7	San Pedro Atocpan	Distrito Federal	Regional landslide event	150	4 June 1935	Lugo and Flores (pers. comm., 1997, based on *Excelsior* and *El Universal* newspapers)
8	Valdivia	Chiapas	Regional landslide event (Tropical Storm 'Earl')	150	6–12 September 1998	*La Jornada* newspaper
9	Atenquique	Jalisco	Regional landslides, mudflows and floods	100	16 October 1955	OFDA–CRED database; Lugo and Flores (pers. comm., 1997, based on *Excelsior* and *El Universal* newspapers)
10	Papantla	Veracruz	Regional landslides, mudflows and floods	60	6 October 1999	DesInventar database according to *La Jornada* newspaper

Alcántara-Ayala 2008.

*Regional landslide events were all triggered by 'heavy' or 'torrential' rains.

Land-use/land cover change

The causes, impacts and implications of the regional landslide event in the Sierra Norte de Puebla in 1999 were investigated subsequently as a consequence of the nature and scale of this disaster (Alcántara-Ayala 2004a, 2004b; Alcántara-Ayala et al. 2004, 2006) and its occurrence just a year after a similar disaster in Chiapas. A particular focus of this research was the specific role of LULCC as a factor controlling the incidence of landslides during the rains of October 1999. Specific influences of vegetation on the stability of individual slopes, such as the beneficial effect of plant transpiration removing water from a hillslope (Greenway 1987), and the ecological implications of landslides removing small patches of vegetation (e.g. Myster et al. 1997) have long been known. Widespread shallow landsliding has been reported to follow deforestation of hillslopes and conversion to farmland in many parts of the world (e.g. the Pyrenees, Spain – García-Ruiz et al. 2010; the Western Ghats, India – Kuriakose et al. 2009; Colombia – López-Rodríguez and Blanco-Libreros 2008). García-Ruiz et al. (2010) noted that the greatly increased incidence of shallow landslides and debris flows continued long after the agricultural activities had been abandoned on the land. Conversely, significant loss of forest cover has also been caused by regional landslide events triggered by rainfall or earthquakes (Table 13.3). Thus, in addition to the potential damage to or loss of homes and infrastructure, there is perhaps a case for including the economic and ecological impacts of landslides resulting from loss of agricultural or forested land in landslide hazard assessments.

Throughout the world, the most widespread and influential type of LULCC is deforestation because wood can provide both shelter and fuel and removal of a forest provides land for agriculture. However, it is often extremely difficult to identify specific causes and implications in any given region because they involve complex interactions between the deforestation (or other vegetation cover change) and biodiversity, climate change, environmental degradation and socio-economic context. For example, removal of forest to create a new pasture may be driven by a significantly increased incidence of inundation of floodplain pastures due to more frequent rainfall and other natural or anthropogenic environmental changes further upstream in the catchment, rather than a direct need to harvest the timber crop or provide additional pasture. Deforestation of the upper slopes of catchments has, in recent years, been identified as a cause of more frequent flooding incidents in Mexico (Mas et al. 2004). It is important to understand the causes if associations between forest removal and increased incidence of floods and landslides are to be identified, in order to inform future policy relating to risk reduction (Velázquez et al. 2010).

Any such work also needs to be set in the context of the wider need to quantify and understand LULCC, including (i) assessment of the effects and implications (adverse or beneficial) with respect to the atmosphere, biosphere and hydrosphere; (ii) assessment of national or regional bioecological resources; and (iii) assessment of natural and environmental hazards resulting from LULCC and development of mitigation measures (Alcántara-Ayala and Dykes 2010a). The apparently increased susceptibility of hillslopes to landslides following a reduction of vegetation cover due to land-use change clearly falls within the third of these actions and is therefore of immediate concern within Mexico. In any case, global accounting of greenhouse gas fluxes under the Kyoto Protocol requires that LULCC generally and losses or gains of forests specifically must be accurately quantified by all non-industrialised countries working on joint projects with industrialised nations. The inadequacies of existing LULCC data are well known and cause difficulties for climate treaty negotiations as well as associated scientific endeavours (Cairns et al. 2000; Velázquez et al. 2003; Höhne et al. 2007), but the collection and utilisation of this data were, until recently, not just expensive but also technically difficult (Rindfuss et al. 2004).

Table 13.3 Impact of landsliding on vegetation cover (after Alcántara-Ayala et al. 2006).

Date	Location	Triggering mechanism	Impact on land cover	References
1935	Papua New Guinea (Torricelli Range)	Earthquake $M=7.9$	130 km² of vegetation removal	Marshall (1937), Simonett (1967) and Garwood et al. (1979)
1960	Chile (Valdivian Andes)	Earthquake $M=9.2$	More than 250 km² of temperate forest lost	Veblen and Ashton (1978)
1970	Papua Guinea (Adelbert Range)	Earthquake $M=7.9$	Vegetation stripped from circa 25% of the surface	Pain and Bowler (1973)
1976	Panama (southeast coast)	Earthquakes $M=6.7$ and $M=7.0$	Removal of about 54 km² of jungle	Garwood et al. (1979)
1982	USA	Mount St. Helens eruption and landslide	550 km² of forest lost	Collins and Dunne (1988)
1987	Ecuador	Reventador earthquakes $M=6.1$ and $M=6.9$	More than 75% of the rainforest of the south-western slopes of Reventador volcano and 230 km² of natural forest lost	Nieto et al. (1991) and Schuster et al. (1996)
1994	Colombia	Paez earthquake $M=6.4$	250 km² of second-growth subtropical brush and forest lost	
1998	Honduras, Nicaragua, El Salvador, Guatemala	Hurricane Mitch (floods and landslides)	>16,000 km² of forest damaged, mostly in protected areas or areas scheduled for protection, 80% of this area in Honduras	ECLAC (1999a,b,c) and CEPAL (1999)
1999	Venezuela	Floods and landslides	33,503 ha of vegetation cover lost	CEPAL (2000)

Reproduced with permission from Alcántara-Ayala et al. (2006). © Elsevier.

Ultimately, adequate monitoring of LULCC on any national or regional basis relies on the collection and accurate classification of appropriate remotely sensed data that can then be analysed and communicated using a GIS framework (e.g. Palacio-Prieto et al. 2000; Read and Lam 2002; Vance and Geohegan 2002; Velázquez et al. 2010). Recent tools such as the International Center for Tropical Agriculture's Terra-i (CIAT 2013), an interactive website providing near real-time monitoring of deforestation in the American tropics, can provide some elements of LULCC data easily and at no cost. However, this resource demonstrates the ongoing difficulty of the trade-off between temporal and spatial resolution (e.g. Brown et al. 2013), and discrimination between vegetation or land cover types, or indeed contexts of particular land covers such as 'grass', remains problematic even at spatial resolutions higher than Terra-i's 250 m pixels (e.g. Wickham et al. 2013). Higher-resolution data are now also freely available such as the interactive 30 m mapping of global forest cover change between 2000 and 2012 (Hansen et al. 2013) although clearly the temporal resolution is very limiting in this case.

The extensive work done in Mexico to develop methodologies for improved assessments of LULCC has highlighted some potentially significant general trends, particularly with respect to changing regional susceptibilities to natural hazards in the tropical states (Alcántara-Ayala and Dykes 2010b). The National Forest Inventory (NFI), first produced in 2000, represents the new standard for ongoing LULCC monitoring in Mexico. Despite some difficulties distinguishing between different forest categories, including a general tendency to underestimate tropical forest losses but overestimate temperate forest losses, the NFI was found to be 64–78% accurate when tested against four sample regions in southern Mexico (Couturier et al. 2010).

Landslide disasters and slope stability

This section examines the details of the two disasters highlighted earlier, that is, events 3 and 6 in Table 13.2. It summarises the nature of the events and the initial research into the causes and characteristics of the natural processes that gave rise to the socio-economic impacts, then attempts to establish the stability conditions of example landslides from each event. In doing so, the difficulties of modelling slope stability and of assessing potential landslide hazards in relatively data-poor locations are examined.

Sierra Norte de Puebla, October 1999
Background

The Sierra Norte de Puebla is the mountainous north-eastern part of Puebla Province, comprising a region of approximately 8000 km² that lies 120–200 km ENE from Mexico City and 70–150 km inland from the Gulf of Mexico. It occupies the transition zone between the Sierra Madre Oriental, that is, the mountain range that separates the coastal plains from the high central plateau, and the TMVB. Altitudes range from around 400 m to a maximum of 4282 m, with local relief of up to 1000 m. Mean annual precipitation varies throughout the region as indicated by the examples in Table 13.4.

The hillslope materials in the Sierra Norte de Puebla have been strongly determined by Mexico's volcanoes. The rocks of Sierra Madre Oriental are dominated by Mesozoic sedimentary formations but increasingly covered with volcanic deposits towards the TMVB. The volcanic belt comprises Late Tertiary and Quaternary stratovolcanoes and associated suites of landforms and deposits. Alcántara-Ayala (2004a) indicated

that a large proportion of the Sierra Norte region is underlain by ignimbrites, often very sandy and poorly consolidated. Exposures we observed in 2003 at Zacapoaxtla and Teziutlán demonstrated considerable heterogeneity within these materials, even at very small scales (e.g. Figure 13.3). Alcántara-Ayala (2004b) highlighted permeability contrasts in these materials between different ash-fall layers and, in some places, interbedded palaeo-soils. Elsewhere, limestones and lutites (argillaceous sedimentary rocks) constitute the bedrock, modified by varying degrees of weathering that result in high clay contents in hillslopes on lutites in particular.

Several hundred landslides occurred across an area of the Sierra Norte exceeding 5000 km² during 4–7 October 1999 as a cold front combined with a tropical depression from the Gulf of Mexico to produce damaging amounts of rainfall over the provinces of Puebla, Veracruz and Hidalgo. Although flooding and landslides occurred throughout these provinces, Puebla Province was affected most severely due to the very high frequency of landslides with 263 people killed, 1.5 million people (30% of the population of the province) affected and total damages exceeding US$450 million (Bitrán and Reyes 2000). Rainfall totals and patterns during the event were analysed by Alcántara-Ayala (2004a). However, to illustrate the scale of the event, at Teziutlán 300 mm of rain fell on 4

Table 13.4 Mean annual rainfall at meteorological stations in the Sierra Norte de Puebla.

Rain gauge	Location of town	Median altitude of town (m)	Mean annual rainfall (mm)
Teziutlán	97°21'40"W, 19°48'50"N	1950	1593
Zacapoaxtla	97°35'20"W, 19°52'20"N	1800	1421
Huauchinango	98°03'30"W, 20°10'20"N	1550	2277

Rainfall data from National Meteorological Service (SMN).

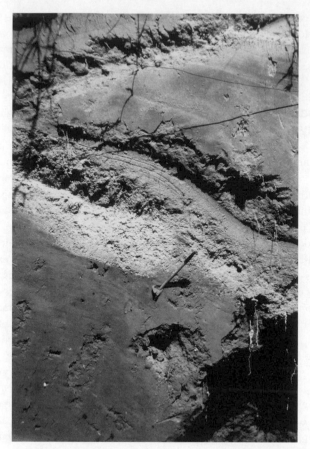

Figure 13.3 Excavated section through an ignimbrite deposit at Zacapoaxtla in 2003 (pen for scale). Two different units of silty–clayey materials are indicated by the smooth cut surfaces of different colours. These units are separated by friable non-cohesive sandy layers. Photo date: October 2003.

October, 360 mm fell on the 5th, and 743 mm fell in total over the four days of the storm (844 mm at Zacapoaxtla). Mean 24 h rainfall intensities of 15.8 mm h^{-1} at Zacapoaxtla and 14.2 mm h^{-1} at Teziutlán were recorded, although intensities over some shorter periods will have been significantly higher.

Most of the landslides were shallow and either involved ignimbrite deposits or occurred in the contact area between sedimentary and volcanic materials (Alcántara-Ayala 2004b). However, some larger, structurally controlled failures also occurred (Alcántara-Ayala 2004a), and at least one landslide dam was formed by a 48 million m^3 failure of pyroclastic rocks that blocked the

Zempoala River (Alcántara-Ayala et al. 2006). The single most damaging landslide occurred below the La Aurora cemetery, Teziutlán (19°49′09″N, 97°21′12″W) (Figure 13.4), which caused around 120 fatalities (figures of 109 or 130 are often cited). This example is analysed in the following text.

Land use

Alcántara-Ayala et al. (2006) used Mexico's NFI, in conjunction with satellite imagery, to analyse patterns of landsliding during the October 1999 rainfall event in the Sierra Norte de Puebla. A bias towards failure of crop-covered slopes is clear, but if 'forested' and 'non-forested' slopes

Figure 13.4 Looking up the site of the La Aurora landslide, Teziutlán. The dashed white line indicates the upper extent of the landslide source area. The white arrow identifies the new CENAPRED monitoring station (see Figure 13.10). Photo date: October 2003.

Table 13.5 Landslide frequencies associated with land cover types (derived from the NFI) in the Sierra Norte de Puebla following the 1999 rainfall event (after Alcántara-Ayala et al. 2006).

Land use	Area (km²)	Area (%)	Landslides (number)	Landslides (%)
Cropland	1213.51	38.9	145	49.2
Temperate forest	1179.52	37.8	98	33.2
Grassland	674.55	21.6	29	9.8
Tropical forest	37.04	1.2	8	2.7
Urban	12.57	0.4	15	5.1
Total	3117.19	100	295	100

are considered, then the frequency of landslides accords almost exactly with the respective land area proportions (36% of the landslides occurred within the 39% of land covered with forest). The categories shown in Table 13.5 hide many variables that may be of significance for slope stability, such as whether the cropland is irrigated, or the elapsed time since the change from forest to grassland. However, further work established that the density of the vegetation cover was much more strongly associated with landslide occurrence, with 72% of landslides occurring on slopes with less than 10% vegetation cover (i.e. effectively bare ground). By comparing satellite imagery from 1989 (Landsat Thematic Mapper, 30 m resolution) and 1999 (Landsat ETM+, 30 m resolution), the proportion of the 5600 km² study

area with <10% cover increased from 48 to 63%. It was inferred that LULCC reduced the density of vegetation cover, most commonly in the form of forest removal that resulted in bare or sparsely vegetated ground (Alcántara-Ayala et al. 2006), and that this gave rise to many of the landslides.

Landslide investigations

Regional assessments of landslide hazards undertaken using GIS techniques can explicitly incorporate quantitative slope stability analyses, but there are limitations to this approach. Some of these limitations were highlighted by Alcántara-Ayala (2004a) with respect to her investigation of the 1999 Sierra Norte event. In particular, the analyses can only be applied to grid cells of a raster GIS, which means that the grid resolution is a critical factor. If the cells are very large (e.g. 100 m), then three problems arise. Firstly, it is generally the case that the larger the cells in a digital elevation model (DEM), the lower the steepest slope gradients associated with each cell. This is because large cells represent the average gradient across each cell, and very steep slope segments will not be represented. Secondly, related to the previous point, a GIS-based regional stability analysis will generally not indicate the stability of individual slope segments. Only if a grid cell falls entirely within a slope of greater extent will a realistic indication of slope stability be obtained. In the Sierra Norte, most critical landslides were of greater extent than a 100 m grid cell so this limitation was not a major constraint (Alcántara-Ayala 2004a). The development and application of new technologies in recent years have been such that greatly improved data resolution and quality are now much more widely available such as the 30 m global DEM released to the public by NASA's Jet Propulsion Laboratory in 2011 (ASTER 2011). Indeed, high-resolution LiDAR data now exist (as of March 2013) for the Sierra Norte de Puebla that will facilitate new research into improved strategies for landslide analyses and hazard assessments in this region.

The third problem is that of adequately representing the geotechnical properties of the slopes. It is not uncommon for investigations of individual landslides, including stability analyses, to remain quantitatively inconclusive because actual conditions within the failed slope (material properties, structural features, pore water pressures) may be highly variable and impossible to know. The potential complexities arising from regional variations and local juxtapositions of these conditions, with respect to their topographic contexts, mean that a regional analysis can only ever provide an indication of relative landslide hazard. In many cases, the use of indicative generalised parameters to represent the necessarily simplified range of materials identified in the GIS may provide useful results

because local variations in the gross material properties will probably average out at larger scales. Hence, Alcántara-Ayala (2004a) found that this approach produced a map of landslide susceptibility that was found to be a good approximation when validated against the field inventory of the 1999 landslides.

The general difficulty is highlighted by an attempt to analyse the landslide that occurred below the La Aurora cemetery, Teziutlán, on 5 October 1999. Similar to many locations in the Sierra Norte, this site comprised a very thick deposit of pyroclastic materials that forms part of a distinct though dissected ridge within a much larger, older valley. The upper surface of the ridge is broad and gently sloping towards steeper (20–40°) lateral slopes. The upper part of the east-facing lateral slope failed over a length of 100 m and at a mean depth of 4.4 m. The high water content of the failed mass caused the initial slide to break down into a rapid mudflow involving 7350 m^3 of material (Alcántara-Ayala 2004a), the relatively small volume having a disproportionate impact because of the high density of poor quality housing that covered the slope. The hillslope materials were described by Alcántara-Ayala (2004a, p. 23) as 'poorly consolidated ignimbrite deposits with a high sand content'. Limited laboratory analyses of ignimbrite samples obtained by Mendoza et al. (2000) provided parameter values for the GIS-based spatial modelling of landslide hazards discussed previously; their data are used in this study. These data and descriptive information are broadly consistent with reports of investigations of other ignimbrite slopes (Calcaterra and Santo 2004; Fiorillo and Wilson 2004; Frattini et al. 2004; Chigira and Yokoyama 2005; Crosta et al. 2005; Table 13.6), although the frictional strength of the Teziutlán material is lower suggesting a more weathered state.

We set out to analyse the La Aurora landslide, in terms of its stability condition at the time of failure, using all of the available (extremely limited) data and making assumptions as necessary. The field evidence indicated that this was a

Table 13.6 Examples of geotechnical characteristics of volcanic deposits.

Location	Source	Sand content	Unit weight, γ (kN m^{-3})	Friction angle, Φ' (°)	Cohesion, c' (kPa)	Porosity	Saturated hydraulic conductivity (m s^{-1})	
Teziutlán, Mexico	Mendoza et al. (2000)	'High'	10.8	20.5	11.7	*	*	'Poorly consolidated ignimbrite'
Pozzano, Campania, Italy	Calcaterra and Santo (2004)	45% sand/gravel	12.5 (7.8 dry)	27–30	20 (0 when saturated)	0.69		'Campanian Ignimbrite formation' ('grey pumice and volcanic ashes')
Western Campania, Italy	Fiorillo and Wilson (2004)	60%	8.7 (dry)			0.67	10^{-7} to 10^{-6}	'Ash level, ignimbrite deposits'
Sarno, Campania, Italy	Frattini et al. (2004)			36–45 (pumice), 38–45 (soil)	0–20 (pumice), 0–34 (soil)	0.67	10^{-8} to 1 (pumice), 10^{-7} to 10^{-5} (soil)	'Pyroclastic soils' – pumice and buried soil
Kagoshima Prefecture, Kyushu Is., Japan	Chigira and Yokoyama (2005)	78–92%	10.8	30–35 (weathered), 38–40 (unweathered)	Up to 10 (weathered), up to 20 (unweathered)	0.55		'Non-welded and unconsolidated ignimbrite'
Santa Tecla, El Salvador	Crosta et al. (2005)	30–63% sand/gravel	14.9 (11–16 dry)	30–35	60–80			'Pyroclastic materials'

*Porosity assumed to be 0.67 and saturated hydraulic conductivity assumed to be 1×10^{-6} m s^{-1}, estimated from Frattini et al. (2004), Fiorillo and Wilson (2004) and Calcaterra and Santo (2004).

dominantly translational landslide, so we initially analysed the hillslope using the infinite slope model (Skempton and DeLory 1957). A mean slope of 27° and failure depth of 4.4 m were used with the shear strength parameters shown in the first row of Table 13.6. The FS was 0.68 if the slope was fully saturated to the surface, but the water table only needed to be 2.2 m below the ground surface for failure to have occurred (FS = 1.0). When back-analysing naturally occurring shallow translational landslides, it is normal practice to assume that failure occurred when the water table was at the ground surface, that is, under the worst-case (hydrostatic) pore water pressure conditions (e.g. Francis 1987). This is because if the slope was not fully saturated to the surface, there arises a difficulty of explaining why it did not fail on a previous occasion when the water table may have been closer to the surface. Alcántara-Ayala (2004a) highlighted the role of permeability contrasts between clay-rich and more sandy layers within the pyroclastic materials, with the higher permeabilities of the sandy layers promoting infiltration and the formation of perched water tables over the clayey layers. It is possible that the permeability of the sandy layers is sufficiently high to prevent the build-up of such perched water tables except in the most extreme rainfall conditions, but the more likely explanation for these FS results is that the slope failure was controlled by 3-dimensional subsurface variations in the heterogeneous properties of the pyroclastic materials and associated groundwater conditions that cannot be represented by an infinite slope analysis.

Our subsequent attempts to model the La Aurora landslide in more detail highlighted the limitations of slope-specific analyses and the potential value of the simpler regional-scale approach to landslide hazard assessments when adequate data are limited or unavailable. We used the commercial SEEP/W software to investigate the subsurface pore water pressure distributions. This is a finite element model that allows the simulation and analysis of subsurface

saturated and unsaturated water flow problems. The outputs from this software can be imported directly into the companion SLOPE/W slope stability model to provide explicit hydrological parameters for stability analyses. In the absence of more appropriate alternatives, hydrological parameters were estimated from the Italian data (Table 13.6). For a finite element seepage analysis, approximate functions are normally sufficient to provide reasonable estimates of likely pore water pressure distributions (Geo-Slope 2009), hence a mean soil moisture characteristic curve of those presented by Frattini et al. (2004) was used with a saturated hydraulic conductivity (K_{sat}) of $1 \times 10^{-6}\,m\,s^{-1}$ (estimated from Frattini et al. (2004) and Fiorillo and Wilson (2004); Table 13.6). Modelling the entire hillslope down to the stream at the bottom, to provide a lower boundary condition, and using rainfall records to specify appropriate infiltration inputs to the slope, we were unable to produce any pore water pressure distributions that would produce a slope failure approximating the observed landslide. This was unsurprising because (i) the K_{sat} order of magnitude was an estimate, (ii) we could not establish antecedent moisture conditions due to the unavailability of adequate ET data and (iii) there is no information regarding the subsurface geometry or even simply the depth of the pyroclastic materials forming the hillslope. Simplifying conditions in order to model them is appropriate and, in most cases, necessary but at the same time is only possible if the constraints and degrees of any uncertainties are known.

At Teziutlán, there were two fundamental difficulties. Firstly, the estimates of either K_{sat} or ET, the latter estimated from field experience related to literature, could be tolerated if the other was better constrained, but estimates of both in combination only multiply the overall uncertainty in the results. Secondly, it is not known if the broad ridge that runs through the town is a ridge of Mesozoic bedrock with a superficial cover of pyroclastic materials up to perhaps a few metres thick or if the entire ridge mass comprises an ignimbrite unit infilling part of a much older valley. This factor will fundamentally influence subsurface flow pathways and distributions and indeed the geometry and type of slope failure that can occur. Hence, in the absence of subsurface information and detailed geotechnical data, future landslide risk planning and management in this region are probably best served by improved regional GIS-based landslide hazard assessments that can incorporate topography, LULCC data, empirical threshold rainfall coefficients and other relevant data as available (Alcántara-Ayala 2004a; Alcántara-Ayala et al. 2006).

Motozintla, Chiapas, September 1998
Background

Motozintla is a town with a population of around 50,000, located 52 km north of Tapachula and 64 km inland from the Pacific coast. The border with Guatemala is 9 km to the SE. It occupies the floor of a deep, steep-sided catchment at an altitude of 1250–1300 m, where three rivers meet. The head of the catchment 9 km to the WNW is up to 2600 m above sea level, and the local relief is 500–800 m (Figure 13.5). Meteorological data comprising monthly mean maximum and minimum temperatures, and monthly mean rainfall, with a few missing entries, were available from two

locations near Motozintla. The Buenos Aires data (station 7333, 15°19′60″N, 92°16′05″W, 1720 m altitude, available records from 1980 to 2004 inclusive) are from a site 4 km SSW from Motozintla, and the Motozintla North data (station 7119, available records from 1922 to 2004 inclusive) actually come from 15°24′58″N, 92°14′31″W, 4 km north of the town at 1728 m altitude in the Chimalapa Valley. Mean monthly data were calculated for both stations, and then the means for the two stations were used for subsequent analyses (Table 13.7).

Most of the catchment surrounding Motozintla is underlain by granites of the Permian to Triassic age Chiapas Batholith and two younger granite stocks of limited spatial extent (Caballero et al. 2006). Consequently, the hillslope materials are dominated by deeply weathered granite with very thin surficial soils (up to 0.2 m) and occasional colluvium deposits (Figure 13.6). Caballero et al. (2006) obtained geotechnical parameters for several samples of weathered rock from the head of the catchment west of Motozintla. They described the three granite samples as follows (pp. 109–110):

MZ54. Granite, highly weathered and partially transformed in saprolite. Very blocky structure (element 5–20 cm) filled with fine matrix

Figure 13.5 The catchment above Motozintla, Chiapas, looking west. The town is located in the bottom of the main valley (beyond the farm) between the black arrows. Photo date: December 2008.

partially cohesive. Blocks can be crumbled with firm blow or perforated with the point of the geological hammer. ... Stable slopes appear to be approximately 34°.

MZ55. Weathered granite. Fine blocky rock mass alternated with some zones of larger elements (5–15 cm). Highly weathered rock element. Discontinuous veins and lenses of quartz (10–15 cm width).

MZ58. Highly weathered granite. Residual elements (5–10 cm), Abundant filling of fine yellow matrix. Fine blocky structure extremely disordered. ...

Exposures that we examined throughout the catchment in December 2008 were consistent with these descriptions, typically comprising grade III–IV weathered granite immediately beneath the surface soil, except on some of the highest ridgetops (e.g. 1.5 km west of El Carrizal) where the rock was almost entirely broken down to sand and gravel (grade V). The soft stone fragments in the grade III–IV materials were not cemented to each other, and where matrix material was found, it was dominantly sandy with little discernible clay. Below about 1 m depth, the rock was typically grade II, with strength largely controlled by the characteristics of the matrix material within the many fractures.

Many landslides occurred in the catchment surrounding Motozintla during 7–8 September 1998 as tropical storm 'Earl' moved across southern Mexico. This event caused landslides and floods throughout southern Chiapas, killed 214 people with over 300 others reported missing,

Table 13.7 Summary climatic data for Motozintla.

Month	Mean minimum temperature (°C)	Mean maximum temperature (°C)	Mean rainfall (mm)
January	11.7	28.9	2
February	12.3	29.8	4
March	13.5	30.9	7
April	14.8	31.3	19
May	15.2	31.1	69
June	15.5	29.4	207
July	14.9	29.5	151
August	14.8	29.5	169
September	14.9	28.8	214
October	14.6	28.6	105
November	13.4	28.5	20
December	12.3	28.6	5
Total			972

Source: National Meteorological Service (SMN).

Figure 13.6 A deep road cut 8 km east from Motozintla showing the upper 8 m of soft, easily excavated weathered granite that has been buried from the south (right of image) by a major colluvium deposit. Note the person (bottom centre) for scale. Photo date: December 2008.

destroyed parts of the Pan-American Highway over a distance of more than 400 km and isolated Motozintla for more than a month afterwards. Motozintla was probably the most severely affected settlement, where most of the deaths were caused by floodwaters laden with sediment from mass movements within the catchment (Caballero et al. 2006). The storm was recorded in daily rainfall data at the two rainfall stations identified previously (Table 13.8). Caballero et al. (2006) used the Motozintla North data to identify a threshold for landsliding according to the mean rainfall intensity since the start of the storm. They showed this to be around $4.5\,mm\,h^{-1}$ ($1.3 \times 10^{-6}\,m\,s^{-1}$). However, the mean rainfall intensity during 7 September alone was $7.3\,mm\,h^{-1}$ at Motozintla North and $11.3\,mm\,h^{-1}$ at Buenos Aires on 6 September. Peak intensities are likely to have been higher than these mean values for short periods of time as observed elsewhere (e.g. Dykes 2000), but they cannot be reliably estimated in the absence of high-resolution rainfall measurements.

Most of the mass movements were associated with cut and fill slopes along roads (Figure 13.7), but some significant landslides occurred on natural slopes as discrete single failures or as sequential failures associated with intense (or prior) gully incision. An example of the latter case was the considerable expansion by mass failure of an antecedent gully on a 32° slope to leave a landslide scar 120 m long, up to 50 m wide and at least 15 m deep involving the total displacement of around 45,000 m³ of material (Figure 13.8). This failure, north of the intersection of Mirador and Avenida Huanacastle in the 'Alianza 2000' district of the town (15°22′20″N, 92°15′04″W), buried a church and several houses though without casualties (according to a resident of a house that the main landslide just missed). An older landslide 1100 m away to the ENE, on a 31° slope at a place identified in Google Earth as San Pablo (15°22′33″N, 92°14′30″W), was wider and deeper (up to 20 m) with a volume exceeding 70,000 m³ (Figure 13.8).

Land use

In Chiapas, analysis of LULCC using the NFI highlighted significant areas of revegetation or 'recovery' of lands originally cleared of forests for agriculture by emigrant peasants and then abandoned (Flamenco-Sandovala et al. 2007). However, the revegetation typically comprises secondary scrub species that resemble forest in satellite imagery. Between 1976 and 2000, five times more land lost 'undisturbed woody vegetation' (i.e. original forest) than regained such forest from disturbed or previously agricultural land (Velázquez et al. 2010). Furthermore, the Grijalva–Usumacinta hydrological region, within which the Motozintla catchment lies, is one of the seven regions that (i) have less than 50% of their original natural forest cover remaining and (ii) suffered the worst flooding during the first decade of the 21st century (Velázquez et al. 2010). The detailed study of deforestation trends in southeast Mexico by Díaz-Gallegos et al. (2010) found that twice as much forest was lost to grasslands and irrigated agriculture in Chiapas state than to slash-and-burn agriculture during the period 1978–2000. However, analysis by hydrological region rather than state sometimes

Table 13.8 Daily rainfall data for the storm event of September 1998.

Date (1998)	Motozintla North	Buenos Aires
1–15 August	42.5	97.9
16–31 August	158.4	44.7
1 September	11	0
2	0	0.5
3	2	10
4	6	0.4
5	5	80
6	18	270
7	175	30
8	130	20
9	27	40
10	10.1	35
11	0	40
12	0	45

Source: National Meteorological Service (SMN).

Figure 13.7 Landslides at the head of the catchment above Motozintla. Two of the landslides are directly related to a main road (white arrows indicate the position of the road), and one is related to a local farm road (black arrow). Photo date: December 2008.

Figure 13.8 View north across the downstream half of Motozintla, showing landslide-dominated hillslope forms including the Mirador (white arrow) and San Pablo (black arrow) landslides discussed in the text. Photo date: December 2008.

revealed different patterns of change. A cursory inspection of Google Earth imagery reveals the Motozintla catchment to have only very limited forest cover (of unknown category: Figure 13.5) compared with the adjacent catchments to the west that fall within the Costa de Chiapas hydrological region. The timing of the deforestation relative to the occurrence of any potentially damaging rainfall events prior to 1998 is not known, but the abundant landslide forms visible on all hillslopes are of ages varying from a few months or years to many decades or centuries. The question relating to the Motozintla catchment is therefore whether LULCC may have altered the frequency, or indeed the dominant type, of landslides relative to the natural 'background' patterns inherent in the long-term landform development context. Findings from the Sierra Norte de Puebla suggest that this may be the case.

Landslide investigations

We used the SEEP/W and SLOPE/W software to investigate the controlling factors that gave rise to some of the more damaging landslides at Motozintla. This allowed the influences of various combinations of site and event conditions to be explored despite the general absence of field data. It was not expected that definitive failure conditions would be identified, but we aimed to develop at least some general ideas that might enhance any future landslide hazard assessments in this part of Mexico. In back-analysing failed slopes, we know that the $FS \leq 1.0$ so the approach used is to adjust the model configuration or parameter values to give $FS = 1.0$.

Our field observations in 2008 and Caballero et al.'s (2006) descriptive information are broadly consistent with reports of investigations of other weathered granite slopes (e.g. Lumb 1975; Ng and Shi 1998; Chen and Lee 2004; Kim et al. 2004). This study therefore uses Caballero et al.'s geotechnical data with hydrological parameters estimated from these other published accounts (Table 13.9). Soil moisture characteristic curves are presented by Ng and Shi (1998) and Kim et al. (2004), and a composite was derived from these using an assumed porosity of 0.4 and a K_{sat} of $1 \times 10^{-5}\,m\,s^{-1}$. The slope that failed in 1998 above Mirador (Figure 13.8) was modelled using a slope profile geometry derived from the 1:50,000 topographic map and cross-checked against aerial imagery (including Google Earth) and field observations and photographs (the gradient measured in the field was 32°). Thus, the failed slope was represented as a planar segment of 30.6° over a horizontal distance of 135 m, with 17° segments above and below the steep critical slope.

Initial hydrology simulations indicated that whatever boundary condition was specified for the outflow from the modelled slope, it had an undue influence on the water conditions throughout the modelled slope. The geometry for the analysis was therefore modified to include the low gradient apron of colluvium

(assumed porosity of 0.3 and a K_{sat} of $1 \times 10^{-4}\,m\,s^{-1}$) between the valley side slope and the river channel, with the water level in the river providing the lower boundary condition for the modelled slope.

Mean monthly data were calculated for both rainfall stations, and then the means for the two stations were used for subsequent analyses (Table 13.7). The mean monthly ET was estimated using the FAO Penman–Monteith (P-M) method (Allen et al. 1998), utilising all of the approximations necessary to accommodate the limited available data identified earlier. However, the annual 'reference ET', that is, from a hypothetical standard grass crop, has been shown to be significantly overestimated if only temperature data are used (Sentelhas et al. 2010). Furthermore, the soil moisture deficit that develops during the dry season will cause the actual ET to be reduced (e.g. Wilske et al. 2010). Quantifying this effect is difficult in the absence of appropriate measurements. The FAO Irrigation and Drainage Paper 56 (Allen et al. 1998) presents a method for estimating the actual 'crop ET' by applying correction factors to the reference value, but the uncertainties in the FAO P-M method combined with the nature of the hillslope modelling meant that such a rigorous approach was not appropriate (or even necessary) for the present study. Hence, the monthly mean 'reference ET' estimated using the FAO P-M method, having been used to provide a mean daily ET for each month, have been arbitrarily reduced to reflect both the overestimation (maximum daily values reduced by 1.0 mm) and the dry season soil moisture deficits (daily values reduced in multiples of 0.5 mm to minima of 1.0–1.5 mm (Giambelluca et al. 2009) to give net monthly infiltrations in the range −20 to −40 mm). Ideally, the effects of land cover change on slope stability would be investigated by applying different ET to the simulated rainfall responses within the modelled hillslope, based on crop correction factors in the FAO P-M method. However, the limitations of the available data (particularly relating to actual

Table 13.9 Examples of geotechnical characteristics of granite slopes.

		Sand content	Unit weight, γ (kN m⁻³)	Friction angle, Φ' (°)	Cohesion, c' (kPa)*	Porosity	Saturated hydraulic conductivity (m s⁻¹)	
Motozintla, Mexico	Caballero et al. (2006)		24	30.6	12	†	†	'Highly weathered granite'
Hong Kong Island and Kowloon Hills, Hong Kong	Lumb (1975)	75% (dominantly coarse sand)		35.6 (30.5–40.6)	3.8 (wet, medium) – 74.1 (dry, dense)	0.33–0.45 (dense–medium)	8×10^{-6}	'Decomposed granite'
Mid-levels, Hong Kong	Ng and Shi (1998)			38	10 (Φ^b=15°)	0.38	4.8×10^{-5}	'Clayey sandy SILT colluvium'
Hong Kong	Chen and Lee (2004)			33–40	0–70,000		4×10^{-7} to 2×10^{-5}	'Decomposed granite'
Seoul, Korea	Kim et al. (2004)	35% 'well-graded sands'	18.64 (16.5 dry)	33	10 (Φ^b=15°)	0.43	7.08×10^{-7}	'Soils formed from in situ weathering of granite and gneiss'

*In unsaturated soils, Φ^b is an additional, separate friction angle representing an additional component of shear strength sometimes referred to as 'apparent cohesion', which arises from the effects of matric suction.

†Porosity assumed to be 0.4 and saturated hydraulic conductivity assumed to be 1×10^{-5} m s⁻¹, estimated from Lumb (1975), Ng and Shi (1998), Kim et al. (2004) and field observations in 2008.

land cover characteristics) cannot support such detailed analyses.

The granite hillslope was initially modelled with uniform hydrological properties throughout, using a finite element mesh comprising approximately 4 m cells except for the uppermost layer which was configured as cells of decreasing thickness upwards to the surface (1.3, 0.4, 0.13 and 0.04 m) to avoid numerical convergence difficulties associated with very steep hydraulic gradients (Geo-Slope 2009). With this configuration, the water table position within the slope did not change its position in response to the changing net monthly infiltration, but its position was controlled by the value of K_{sat} that was used. These observations are consistent with theoretical expectations. Despite being unsaturated throughout the thickness of the failure zone above the water table, a stability analysis was undertaken to investigate the likely form of any mass movement. This form is determined by the least stable geometry if a given set of geotechnical properties is assumed to apply uniformly throughout a slope. The modelled failure geometry with the lowest FS (FS = 1.70) corresponded almost exactly with the observed landslide scar in the field, having the same length and maximum depth and occupying the same position on the slope (Figure 13.9). Thus, there was no need to

define a separate zone of higher strength material at the failure depth observed in the field to constrain the landslide geometry, as may be necessary in other settings (Dykes 2002). However, if the granite was modelled with a lower K_{sat} of $1 \times 10^{-7} \, \mathrm{m \, s^{-1}}$, which brought the water table within 20 m of the ground surface beneath the steepest part of the slope, then a failure surface much deeper than observed was generated with an FS of around 1.1.

Average 24 h rainfall intensities of at least 11.3 mm h^{-1} were recorded on 6 September 1998 (Table 13.8). Assuming that all of the rainwater infiltrated into the slope, none of the intensities indicated by the available data (Table 13.8) were sufficient to produce any saturation within the upper 15–20 m of the hillslope if the K_{sat} is $1 \times 10^{-5} \, \mathrm{m \, s^{-1}}$, even if a separate 'unweathered' zone >15 m deep was defined using $K_{sat} = 1 \times 10^{-7} \, \mathrm{m \, s^{-1}}$. However, the K_{sat} values used for the modelling are based on limited published data from other weathered granites in (sub)tropical environments (Table 13.9). The actual hydrological properties of the Motozintla granite may be significantly different from those values, not least with respect to depth variations within the upper 15–20(+) m thick weathered zone. Field observations indicated that the assumption of

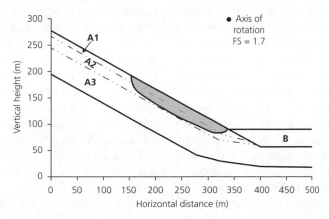

Figure 13.9 Representation of the back-analysis of the landslide above Mirador, Motozintla, showing hypothetical geological units: A1 = highly weathered surficial layer, A2 = less weathered (grade II) zone, A3 = effectively unweathered bedrock, B = colluvium (extends to the river).

$K_{sat} = 1 \times 10^{-5} \, \mathrm{m\,s^{-1}}$ within 1–2 m of the ground surface was probably reasonable, but there is no information to indicate properties at greater depths. The modelling analyses suggest that much of the upper 15 m of the bedrock would need $K_{sat} \leq 1 \times 10^{-6} \, \mathrm{m\,s^{-1}}$ for any saturation to develop within this zone, an essential requirement for failure to 15 m depth with FS < 1.0. In the absence of more detailed subsurface information and event rainfall data (e.g. hourly totals), we were unable to model changing stability patterns due to infiltrating rainwater throughout the storm.

In the light of the results obtained and taking account of the limitations identified previously, we interpret the Motozintla area as comprising steep hillslopes mostly underlain by granite, with the many recent and older landslide scars indicating a typical depth of significant weathering penetration of 15–20 m. Although the near-surface weathered granite is mostly uncemented sand and soft gravel, the saturated hydraulic conductivity possibly decreases to substantially less than the assumed mean value derived from literature accounts $(1 \times 10^{-5} \, \mathrm{m\,s^{-1}})$ in the less weathered (grade II) zone between 1–2 m and 15–20 m.

The landslides are therefore associated with prolonged rainfall of mean intensity greater than or equal to K_{sat} of the less weathered zone, causing a water table to rise up into this zone sufficiently to reduce the FS of the steep slope segments below 1.0 and trigger failure. If rainfall intensities are too high, the rainwater will be unable to penetrate the grade II zone and will instead form downslope throughflow within the thin surficial cover of more highly weathered granite. This may trigger shallow translational landslides up to 1–2 m deep, of which many were seen in the upper parts of the catchment and particularly above the cut slopes of the mountain roads. The uncertainties in the controlling hydrological parameters preclude the determination of meaningful rainfall thresholds for landslide triggers. The high permeability of the surficial materials and the thickness of the

grade II zone are such that large depths of rainfall appear to be needed to trigger either type of landslide. It is therefore suggested that in this location, any differences in antecedent hillslope water contents arising from different land uses probably have little influence over the threshold rainfall event magnitude.

Discussion

Many regions of Mexico are highly susceptible to hazards, because of the combination of the country's tectonic setting, which has produced a variety of topographic landscapes inherently susceptible to mass movements, and its geographical location between two oceans at latitudes where tropical storms readily develop. Some of the data relevant for landslide hazard assessments such as the topographic surface, rainfall, land use/land cover and surficial geology are often available and are in any case improving rapidly. However, other knowledge that could lead to improved understanding of past landslide events and, thus, improved assessments of future risks, such as subsurface geology and regional groundwater characteristics, has not yet been obtained. A modelling approach can often be used to investigate the controls on processes and patterns of change resulting from those processes in situations where data are limited. Such an approach necessarily requires reasonable assumptions to be made about many of the system parameters and processes, based on at least some relevant data, as well as sufficient information about example sites against which a model's outputs can be compared for reasons of validation. However, the scale of the model's conceptual framework, and hence its degree of generality, will often determine the effectiveness of this type of approach.

Alcántara-Ayala et al.'s (2006) GIS-based study of the LULCC influence on landsliding in the Sierra Norte de Puebla utilised all of the readily available types of data at a very general, regional scale and obtained a degree of association between LULCC and landslides that corresponded reasonably well with the inventory of the 1999 event. We used the other end of the modelling scale in this chapter, that is, focusing on modelling changing patterns of the hydrology/hydrogeology and stability of individual hillslopes in response to rainfall, so as to enhance our understanding of the relevant thresholds and controls that determine whether a slope will fail (or not). We have shown that even for the relatively simple case of the weathered granite hillslopes at Motozintla, although enough assumptions could be made to support exploratory analyses that we were able to relate to the observed reality, there were insufficient data to constrain all of the assumptions needed to (i) generate model outputs that could provide definitive explanations for the observed landslides or (ii) investigate the critical threshold rainfall conditions for landsliding in this region. The situation at Teziutlán was even more difficult because of the unknown 3-dimensional extent of the pyroclastic sediments that constitute the failed slope.

The reasons for the demonstrated general association between LULCC and landsliding for the Sierra Norte de Puebla (Alcántara-Ayala et al. 2006) can be understood in a theoretical sense. However, if the landslide hazard is to be reliably assessed for other parts of the country, then it will be important to establish whether the same patterns of association arise elsewhere and to investigate the detailed causal mechanisms. The latter can be achieved with respect to shallow failures of the soil layer on a slope by means of monitoring and modelling the soil moisture variations, that is, more of the work that John Thornes undertook in Mexico. More problematic is the fact that many of the observed landslides in the disasters at Motozintla and the Sierra Norte de Puebla were much more than shallow soil slides. Future attempts to model the changing stability of slopes mantled with significant thicknesses of colluvium, pyroclastic materials or in situ weathered profiles will require

more extensive field and laboratory investigations of the respective materials and their 3-dimensional distributions on the slopes. In addition, more detailed meteorological data are needed in order to estimate more reliably the contribution of ET losses from hillslopes under different vegetation covers. The latter requirement is especially important given the potential for climate change to enhance the effect of LULCC in giving rise to landslides.

The Intergovernmental Panel on Climate Change identified several fundamental issues for landsliding worldwide (IPCC 2007), of which two are of particular relevance to Mexico: (i) 'Trends from 1900 to 2005 have been observed in precipitation amounts in many large regions. Over this period, precipitation increased significantly in eastern parts of North and South America, …'; and (ii) 'It is likely that the frequency of heavy precipitation events (or the proportion of total rainfall from heavy falls) has increased over most areas'. Climate change modelling has been performed at both global (Trenberth 1996; Gates et al. 1996) and regional scales (Buma and Dehn 1998; Dehn and Buma 1999) with respect to its impact on first time landslides (Corominas 2001) and reactivation of mass failures (Dixon and Brook 2007). The increase of extreme precipitation resulting from climate change is one of the most significant factors in landslide occurrence (Buma and Dehn 1998; Corominas and Moya 1999; Dikau and Schrott 1999; Schmidt and Dikau 2004). Climate changes could potentially alter the stability conditions of slopes, in accordance with the geomorphological environments, as most landslides are caused by climatically controlled processes such as intense and/or prolonged rainfall events, fast snowmelt, permafrost degradation or river migration (Borgatti and Soldati 2010).

Besides greenhouse gas emissions, land-use changes and primarily deforestation have produced shifts in Mexico's climate. Based on the analysis of precipitation tendencies, rainfall will increase in most regions of the country. This will involve an increment in the number and intensity of storms and hurricanes that will give rise to shorter cycles of rainfall (Magaña et al. 2004) that are likely to trigger landslides in short and medium terms. Additionally, regions where urban growth and sprawling is not planned or regulated will be severely affected as they are frequently located on unstable slopes. This is now of serious concern. The National Centre for Disaster Prevention (CENAPRED) made major efforts to develop such monitoring systems for volcanic hazards, in association with the USGS, for example, by establishing an automated lahar-detection system on the northern slope of Popocatépetl volcano, but landslide monitoring approaches in Mexico have not been widely developed. There is progress, however: In response to the Teziutlán landslide in 1999, a programme of instrumentation and monitoring on the La Aurora slope was proposed. Subsequently, and supported by local authorities, a data registration stall for landslide instrumentation and monitoring (known as 'Casa CENAPRED') was officially established on the same hillslope (Table 13.10; Figure 13.10).

More generally, strategies for landslide hazard management and mitigation in Mexico ultimately depend on continuing research into the controlling site factors and mechanisms of landslide occurrence, which must necessarily

include the extent to which 'recent' LULCC may be a more – or less – important spatial factor than the actual vegetation cover and site (e.g. geological) factors at any given time. The principle of integrating LULCC data with other spatial data has been demonstrated but only for one region so far. Efforts to further improve the accuracy and reliability of land cover assessments from remotely sensed data, for the primary purpose of maintaining the NFI, have been continuing (e.g. Couturier et al. 2010) and may already be adequate for more detailed spatial modelling of landslide susceptibility. Likewise, the quality and resolution of topographic data for the DTM layer in a GIS have been rapidly improving with the acquisition of LiDAR data for regions previously identified as susceptible to natural hazards, particularly landslides (CENAPRED, pers. comm., 2013). The most susceptible regions are known; the difficulty is to know whether and if so, when and how, warnings should be issued to local populations. Such actions need to be based on accurate rainfall forecasts and better-constrained estimates of rainfall thresholds likely to trigger landslides. This chapter has shown that these rainfall thresholds cannot be estimated from hillslope modelling in the absence of more detailed climate/weather monitoring and geological information – but when these data

Table 13.10 Instrumentation developed by the National Centre for Disaster Prevention (CENAPRED) at La Aurora, Teziutlán, Puebla (Domínguez and Mendoza 2008).

Item	Equipment	Quantity	Comments
1	Automatic data acquisition system	1	1 data logger (channels) and 2 multiplexors (16 channels)
2	Surface piezometers	3	0–35 kPa, 60 m cable
3	Deep piezometers	3	5, 7, 10 m depth, 0–170 kPa, 80 m cable
4	Electronic surface tensiometers	3	0–999 mb, 40 m cable
5	Electronic tensiometers	3	0–999 mb, 60 m cable
6	Electronic deep tensiometers	3	0–999 mb, 80 m cable
7	Rain gauge	1	Digital
8	Surface extensometers	2	Situated at the crown and base of the slope
9	Deep extensometers	2	Situated at the crown and base of the slope; 0.5, 2.0, 3.5, 5.0, 6.5 m deep
10	Triaxial accelerometer	1	Situated next to the registration stall

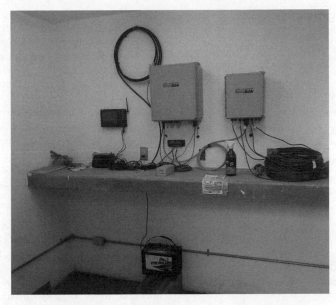

Figure 13.10 The new CENAPRED monitoring station at La Aurora landslide, Teziutlán (from Domínguez and Mendoza 2008).

become available, the assessment and management of landslide hazards should improve greatly as ongoing research, identified previously, delivers new findings and outputs.

In consequence, effective landslide warning systems ought to be based on a multi-actor structure on which communities, local and national governments, stakeholders, regional institutions and organisations, international bodies and the scientific and technological community and communication experts could contribute to the definition and implementation of appropriate disaster risk management strategies (Alcántara-Ayala and Garnica 2013), that include a thorough understanding of hazards, but also efficient and realistic practices for reducing social vulnerability.

Conclusions

Environmental change and landslide hazards in Mexico are both known to be significant, but very little research has been done to improve understanding of the links between these two issues and, as a result, to develop strategies to assess the hazards and manage the associated risks. This chapter has reviewed the state of knowledge and has used recent field studies to show that although monitoring of LULCC has been dramatically improved in recent years, there are – as yet – insufficient relevant data to facilitate detailed modelling of the stability of slopes, particularly in terms of demonstrating the hydrological role of LULCC in the occurrence of rainfall-triggered landslides. Field observations throughout the two study regions revealed the significance, if not dominance, of mass movements as the long-term geomorphological mechanisms of landform development, and as such landslides are common and inevitable occurrences. However, in the absence of more comprehensive geological and meteorological data, the simple generalised GIS-based approach is probably still the most effective means of assessing the regional susceptibility to rainfall-triggered landslides at the present time.

Acknowledgements

Special thanks are due to CONACYT and UNAM for the support kindly provided through the research projects 156242 and IN307410, respectively. We are both deeply grateful for the time and energy that John Thornes put into supervising our respective PhD projects (we worked in the same office at King's College London for one of those years) and for his seemingly endless encouragement and support before, during and through the years following that time.

References

Alcántara-Ayala, I.(2004a) Hazard assessment of rainfall-induced landsliding in Mexico. *Geomorphology* **61**, 19–40.

Alcántara-Ayala, I.(2004b) Flowing mountains in Mexico: incorporating local knowledge and initiatives to confront disaster and promote prevention. *Mountain Research and Development* **24**, 10–13.

Alcántara-Ayala, I.(2008) On the historical account of disastrous landslides in Mexico: the challenge of risk management and disaster prevention. *Advances in Geosciences* **14**, 159–164.

Alcántara-Ayala, I., Dykes, A. P. (2010a) Introduction – land use change in the tropics: causes, consequences and monitoring in Mexico. *Singapore Journal of Tropical Geography* **31**, 143–151.

Alcántara-Ayala, I., Dykes, A. P. (eds.) (2010b) Environmental dimensions of the Mexican tropics. Special Theme (6 papers), *Singapore Journal of Tropical Geography* **31**, 143–223.

Alcántara-Ayala, I., Garnica, R. J. (2013) Landslide monitoring and warning systems in Mexico. In: Sassa, K., Rouhban, B., Briceno, S., McSaveney, M., He, B. (eds.), *Landslides: Global Risk Preparedness*. Springer-Verlag, Berlin/Heidelberg, pp. 301–316.

Alcántara-Ayala, I., López-Mendoza, M., Melgarejo-Palafox, G., Borja-Baeza, R. C., Acevo-Zarate, R. (2004) Natural hazards and risk communication strategies among indigenous communities. *Mountain Research and Development* **24**, 298–302.

Alcántara-Ayala, I., Esteban-Chávez, O., Parrot, J. F. (2006) Landsliding related to land-cover change: a diachronic analysis of hillslope instability distribution in the Sierra Norte, Puebla, Mexico. *Catena* **65**, 152–165.

Allen, R. G., Pereira, L. S., Raes, D., Smith, M. (1998) Crop evapotranspiration – guidelines for computing crop water requirements – FAO irrigation and drainage paper 56. Food and Agriculture Organisation of the United Nations, Rome. Available at: www.fao. org/docrep/x0490e/x0490e00.htm#Contents (accessed 14 July 2014).

ANSS (2010) Composite earthquake catalog. Advanced national seismic system. Northern California Earthquake Data Center. Available at: www.ncedc.org/anss/ (accessed 14 July 2014).

ASTER (2011) ASTER global digital elevation map announcement. Available at: http://asterweb.jpl. nasa.gov/gdem.asp (accessed 14 July 2014).

Atkinson, P. M., Massari, R. (2011) Autologistic modelling of susceptibility to landsliding in the Central Apennines, Italy. *Geomorphology* **130**, 55–64.

Bălteanu, D., Chendeş, V., Sima, M., Enciu, P. (2010) A country-wide spatial assessment of landslide susceptibility in Romania. *Geomorphology* **124**, 102–112.

Bitrán, D., Reyes, C. (2000) Evaluación del impacto económico de las inundaciones ocurridas en octubre de 1999 en el estado de Puebla. In: Bitrán, D. (ed.) *Evaluación del impacto socioeconómico de los principales desastres naturales ocurridos en la República Mexicana durante 1999*. Cuadernos de Investigación 50. CENAPRED, México City, pp. 161–194.

Bonasia, R., Capra, L., Costa, A., Macedonio, G., Saucedo, R. (2011) Tephra fallout hazard assessment for a Plinian eruption at Volcán de Colima (Mexico). *Journal of Volcanology and Geothermal Research* **203**, 12–22.

Borgatti, L., Soldati, M. (2010) Landslides as a geomorphological proxy for climate change: a record from the Dolomites (northern Italy). *Geomorphology* **120**, 56–64.

Brown, J. C., Kastens, J. H., Coutinho, A. C., Victoria, D. de C., Bishop, C. R.(2013) Classifying multiyear agricultural land use data from Mato Grosso using time-series MODIS vegetation index data. *Remote Sensing of Environment* **130**, 39–50.

Brunsden, D., Thornes, J. B. (1979) Landscape sensitivity and change. *Transactions of the Institute of British Geographers (New Series)* **4**, 463–484.

Buma, J., Dehn, M. (1998) A method for predicting the impact of climate change on slope stability. *Environmental Geology* **35**, 190–196.

Caballero, L., Macías, J. L., García-Palomo, A., Saucedo, G. R., Borselli, L., Sarocchi, D., Sánchez, J. M. (2006) The September 8–9, 1998 rain-triggered flood events at Motozintla, Chiapas, Mexico. *Natural Hazards* **39**, 103–126.

Cairns, M. A., Haggerty, P. K., Alvarez, R., de Jong, B. H. J., Olmsted, I. (2000) Tropical Mexico's recent land-use change: a region's contribution to the global carbon cycle. *Ecological Applications* **10**, 1426–1441.

Calcaterra, D., Santo, A. (2004) The January 10, 1997 Pozzano landslide, Sorrento Peninsula, Italy. *Engineering Geology* **75**, 181–200.

CENAPRED (1997) Daños ocurridos en Acapulco por el huracán Pauline. Informe Preliminar, Revista Prevención (Órgano Informativo del Sistema Nacional de Protección Civil), 19, México. agosto-diciembre, pp. 2–7.

CEPAL (1999) Honduras: evaluación de los daños ocasionados por el Huracán Mitch, 1998. Sus implicaciones para el desarrollo económico y social y el medio ambiente. Economic Commission for Latin America and the Caribbean, Santiago. Available at: www.eclac.cl/cgi-bin/getProd.asp?xml=/publicaciones/xml/1/15501/P15501.xml&xsl=/mexico/tpl/p9f.xsl&base=/mexico/tpl/top-bottom.xsl (accessed 14 July 2014).

CEPAL (2000) Los efectos socioeconómicos de las inundaciones y deslizamientos en Venezuela en 1999: Perfiles de proyectos. Economic Commission for Latin America and the Caribbean, Santiago Available at: www.eclac.cl/cgi-bin/getProd.asp?xml=/publicaciones/xml/5/10135/P10135.xml&xsl=/mexico/tpl/p9f.xsl&base=/mexico/tpl/top-bottom.xsl (accessed 14 July 2014).

Chen, H., Lee, C. F. (2004) Geohazards of slope mass movement and its prevention in Hong Kong. *Engineering Geology* **76**, 3–25.

Chigira, M., Yokoyama, O. (2005) Weathering profile of non-welded ignimbrite and the water infiltration behavior within it in relation to the generation of shallow landslides. *Engineering Geology* **78**, 187–207.

CIAT (2013) Terra-i. International center for tropical agriculture, Palmira. Available at: www.terra-i.org/terra-i.html (accessed 14 July 2014).

Collins, B. D., Dunne, T. (1988) Effects of forest land management on erosion and revegetation after the eruption of Mount St. Helens. *Earth Surface Processes and Landforms* **13**, 193–205.

Corominas, J. (2001) Landslides and climate. In: Bromhead, E., Dixon, N., Ibsen, M. (eds.) *Landslides in research, theory and practice*. Proceedings of the 8th International Symposium on Landslides. Thomas Telford, London, CD-ROM: Keynote Lectures, 26–30 June 2000, Cardiff.

Corominas, J., Moya, J. (1999) Reconstructing recent landslide activity in relation to rainfall in the Llobregat River basin, Eastern Pyrenees, Spain. *Geomorphology* **30**, 79–93.

Couturier, S., Mas, J. F., López-Granados, E., Benítez, J., Coria-Tapia, V., Vega-Guzmán, Á. (2010) Accuracy assessment of the Mexican national forest inventory map: a study in four ecogeographical areas. *Singapore Journal of Tropical Geography* **31**, 163–179.

Covaleda, S., Gallardo, J. F., García-Oliva, F., Kirchmann, H., Prat, C., Bravo, M., Etchevers, J. D. (2011) Land-use effects on the distribution of soil organic carbon within particle-size fractions of volcanic soils in the Transmexican Volcanic Belt (Mexico). *Soil Use and Management* **27**, 186–194.

Crosta, G. B., Imposimato, S., Roddeman, D., Chiesa, S., Moia, F. (2005) Small fast-moving flow-like landslides in volcanic deposits: the 2001 Las Colinas Landslide (El Salvador). *Engineering Geology* **79**, 185–214.

Crozier, M. J. (1986) *Landslides: causes, Consequences and Environment*. Croom Helm, London.

Dehn, M., Buma, J. (1999) Modelling future landslide activity based on general circulation models. *Geomorphology* **30**, 175–187.

Díaz-Gallegos, J. R., Mas, J. F., Velázquez, A. (2010) Trends of tropical deforestation in Southeast Mexico. *Singapore Journal of Tropical Geography* **31**, 180–196.

Dikau, R., Schrott, L. (1999) The temporal stability and activity of landslides in Europe with respect to climatic change (TESLEC): main objectives and results. *Geomorphology* **30**, 1–12.

Dixon, N., Brook, E. (2007) Impact of predicted climate change on landslide reactivation: case study of Mam Tor, UK, *Landslides* **4**, 137–147.

Domínguez, L., Mendoza, M. J. (2008) Instrumentación geotécnica de una ladera en la colonia La Aurora, municipio de Teziutlán, Puebla. Informe Técnico, CENAPRED SEGOB, Mexico City, 21pp.

Dykes, A. P. (2000) Climatic patterns in a tropical rainforest in Brunei. *Geographical Journal* **166**, 63–80.

Dykes, A. P. (2002) The role of bedrock failures driven by uplift in long-term landform development in Brunei, northwest Borneo. *Transactions of the Japanese Geomorphological Union* **23**, 201–222.

ECLAC (1999a) El Salvador: assessment of the damage caused by hurricane Mitch, 1998. Implications for Economic and Social Development and for the Environment. Economic Commission for Latin America and the Caribbean, Santiago. Available at: www.eclac.cl/cgi-bin/getProd.asp?xml=/mexico/noticias/paginas/0/15510/P15510.xml&xsl=/mexico/

tpl/p18f.xsl&base=/mexico/tpl/top-bottom.xslt (accessed 14 July 2014).

ECLAC (1999b) Guatemala: assessment of the damage caused by hurricane Mitch, 1998. Implications for Economic and Social Development and for the Environment. Economic Commission for Latin America and the Caribbean, Santiago. Available at: www.eclac.cl/cgi-bin/getProd.asp?xml=/mexico/noticias/paginas/0/15510/P15510.xml&xsl=/mexico/tpl/p18f.xsl&base=/mexico/tpl/top-bottom.xslt (accessed 14 July 2014).

ECLAC (1999c) Nicaragua: assessment of the damage caused by hurricane Mitch, 1998. Implications for Economic and Social Development and for the Environment. Economic Commission for Latin America and the Caribbean, Santiago. Available at: www.eclac.cl/cgi-bin/getProd.asp?xml=/mexico/noticias/paginas/0/15510/P15510.xml&xsl=/mexico/tpl/p18f.xsl&base=/mexico/tpl/top-bottom.xslt (accessed 14 July 2014).

EM-DAT (undated).The OFDA/CRED International disaster database. Université catholique de Louvain, Brussels. Available at: www.emdat.be (accessed 14 July 2014).

Fiorillo, F., Wilson, R. C. (2004) Rainfall induced debris flows in pyroclastic deposits, Campania (southern Italy). *Engineering Geology* **75**, 263–289.

Flamenco-Sandovala, A., Martínez-Ramos, M., Masera, O. M. (2007) Assessing implications of land-use and land-cover change dynamics for conservation of a highly diverse tropical rain forest. *Biological Conservation* **138**, 131–145.

Francis, S. C. (1987) Slope development through the threshold slope concept. In: Anderson, M. G., Richards, K. S. (eds.) *Slope Stability*. John Wiley & Sons, Ltd, Chichester, pp. 601–624.

Frattini, P., Crosta, G. B., Fusi, N., Dal Negro, P. (2004) Shallow landslides in pyroclastic soils: a distributed modelling approach for hazard assessment. *Engineering Geology* **73**, 277–295.

García-Ruiz, J. M., Beguería, S., Alatorre, L. C., Puigdefábregas, J. (2010) Land cover changes and shallow landsliding in the flysch sector of the Spanish Pyrenees. *Geomorphology* **124**, 250–259.

Garwood, N. C., Janos, D. P., Brokaw, N. (1979) Earthquake-induced landslides: a major disturbance to tropical soils. *Science* **205**, 997–999.

Gates, W. L., Henderson-Sellers, A., Boer, G. J., Folland, C. K., Kitoh, A., McAvaney, B. J., Semazzi, F., Smith, N., Weaver, A. J., Zeng, Q. C. (1996) Climate models – evaluation. In: Houghton, J. T., Meira Filho, L. G., Callander, B. A., Harris, N.,

Kattenberg, A., Maskell, K. (eds.) *Climate Change 1995: The Science of Climate Change*. Cambridge University Press, Cambridge, UK, pp. 229–284.

Geo-Slope (2009) *Seepage Modeling with SEEP/W© 2007. An Engineering Methodology* (4th ed.). GEO-SLOPE International Ltd., Calgary.

Giambelluca, T. W., Scholz, F. G., Bucci, S. J., Meinzer, F. C., Goldstein, G., Hoffmann, W. A., Franco, A. C., Buchert, M. P. (2009) Evapotranspiration and energy balance of Brazilian savannas with contrasting tree density. *Agricultural and Forest Meteorology* **149**, 1365–1376.

Glade, T., Crozier, M. J. (2005) A review of scale dependency in landslide hazard and risk analysis. In: Glade, T., Anderson, M., Crozier, M. J. (eds.) *Landslide Hazard and Risk*. John Wiley & Sons, Ltd, Chichester, pp. 75–138.

Greenway, D. R. (1987) Vegetation and slope stability. In: Anderson, M. G., Richards, K. S. (eds.) *Slope Stability*. John Wiley & Sons, Ltd, Chichester, pp. 187–230.

Hansen, M. C., Potapov, P. V., Moore, R., Hancher, M., Turubanova, S. A., Tyukavina, A., Thau, D., Stehman, S. V., Goetz, S. J., Loveland, T. R., Kommareddy, A., Egorov, A., Chini, L., Justice, C. O., Townshend, J. R. G. (2013) High-resolution global maps of 21st-century forest cover change. *Science* **342**(6160), 850–853. Interactive map available at: http://earthenginepartners.appspot.com/science-2013-global-forest (accessed 14 July 2014).

Höhne, N., Wartmann, S., Herold, A., Freibauer, A. (2007) The rules for land use, land use change and forestry under the Kyoto Protocol – lessons learned for the future climate negotiations. *Environmental Science & Policy* **10**, 353–369.

Huber-Sannwald, E., Maestre, F. T., Herrick, J. E., Reynolds, J. F. (2006) Ecohydrological feedbacks and linkages associated with land degradation: a case study from Mexico. *Hydrological Processes* **20**, 3395–3411.

IPCC (2007) Climate change 2007: synthesis report. Contribution of Working Groups I, II and III to the Fourth Assessment Report of the Intergovernmental Panel on Climate Change [Core Writing Team: Pachauri, R. K., Reisinger, A. (eds.)]. IPCC, Geneva, 104pp.

Kim, J., Jeong, S., Park, S., Sharma, J. (2004) Influence of rainfall-induced wetting on the stability of slopes in weathered soils. *Engineering Geology* **75**, 251–262.

Klemetti, E. (2012) Looking back at the 1982 eruption of El Chichón in Mexico. *Wired Science Blogs*, Condé Nast Publications. Available at: www.wired.com/

wiredscience/2012/03/looking-back-at-the-1982-eruption-of-el-chichon-in-mexico/ (accessed 14 July 2014).

Kuriakose, S. L., Sankar, G., Muraleedharan, C. (2009) History of landslide susceptibility and a chorology of landslide-prone areas in the Western Ghats of Kerala, India. *Environmental Geology* **57**, 1553–1568.

López-Rodríguez, S. R., Blanco-Libreros, J. F. (2008) Illicit crops in Tropical America: deforestation, landslides, and the terrestrial carbon stocks. *Ambio* **37**, 141–143.

Lumb, P. (1975) Slope failures in Hong Kong. *Quarterly Journal of Engineering Geology* **8**, 31–65.

Magaña, V., Mendez, J. M., Morales, R., Millán, C. (2004) Consecuencias presentes y futuras de la variabilidad y el cambio climático. In: Martínez, J., Fernández Bremauntz, A. (eds.) *Cambio Climático: una visión desde México*. SEMARNATe INE, México City.

Marshall, A. J. (1937) Northern New Guinea (1936). *Geographical Journal* **89**, 489–506.

Mas, J. F., Velázquez, A., Díaz-Gallegos, J. R., Mayorga-Saucedo, R., Alcántara, C., Bocco, G., Castro, R., Fernández, T., Pérez-Vega, A. (2004) Assessing land use/cover changes: a nationwide multidate spatial database for Mexico. *International Journal of Applied Earth Observation and Geoinformation* **5**, 249–264.

Mendoza, M. J., Noriega, I., Domínguez, L. (2000) Deslizamientos de laderas en Teziutlán, pue., provocados por las lluvias intensas de octubre de 1999. SEGOB, CENAPRED, México City.

Montrasio, L., Valentino, R., Losi, G. L. (2011) Towards a real-time susceptibility assessment of rainfall-induced shallow landslides on a regional scale. *Natural Hazards and Earth System Sciences* **11**, 1927–1947.

Moreno Murillo, J. M. (1995) The 1985 Mexico Earchquake. *Geofisica Coumbia (Universidad Nacional de Colombia)* **3**, 5–19.

Myster, R. W., Thomlinson, J. R., Larsen, M. C. (1997) Predicting landslide vegetation in patches on landscape gradients in Puerto Rico. *Landscape Ecology* **12**, 299–307.

Ng, C. W. W., Shi, Q. (1998) Influence of rainfall intensity and duration on slope stability in unsaturated soils. *Quarterly Journal of Engineering Geology* **31**, 105–113.

Nieto, A. S., Schuster, R. L., Plaza-Nieto, G. (1991) Mass wasting and flooding. In: Schuster, R. L. (ed.) *The March 5, 1987, Ecuador Earthquakes – Mass Wasting and Socioeconomic Effects* (Vol. **5**). Natural Disaster Studies, Natural Research Council, Washington, DC, pp. 51–82.

Orellana, C. (2005) Mexico addresses desertification. *Frontiers in Ecology and the Environment* **3**, 240.

Pain, C. F., Bowler, J. M. (1973) Denudation following the November 1970 earthquake at Madang, Papua New Guinea. *Zeitschrift für Geomorphologie, Supplement bd.* **18**, 92–104.

Palacio-Prieto, J. L., Bocco, G., Velázquez, A. et al. (2000) La condición actual de los recursos forestales en México: resultados del Inventario Forestal Nacional 2000. *Investigaciones Geográficas* **43**, 183–203.

Pidgeon, N. (2012) Public understanding of, and attitudes to, climate change: UK and international perspectives and policy. *Climate Policy* **12**(sup01), S85–S106.

Ratter, B. M. W., Philipp, K. H. I., von Storch, H. (2012) Between hype and decline: recent trends in public perception of climate change. *Environmental Science and Policy* **18**, 3–8.

Read, J. M., Lam, N. S.-N. (2002) Spatial methods for characterizing land cover and detecting land-cover changes for the tropics. *International Journal of Remote Sensing* **23**, 2457–2474.

Rindfuss, R. R., Walsh, S. J., Turner, B. L., Fox, J., Mishra, V. (2004) Developing a science of land change: challenges and methodological issues. *Proceedings of the National Academy of Sciences* **101**, 13976–13981.

Ronchetti, F., Borgatti, L., Cervi, F., Lucente, C. C., Veneziano, M., Corsini, A. (2007) The Valoria landslide reactivation in 2005–2006 (Northern Apennines, Italy). *Landslides* **4**, 189–195.

Saldaña-Zorrilla, S. O. (2007) *Socioeconomic Vulnerability to Natural Disasters in Mexico: Rural Poor, Trade and Public Response*. Estudios y perspectivas series, No. 92. Economic Commission for Latin America and the Caribbean (ECLAC), Disaster Evaluation Unit, México City. Available at: www.eclac.org/publicaciones/xml/4/31414/Serie_92.pdf (accessed 14 July 2014).

Schmidt, J., Dikau, R. (2004) Modelling historical climate variability and slope stability. *Geomorphology* **60**, 433–447.

Schuster, R. L., Nieto, A. S., O'Rourke, T. D., Crespo, E. (1996) Mass wasting triggered by the 5 March 1987 Ecuador earthquakes. *Engineering Geology* **42**, 1–23.

Seach, J. (2010) Volcano Eruption Fatalities. *Volcano Live*, John Seach, New South Wales. Available at: www.volcanolive.com/fatalities.html (accessed 14 July 2014).

Sentelhas, P. C., Gillespie, T. J., Santos, E. A. (2010) Evaluation of FAO Penman–Monteith and alternative methods for estimating reference evapotranspiration with missing data in Southern Ontario, Canada. *Agricultural Water Management* **97**, 635–644.

Simonett, D. S. (1967) Landslide distribution and earthquakes in the Bewani and Torricelli Mountains. In: Jennings, J. N., Mabbutt, J. A. (eds.) *Landform Studies from Australia and New Guinea*. National University Press, Canberra, pp. 64–84.

Skempton, A. W., DeLory, F. A. (1957) Stability of natural slopes in London clay. Proceedings of the 4th International Conference on Soil Mechanics and Foundation Engineering, 12–24 August 1957, London, 2, 378–381.

Thiery, Y., Malet, J.-P., Sterlacchini, S., Puissant, A., Maquaire, O. (2007) Landslide susceptibility assessment by bivariate methods at large scales: application to a complex mountainous environment. *Geomorphology* **92**, 38–59.

Thornes, J. B., Brunsden, D. (1977) *Geomorphology and Time*. Methuen, London.

Trenberth, K. E. (1996) Coupled climate system modelling. In: Giambelluca, T. W., Henderson-Sellers, A. (eds.) *Climate Change: Developing Southern Hemisphere Perspectives*. John Wiley & Sons, Ltd, Chichester, pp. 63–88.

van Asch, T. W. J., Malet, J.-P. (2009) Flow-type failures in fine-grained soils: an important aspect in landslide hazard analysis. *Natural Hazards and Earth System Sciences* **9**, 1703–1711.

Vance, C., Geohegan, J. (2002) Temporal and spatial modelling of tropical deforestation: a survival analysis linking satellite and household survey data. *Agricultural Economics* **27**,317–332.

Veblen, T. T., Ashton, D. H. (1978) Catastrophic influences on the vegetation of the Valdivian Andes, Chile. *Vegetation* **36**, 149–167.

Velázquez, A., Durán, E., Ramírez, I., Mas, J.-F., Bocco, G., Ramírez, G., Palacio-Prieto, J.-L. (2003) Land use-cover change processes in highly biodiverse areas: the case of Oaxaca, Mexico. *Global Environmental Change* **13**, 175–184.

Velázquez, A., Mas, J. F., Bocco, G., Palacio-Prieto, J. L. (2010) Mapping land cover changes in Mexico, 1976–2000 and applications for guiding environmental management policy. *Singapore Journal of Tropical Geography* **31**, 152–162.

Wickham, J. D., Stehman, S. V., Gass, L., Dewitz, J., Fry, J. A., Wade, T. G. (2013) Accuracy assessment of NLCD 2006 land cover and impervious surface. *Remote Sensing of Environment* **130**, 294–304.

Wilske, B., Kwon, H., Wei, L., Chen, S., Lu, N., Lin, G., Xie, J., Guan, W., Pendall, E., Ewers, B. E., Chen, J. (2010) Evapotranspiration (ET) and regulating mechanisms in two semiarid Artemisia-dominated shrub steppes at opposite sides of the globe. *Journal of Arid Environments* **74**, 1461–1470.

PART B

CHAPTER 14

John Thornes and palaeohydrology

Ken Gregory[1] and Leszek Starkel[2]

[1] *Department of Geography, University of Southampton, Southampton, UK*
[2] *Department of Geomorphology and Hydrology, Institute of Geography, Polish Academy of Sciences, Krakow, Poland*

Palaeohydrology was first formally defined by Leopold and Miller (1954). Quaternary palaeohydrology was described by Schumm (1965) and applied to geologic timescales (Schumm 1968, 1977). The subject was progressed by individual research investigations but required international research collaboration to establish a substantial research foundation. As a result of an enterprising initiative by Leszek Starkel and Björn Berglund, conceived in the Eurosiberian Holocene Subcommission of INQUA, an international research programme was proposed in the mid-1970s. The programme suggested parallel studies on the evolution of fluvial systems (sub-project A) and lakes and bogs (sub-project B) during the late glacial and Holocene periods. This project was accepted as International Geological Correlation Programme (IGCP) Project 158 which operated from 1977 to 1988. When that project concluded, a subsequent international project was initiated under the auspices of INQUA, was styled the Global Continental Palaeohydrology Commission (GLOCOPH) and as such operated for 12 successful years from 1991 to 2003. Under a slightly different INQUA structure, the palaeohydrology research achieved by GLOCOPH continues, with the most recent meeting held in Israel in 2009. Whereas IGCP Project 158 research was focused on the temperate zone, subsequent GLOCOPH research was directed towards all world environments. These research endeavours created a body of information about palaeohydrology, involved international scientists from a range of disciplines and made considerable progress in establishing publications about palaeohydrology in general and applications in specific areas (e.g. Starkel et al. 1991; Gregory et al. 1995; Benito et al. 1998; Gregory and Benito 2003).

Regular meetings held from 1977, first in the temperate zone and subsequently worldwide, achieved considerable scientific advances. The fluvial sub-project A involved mainly geomorphologists, sedimentologists and hydrologists, whereas sub-project B included palaeobotanists and palaeolimnologists. Although the two sub-projects worked closely together initially (Figure 14.1), as the research progressed, they worked more independently. John Thornes was heavily involved in the early years, and his fundamental ideas were extremely significant in creating the foundation upon which international palaeohydrological research prospered and flourished. His clear thinking and imaginative

Monitoring and Modelling Dynamic Environments, First Edition. Edited by Alan P. Dykes, Mark Mulligan and John Wainwright.

Figure 14.1 Professor John Thornes (rear) during a field excursion of the IGCP (sub-projects 158 A and B) meeting in Finland in 1978, dynamically providing physical support for the Arctic Circle.

contributions were vital inputs to the way in which the scientific papers were discussed, to the way in which international projects were organised and to the way in which scientific evidence was evaluated in the field. Appreciation of his influence on many individuals was greatly evident at meetings in the Severn Basin and in Finland (Figure 14.1), Poland (Figure 14.2) and France. Although it is not easy to isolate John's impact, because he was always so generous in informal discussions and did not guard his ideas, we can identify at least three ways in which he made very significant contributions fundamental to the progress of palaeohydrology.

First was his contribution to methodology and to the preparation of guidelines published by the British Geomorphological Research Group

(BGRG). Once IGCP Project 158 was accepted in May 1977 for a 5-year period, it was decided that there would be two sub-projects – one (158 A) concerned with the fluvial environment and one (158 B) devoted to lake and mire environments. As the main objective of sub-project A was to understand changes in the fluvial environment on the basis of multidisciplinary studies of fluvial sediments and forms created in the last 15,000 years, it was necessary to produce a guide for international participants. This guide was produced after discussion meetings in Szymbark, Poland, in October 1977 and in London in March 1978, and the volume edited by Starkel and Thornes (1981) included contributions by K.J. Gregory, F. Gullentops, E. Paulissen and M. Church as well as by the editors. Subsequently,

Figure 14.2 Professor John Thornes in the Vistula Valley in 1981 with a subfossil oak resulting from deforestation in the Late Roman period.

in order to coordinate early progress in palaeohydrology, an international conference was organised at Attingham Park, Shrewsbury, United Kingdom, in 1983, and a volume was produced before the meeting (Gregory 1983). That volume, designed to provide the theoretical background for the regional research investigations, included John Thornes' contribution on discharge (Thornes 1983) demonstrating his ability to place contemporary discharge data in a temporal context. He indicated how and why palaeohydrological analyses are data deficient and how this might be remedied, suggesting that 'as the dynamics of the process become better understood it should be easier to unravel the real world complexities which generate historical series' (Thornes 1983, 61). This suggestion anticipated the progress that was later made by palaeoflood analysis (e.g. Baker 2003).

Second were his modelling contributions. The approach to temperate palaeohydrology was agreed to be based on results from type basins, the Severn Basin being selected in the United Kingdom and referred to in the final part of the

paper in 1983 (Thornes 1983). A series of field meetings in the Severn Basin were the basis for the collection of research results, and at all of these, John Thornes' contribution was incisive and particularly informed by the need to link contemporary discharge and sediment production to past behaviour in the modelling context (Figure 14.3). He pointed out that the most productive and widespread approach to establishing palaeohydrological responses to various forcing agents had been inductive but also explored whether deductive models could be employed (Thornes 1987, 18). He concluded that, although we will probably never reach a general model of river basin response, palaeohydrologists can, and should, be making even more use of approximate and deductive models (Thornes 1987, 33), suggesting that 'this will eventually shift palaeohydrology away from a preoccupation with pollen diagrams and fluvial sediments to dynamic modelling of the catchment ecological systems in a way that provides a better understanding of water and sediment yield' (Thornes 1987, 33). This objective typically involved the

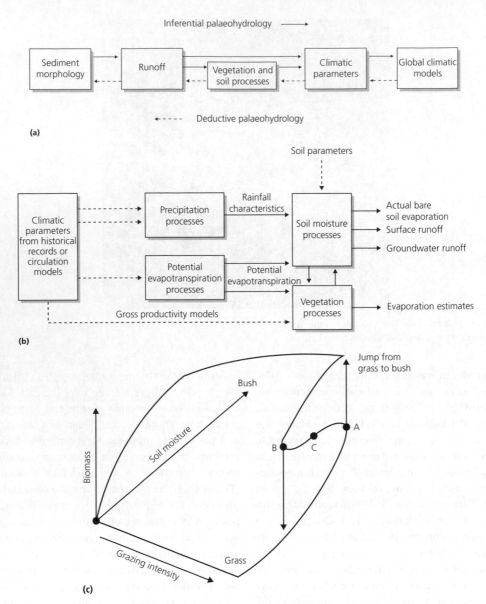

Figure 14.3 Characteristic specifications: (a) relationship between elements in conventional and modelling approaches to palaeohydrology; (b) components of an interactive soil–vegetation model. (Gregory et al. 1987. Reproduced with permission of John Wiley & Sons.) (c) a catastrophe theory interpretation of bush–grass transition used by Thornes (2003, 301) to illustrate multiple stable states in palaeohydrological modelling (Gregory and Benito 2003. Reproduced with permission of John Wiley & Sons.)

distinction between inferential and deductive palaeohydrology (Figure 14.3a) and required specification of the parameters of an interactive soil–vegetation model in a format appropriate for palaeohydrology (Figure 14.3b). In the final volume produced after the end of the second phase of IGCP Project 158 in 1989, research results from the contributing basin investigations were collated (Starkel et al. 1991). In a final chapter, pointing to a continuing agenda

(Thornes and Gregory 1991), John extended the modelling opportunities by focusing on stability and change, elaborated by stability indicated by dominance of a particular state over a long period of time, by rapid change indicated by switches over relatively short time periods and by oscillations between states (Thornes and Gregory 1991, 527). This provided a perspective which could be pursued subsequently because John contrasted the classical stability analysis approach with multiple stable states, catastrophic transitions and instability thresholds. Although he noted that at that time none of the models had been subject to significant testing, each could claim some support from the literature and offered further potential. These ideas of multiple stable states have been demonstrated in more recent research demonstrating John's ability to anticipate developments necessary for further research.

Third was his expertise in relation to ecological influences. At that time, John was one of few who were exploring relationships between water circulation (especially runoff) and biotic elements (Thornes 1990), building bridges with botanists and ecologists investigating the palaeohydrological impacts of historical land use changes caused by human and climatic factors. International collaboration subsequent to IGCP Project 158 in palaeohydrology research was under the auspices of the INQUA commission of GLOCOPH (1991–) for which the prospectus was outlined (Gregory et al. 1995). John was less frequently involved in this stage of the development of palaeohydrology particularly because his research in Spain, involving modelling in relation to vegetation, was taking much of his time – all extremely productive as testified by other chapters in this volume. However, his research in the Mediterranean region, with a long history of human occupation as well as recent effects of climatically controlled desertification, provided results of significance for palaeohydrology (e.g. Thornes 2003). Indeed, one consequence of that research is that John has now been identified as one of the key researchers who provided the inspiration for

biogeomorphology by kick-starting a reassessment of the role of vegetation in erosion (Viles 2011; see Thornes 1985). However, his contributions continued to affect palaeohydrology by developing modelling with particular reference to vegetation stability, placing his visionary ideas in the context of what he identified as the history of palaeohydrological modelling in six main phases: historic, empirical, analytical, evolutionary, global and contemporary (Thornes 2003). His clear specification of the modelling framework allowed him to apply his research on the stability of plant communities to the way in which vegetation changes, characterised by slow growth at first followed by steeply rising growth until it is checked by limited resources (e.g. water stress) when the rate of biomass increase levels off or declines. This demonstrated how shift in vegetation types prompted by climate change is non-linear, more complicated than formerly thought, inexplicable without recourse to human factors (including grazing) and not accountable simply in terms of regression models of biomass productivity by rainfall. The complexity was illustrated by a catastrophe theory interpretation of bush–grass transition (Figure 14.3c). In exposing the limitations of modelling based on regression procedures, John (Thornes 2003, 301–303) contended that 'The complexity of vegetation response calls for an increased effort in this area from hydrologists and palaeohydrologists to better understand the multiple stable outcomes of global climate change' (Thornes 2003, 303). This not only demonstrated the importance of understanding and modelling land cover behaviour during past environmental changes but also signalled the ways in which modelling may relate to future environments.

In an evaluation of John Thornes' contribution through 'The ecology of erosion', his enthusiasm for process-based understanding of landscape evolution as related to palaeohydrology was stressed and it was noted how his interests came full circle with the evaluation of long-term evolution of Mediterranean landscapes (Wainwright and Parsons 2010). Much of the contribution that John Thornes made to the progress of palaeohydrology concerns what is now referred to as memory; but we suspect that he would not have been content to see landscape memory visualised simply in terms of geologic, climatic and anthropogenic (Brierley 2010) because his research demonstrated how land cover is intrinsically associated with regulation of environmental systems and with palaeohydrological scenarios.

Contributions by John Thornes to the development of palaeohydrological research were effected particularly through his contributions to the organisation of the international project, by his ability to build bridges between studies of present-day processes and their reflection in the sediments and forms inherited from the past, by his modelling prowess and by his understanding of land cover behaviour. Therefore, his research (especially in the Mediterranean region) concerning the impact of land use changes on acceleration of runoff and gradual desertification could be employed with great advantage in studies of palaeohydrology. However, his contributions added up to much more than the sum of these inputs and are perhaps encapsulated by the fact that he was able to visualise the bigger picture, to see what was required to elucidate it and then to establish the foundations to achieve it. That was why John Thornes' inspirational ideas were always sought during informal discussions at meetings, why his proposals and papers were eagerly awaited and why his research initiatives developed and continued so successfully, as demonstrated by earlier chapters in this volume.

References

Baker, V. R. (2003) Palaeofloods and extended discharge records. In: Gregory, K. J., Benito, G. (eds.) *Palaeohydrology: Understanding Global Change*. John Wiley & Sons, Ltd, Chichester, pp. 307–323.

Benito, G., Baker, V. R., Gregory, K. J. (eds.) (1998) *Palaeohydrology and Environmental Change*. John Wiley & Sons, Ltd, Chichester.

Brierley, G. J. (2010) Landscape memory: The imprint of the past on contemporary landscape forms and processes. *Area* **42**, 76–85.

Gregory, K. J. (ed.) (1983) *Background to Palaeohydrology*. John Wiley & Sons, Ltd, Chichester.

Gregory, K. J., Benito, G. (eds.) (2003) *Palaeohydrology: Understanding Global Change*. John Wiley & Sons, Ltd, Chichester.

Gregory, K. J., Starkel, L., Baker, V. R. (eds.) (1995) *Global Continental Palaeohydrology*. John Wiley & Sons, Ltd, Chichester.

Leopold, L. B., Miller, J. P. (1954) *Postglacial Chronology for Alluvial Valleys in Wyoming*. U.S. Geological Survey Water Supply Paper 1261. Wyoming Water Science Center, Cheyenne, pp. 61–85.

Schumm, S. A. (1965) Quaternary palaeohydrology. In: Wright, H. E. Jr., Frey, D. G. (eds.) *The Quaternary of the United States*. Princeton University Press, Princeton, pp. 783–794.

Schumm, S. A. (1968) Speculations concerning palaeohydrologic controls of terrestrial sedimentation. *Geological Society of America Bulletin* **79**, 1573–1588.

Schumm, S. A. (1977) *The Fluvial System*. John Wiley & Sons, Inc, New York.

Starkel, L., Thornes, J. B. (eds.) (1981) *Palaeohydrology of River Basins. Guide to the Sub-project A on Palaeohydrological Changes in the Temperate Zone in the last 15,000 years*. British Geomorphological Research Group Technical Bulletin No. 28. Geo Books, Norwich.

Starkel, L., Gregory, K. J., Thornes, J. B. (eds.) (1991) *Temperate Palaeohydrology: Fluvial Processes in the Temperate Zone During the Last 15,000 Years*. John Wiley & Sons, Ltd, Chichester.

Thornes, J. B. (1983) Discharge: Empirical observations and statistical models of change. In: Gregory, K. J. (ed.) *Background to Palaeohydrology*. John Wiley & Sons, Ltd, Chichester, pp. 51–67.

Thornes, J. B. (1985) The ecology of erosion. *Geography* **70**, 222–236.

Thornes, J. B. (1987) Models for palaeohydrology in practice. In: Gregory, K. J., Lewin J., Thornes, J. B. (eds.) *Palaeohydrology in Practice: A River Basin Analysis*. John Wiley & Sons, Ltd, Chichester, pp. 17–36.

Thornes, J. B. (ed.) (1990) *Vegetation and Erosion: Processes and Environments*. John Wiley & Sons, Ltd, Chichester.

Thornes, J. B. (2003) Palaeohydrological modelling: From palaeohydraulics to palaeohydrology. In: Gregory, K. J., Benito, G. (eds.) *Palaeohydrology:*

Understanding Global Change. John Wiley & Sons, Ltd, Chichester, pp. 291–305.

Thornes, J. B., Gregory, K. J. (1991) Unfinished business: A continuing agenda. In: Starkel, L., Gregory, K. J., Thornes, J. B. (eds.) *Temperate Palaeohydrology: Fluvial Processes in the Temperate Zone during the Last 15,000 Years.* John Wiley & Sons, Ltd, Chichester, pp. 521–536.

Viles, H. M. (2011) Biogeomorphology. In: Gregory, K. J., Goudie, A. S. (eds.) *Handbook of Geomorphology*. Sage, London.

Wainwright, J., Parsons, A. J. (2010) Classics in physical geography revisited. 'Thornes, J.B. 1985: The ecology of erosion. Geography 70, 222–235'. *Progress in Physical Geography* **34**(3), 399–408.

CHAPTER 15

John Thornes: Landscape sensitivity and landform evolution

Tim Burt

Department of Geography, Durham University, Durham, UK

Introduction

Three papers were presented at a meeting of the *Institute of British Geographers* at Manchester University in January 1979 which, on the face of it, seem to come from different eras – indeed paradigms – of geomorphological research. On the one hand, Eric Brown's presidential address dealt with his lifelong interest in erosion surfaces, poorly dissected areas of great antiquity and evidence of several cycles of denudation (Brown 1979). On the other hand, Stan Schumm provided excerpts from his recent book – *The Fluvial System* (Schumm 1977) – to illustrate the concept and application of thresholds in geomorphology. In between those two papers (at the conference and in the subsequent publication), Denys Brunsden and John Thornes attempted to assess the 'general concepts generated by modern studies of geomorphological processes – in terms of their utility for models of long-term landform evolution'. The work of Brunsden and Thornes (1979) is one of the most significant publications in the science of geomorphology in the second half of the 20th century. Of course, it is the product of its

time, rooted in process studies, but in best Kuhnian tradition, it integrates the new paradigm with the old, offering novel approaches to dealing with the inadequacies of the Davisian approach, and thereby, the subject matter covered by Brown and Schumm is in effect incorporated into a single analysis of long-term landscape change.

Landscape sensitivity and change

Brunsden and Thornes (1979) summarised their ideas via four fundamental propositions of landform genesis:

1 *For any given set of environmental conditions, through the operation of a constant set of processes, there will be a tendency over time to produce a set of characteristic landforms.*

2 *Geomorphological systems are continually subject to perturbations which may arise from changes in the environmental conditions of the system or from structural instabilities within. These may or may not lead to a marked unsteadiness or transient behaviour of the system over a period of 10^2 to 10^5 years.*

Monitoring and Modelling Dynamic Environments, First Edition. Edited by Alan P. Dykes, Mark Mulligan and John Wainwright.

3 *The response to perturbing displacement away from equilibrium is likely to be temporally and spatially complex and may lead to considerable diversity of landform.*

4 *Landscape sensitivity is a function of the temporal and spatial distributions of the resisting and disturbing forces and may be described by the landscape change safety factor here considered to be the ratio of the magnitude of barriers to change to the magnitude of the disturbing forces.*

Brunsden and Thornes started from the idea that process domains exist within which the constant operation of process leads eventually to landforms characteristic of the domain. They distinguished between relaxation time, the period required to attain characteristic form and the length of time over which the characteristic form persists. Of course, they recognised that conditions were unlikely to remain steady over very long periods of time but argued that the characteristic form concept remained a valid and applicable position to adopt for less resistant systems where the time needed for adjustment was of a similar order of magnitude to that for changes to the external controlling variable, as well as a useful concept for theoretical modelling. Recognising that geomorphological controls rarely remain constant for long enough to enable the characteristic form to evolve, their second proposition focused on transient behaviour, unsteady landform response to changing external controls. They discussed the impact of both low-frequency–high-magnitude events (pulsed inputs) and gradual changes in environmental conditions (ramped inputs). Acknowledging Schumm's work, they identified the particular significance of thresholds; in simple terms, a threshold is the point at which a system's behaviour changes (Phillips 2006). This in turn led on to the third proposition – that landscape response to change is likely to be complex in time and space. While the most obvious type of response was the simple, self-limiting, stabilising response in which an impulse was damped out and the previous state restored (negative feedback), some impulses yielded a sustained response at a new level of geomorphological activity, often in relation to the existence of a threshold, while in still other cases, there was reinforcement of initial change through positive feedback. It followed that landscapes were likely to vary in their sensitivity to change, the fourth basic proposition. They borrowed the idea of a factor of safety from engineering geology to argue that landscape stability was a balance between the magnitude of disturbing forces and the magnitude of barriers to change. Once change had been initiated, the rate of change determined the time of attainment of a new characteristic form or, conversely, the persistence of the characteristics of the former state. They identified two end-member cases: mobile, sensitive systems which quickly relaxed and slowly responding, insensitive areas. Complexity of landscape response to change thus varied in time and space, with change most likely being episodic, and both transient and characteristic forms normally coexisting in the same landscape.

Brunsden and Thornes' synthesis drew on much work from the previous three decades, Hack (1960), Chorley (1962) and Schumm (1977, 1979) in particular. They saw the need to integrate the cycle of erosion as propounded by Davis (1899), in which time was the key variable, with the theory of dynamic equilibrium championed by Hack and Chorley, where landforms were independent of time; such views were not mutually exclusive. In many ways, their division of landforms into 'adjusting' and 'characteristic' complemented those of Davis (1899) and Schumm and Lichty (1965) but placed more emphasis on 'young', rapidly adjusting systems than on the graded condition where negative feedback dominated. They concluded that landscape change (in time *and* space) would be episodic and not continuous and, interestingly, quoted Derek Ager (1973) who had opined that evolution 'has been a very episodic affair, with short happenings, interrupting long periods of nothing much in particular'. The diffusion of change across a geomorphological system (again, in time *and* space) meant the location of

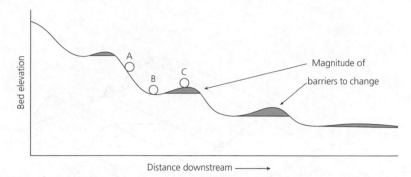

Figure 15.1 The two-dimensional analogy of states in a dynamic system showing unstable (A) and metastable (B) conditions and the notion of barriers to change (C). Redrawn from Brunsden and Thornes 1979; the caption is their original one.

transient behaviour moved – on either side, not much happens. The ultimate importance of the paper was therefore the emphasis on transient behaviour, recognition that under apparently constant conditions, very small changes in the environment might neverthe-less produce dramatic changes in behaviour, a theme discussed later.

Denys Brunsden has told me something about the way in which *Landscape Sensitivity and Change* was devised and written:

'We had endless discussions in the field, as you can imagine, and I am sure that the ideas emerged together but differently in both of us. In the end I went down to Dorset for a weekend and wrote a first draft. I brought it back a little unconfident to go ahead and publish it. So, I gave it to John for comment. He read it and made the comment 'I think we are on to something here'. He reorgan-ised, cut out bits and added all the stuff with balls and hollows, etc. It is fairly easy to pick out him and me – I am distinctly more on the traditional, less modelling side. Unlike *Geomorphology and Time* which was his first draft, the paper was mine. Nevertheless, so much of what we did was a prod-uct of endless field trips together in Spain 1969–88 so who knows who originated what'. (pers. comm.)

Indeed, John's interest in theory and concep-tual ideas was clearly evident, matters he had already addressed in Thornes (1978) and in *Geomorphology and Time* (Thornes and Brunsden

1977). The 'balls and hollows' to which Denys refers are the diagrammatic representations of stability (their Figure 15.7, reproduced here as Figure 15.1) and the way in which systems become more entrenched over time (their Figure 15.8, reproduced here as Figure 15.2). John's growing fascination with stability and equilibrium was further developed in a number of later articles including Thornes (1982).

Landscape Sensitivity and Change has had wide-spread influence across the field of geomorphol-ogy and beyond, and it is highly quoted. For example, Michael Thomas organised a special meeting on the theme, leading to a special issue of *Catena*. The abstract to his introduction (Thomas 2001) is worth quoting at length since it both neatly re-expresses the original ideas and illustrates the way in which the paper's influ-ence has spread right across the subject, both pure and applied:

Landscape sensitivity may be discussed in terms of the response of landscape systems to perturbation on different time and spatial scales. Unstable systems behave chaotically but may show self-organised criticality, while stable systems resist change until threshold values of system parameters are exceeded. Spatial sensitivity is expressed in dif-ferent rates of change, between landscape compo-nents or elements. This leads to divergence between landscape elements, and the inheritance of palaeoforms in present-day landscape mosaics. Temporal sensitivity reflects the magnitude and

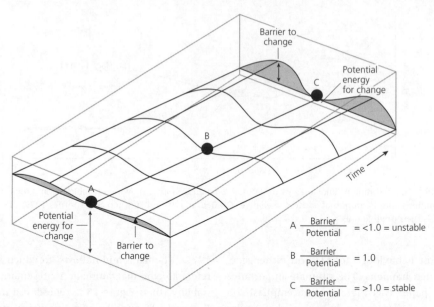

Figure 15.2 Through time, systems become more entrenched in furrows in the space–time manifold. The barriers to change become greater and more energy is required for a change from one equilibrium state to another. Redrawn from Brunsden and Thornes 1979; the caption is their original one.

frequency of individual events nested within patterns of longer term environmental changes occurring on different timescales. The resulting landscape complexity reflects the spatio-temporal sensitivity of earth surface systems over ten orders of scale magnitude. The connectivity within landscapes ensures that site instabilities can be propagated within multi-event feedback systems. Landscapes record their own histories in sediments and soils, but interpretation of event stratigraphy may not be straightforward, while soil profiles can absorb individual events without erosion. Although we are increasingly able to model the present, environmental management is dominantly about conserving inherited properties of landscapes: forests, soils, floodplains, coastlines. Landscape sensitivity for landscape management must, therefore, address not only active, largely non-linear, environmental systems, but also the mosaics and palimpsests that are the inheritance from past environments. (Thomas 2001, p. 83)

In my own contribution to the *Catena* special issue (Burt 2001), I began to explore the notion of management of catchment systems, identifying sensitivities in both time (e.g. periods of bare soil before crops can establish cover) and space (e.g.

the role of riparian zones providing a protective barrier between farmland and river); in doing so, of course, I inadvertently ventured into one of John Thornes' pet interests, his fascination with the role of vegetation (see Kirkby et al., 2015). Some of the spatial themes are further explored in Burt and Pinay (2005). I also included consideration of long water quality time series and might well have profited from making reference to Thornes (1982). One of John's earlier papers with Tony Edwards (Edwards and Thornes 1973) was on the same theme, and it remains relevant today, with ongoing concerns about long-term change and the dynamic metastability of catchment systems. Burt et al. (2008) used data from the same river (Figure 15.3), showing what can happen to water quality in response to large-scale land conversion (ploughing of grassland) and warning of the long-term consequences (water pollution) of short-term gain (crop yields). John's key question (Thornes 1982), whether we have enough data for meaningful systems analysis (we rarely have too much data), also remains very relevant in times of cost-cutting by monitoring

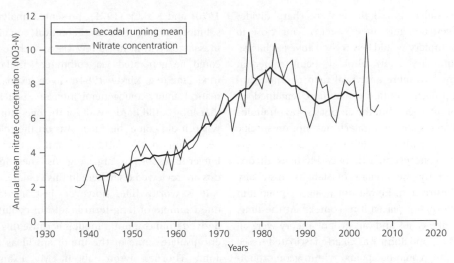

Figure 15.3 Nitrate concentrations in the River Stour, Essex, England, since 1937 (data from Burt et al. 2008).

agencies putting the maintenance of long bench-mark records under threat (Burt et al. 2010).

Evolutionary geomorphology

Notwithstanding the emphasis on transient behaviour, John's vision of geomorphology was ultimately concerned with the long-term behaviour of landforms. In his 1983 paper in *Geography* (Thornes 1983), he laid out what Jonathan Phillips (2006) describes a 'blueprint for evolutionary geomorphology' based on complex dynamical systems. This synthesised a number of strands of enquiry that John had followed over the recent years. In noting a renewed interest in long-term landform change, he argued that such studies should be soundly based in theory rather than on historical inference. This echoes an earlier book chapter (Thornes 1978) where he reviewed the character of geomorphological theory and, in so doing, attempted to bridge between earlier work on denudation chronology and modern process studies. Casting aside 'weaker' forms of theory, John called for geomorphological theory to be as well grounded as possible in the basic laws of physics and chemistry (he might

well have added 'biology' – see Thornes 1985). It is interesting that, in order to avoid the problem of specifying all the processes in operation, he proposed a coarser scale of resolution which avoided considering the detailed mechanics. This aligned nicely with his call for the developments of macroscale theoretical models, including the development of computer simulation models (in 1978, requiring 'formidable' amounts of computing but soon to be made much easier).

In Thornes (1983), John returned to the first proposition of *Landscape Sensitivity and Change*, the idea of process domains. Defining an area dominated by a particular landform or process as a domain, process geomorphology was primarily concerned with behaviour within the domain. Evolutionary geomorphology had a wider scope, concerned with 'the initiation and development of the structure giving rise to the domains' (Thornes 1983, p.227). Phillips (2006, p.368) interpreted 'structure' as the structural relationships among processes, geological controls, climate, relief and other factors rather than geological structure per se; in other words, the broad combination of environmental conditions within which process determines form. At the heart of a domain, systems would be expected to be stable, but if there were sharp divides between domains, then systems lying close to the boundary would be very sensitive to change. Evolutionary geomorphology, being rooted in theory, was more analytical than chronological in approach: landform evolution remained an important focus, but the emphasis was on multiple stable states and trajectories, not on specific historical pathways. John went on to review recent concepts in dynamical systems theory, particularly the notions of stability, instability and bifurcation; he did not neglect equilibrium therefore, but placed it in context. At the time, ideas like chaos and catastrophe theory were all the rage, and John was clearly fascinated to see how they could be applied within geomorphology. In particular, he spent some time describing the cusp catastrophe and its possible application to the sediment transport process. He quotes the work of Will Graf, and indeed, it is no surprise to find that Graf (1982) is a fellow contributor to the same Binghamton Symposium volume in which Thornes (1982) appears.

These themes, and many more, were presaged in *Geomorphology and Time* (Thornes and Brunsden 1977). There, John Thornes and Denys Brunsden 'attempt(ed) to provide and overall framework within which to consider and compare old and new approach to the subject: the qualitative model-building approach of the pre-1950s, the empirical, observation, process measurement and field measurement technique of the 1960s and currently fashionable analytical modelling'. As noted previously, John produced the first draft, and his influence was particularly evident in the final chapters dealing with quantitative deterministic models of temporal change and stochastic models of form evolution. Of course, the book reflected fully the quantitative revolution in geomorphology, but as with many of the articles reviewed here, the timing was effectively pre-computer – computing capacity was very limited in the mid-1970s compared to now. Thus, the review of hillslope process-response models in Chapter 7, built on the work of Culling (1960, 1963), Scheidegger (1961,

1970) and Kirkby (1971), presented analytical solutions to the general theoretical equations presented. Within just a few years, such results could be generated via computer simulation modelling (e.g. Kirkby 1984). Given his mathematical ability and general interests, one feels that John could hardly wait for the opportunity; when it did come, his focus was on the role of vegetation in the landscape, rather than on longer timescales or larger regions. These interests are reviewed elsewhere in this volume.

It is worthwhile, however, just including one example of long-term landform evolution derived from computer models because this can encapsulate some of the important ideas that John Thornes wrote about. My example (Figure 15.4) comes via Mike Kirkby, using a later version of a computer simulation model he had developed (Kirkby 1984) based on his process equations for characteristic slope forms under both transport-limited and weathering-limited removal (Kirkby 1971) to model a series of slope profiles in South Wales originally described by Savigear (1952). The Savigear 'sequence' is one of the most frequently cited examples of space–time substitution: that is, the set of slope profiles in space may be regarded as demonstrating the pattern of slope evolution over time. The processes included in the Kirkby model were creep/solifluction, wash and mass movement. Kirkby's main purpose was to produce a version of his model that would satisfactorily reproduce the Savigear sequence, but in so doing, he was able to speculate more generally and to relate his findings to the various models of slope development available in the literature. The relevance of the example here is that several of the concepts presented in *Landscape Sensitivity and Change* and elsewhere could be briefly illustrated. The role of thresholds was shown during the early phase of valley incision when undercutting of slopes generated mass movements with slope angles at the maximum angle of stability. Once undercutting ceased, creep and wash reduced slope angles below this angle, and a

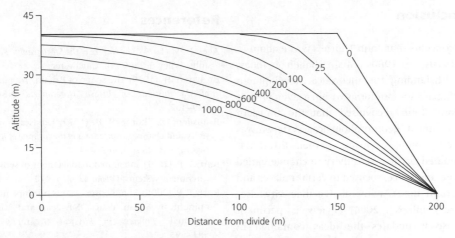

Figure 15.4 The pattern of slope evolution over time (see text for explanation).

characteristic convexo-concave slope form gradually developed. Note that the characteristic upper convexity developed while the transient straight slope form was still evident; as noted earlier, both transient and characteristic forms normally coexist in the same landscape. Insensitivity to change was illustrated via the interfluves, well away from the valley: this is where change was slowest to occur and thus where evidence of previous process domains remained longest – Eric Brown's erosion surfaces in effect. Of course, the simulation results shown on Figure 15.4 are just one set of very many possibilities: varying one or more of the model parameters can yield virtually innumerable sequences (a few are provided in Burt 2005). While in one sense the original Kirkby model was designed to reproduce the Savigear sequence, there was a much wider purpose: to speculate more generally about the multiple pathways along which slopes might evolve. Small changes could have dramatic effects: for example, if the rate of valley incision was too slow, steep, straight valley-side slopes never appeared in the landscape. One of John Thornes' abiding themes in evolutionary geomorphology was the possibility of multiple stable states, the bifurcations between them and the trajectories connecting them. It is clear that

he was dissatisfied with the 'equilibrium approach' and sought to understand the movement of systems away from equilibrium and, in due course, back towards equilibrium again.

Since John's long-lasting interest in the role of vegetation within geomorphological systems is being reviewed elsewhere in this volume, little needs to be added here. However, it is worth making reference to his introduction to *Vegetation and Erosion* (Thornes 1990), given the way he linked process studies through to landform evolution. He saw three major areas of study where vegetation was important in geomorphology: first and the most obvious, its role in controlling the nature and rate of operation of processes. Second, he recognised the interaction between plants and geomorphological processes and implicitly linked this to the notion of process domains and landscape stability. Third and the most complex in his opinion, he identified relationships between climate, vegetation and landform evolution. Again, it was clear that he was interested in both equilibria and the transitional states between them. Notwithstanding imperfect understanding of the impact of vegetation cover on geomorphological processes, he characteristically advocated making a start on the work of modelling landform change.

Conclusion

The concepts that John Thornes wrote about in the 1970s and 1980s remain challenges today. Notwithstanding the theoretical development of characteristic landforms within a given process domain, John recognised that the transient behaviour of geomorphological systems was typically complex, owing to their threshold-dominated nature. Sensitivity to change varied in time and space, focused in certain places and at certain times, with relative inactivity elsewhere. Phillips' (2006) review is pertinent because it updates the ideas contained in Brunsden and Thornes (1979) and Thornes (1983), stressing in particular the importance of contingency, both geographical and historical. Geography mattered because local variations and disturbances resulted in increasing divergence over time. History mattered because geomorphological systems remained sensitive to initial conditions and perturbations. John's later research illustrated in particular complex response and the sensitive nature of vegetation cover in controlling process and change.

In the late 1970s, John noted that geomorphology seemed 'poised on the verge of considerable advances in the theory of processes which should lead to understanding and perhaps even some answers to the classical problems of landform evolution' (Thornes 1978, p. 21). At the same time, he expressed the thought that 'it would be a great pity if the flowering of theory was nipped in the bud by a complete shift in emphasis to "relevant" or "applied" research' (Thornes 1978, p.19). To some extent, 'applied research' may drive current research focus more than it used to, but not apparently at the expense of theoretical developments. Moreover, parallel developments in cosmogenic dating of ancient surfaces have allowed a revival of studies of long-term landform evolution. If anything, the science of geomorphology is more varied and challenging than ever. John Thornes' contribution was to help lay the foundations for the study of complex landform systems.

References

Ager, D. V. (1973) *The Nature of the Stratigraphic Record*. John Wiley & Sons, Ltd, Chichester.

Brown, E. H. (1979) The shape of Britain. *Transactions of the Institute of British Geographers* **NS4**(4), 449–462.

Brunsden, D., Thornes J. B. (1979) Landscape sensitivity and change. *Transactions of the Institute of British Geographers* **NS4**(4), 463–484.

Burt, T. P. (2001) Integrated management of sensitive catchment systems. *Catena* **42**, 275–290.

Burt, T. P. (2005) Some observations on slope development in South Wales: Savigear and Kirkby revisited. *Progress in Physical Geography* **27**, 581–595.

Burt, T. P., Pinay, G. (2005) Linking hydrology and biogeochemistry in complex landscapes. *Progress in Physical Geography* **29**, 297–316.

Burt, T. P., Howden, N. J. K., Worrall, F., Whelan, M. J. (2008) Importance of long-term monitoring for detecting environmental change: lessons from a lowland river in south east England. *Biogeosciences* **5**, 1529–1535.

Burt, T. P., Howden, N. J. K., Worrall, F., Whelan, M. J. (2010) Long-term monitoring of river water nitrate: how much data do we need? *Journal of Environmental Monitoring* **12**, 71–79.

Chorley, R. J. (1962) Geomorphology and general systems theory. United States Geological Survey Professional Paper 500(B). US Government Printing Office, Washington, DC.

Culling, W. E. H. (1960) Analytical theory of erosion. *Journal of Geology* **69**, 336–344.

Culling, W. E. H. (1963) Soil creep and the development of hillside slopes. *Journal of Geology* **71**, 127–161.

Davis, W. M. (1899) The geographical cycle. *The Geographical Journal* **14**, 481–504.

Edwards, A. M. C., Thornes, J. B. (1973) Annual cycle in river water quality: a time series approach. *Water Resources Research* **9**, 1286–1295.

Graf, W. L. (1982) Spatial variation of fluvial processes in semi-arid lands. In: Thorne, C. E. (ed.) *Space and Time in Geomorphology*. George Allen & Unwin, London, pp.193–217.

Hack, J. T. (1960) Interpretation of erosional topography in humid temperate regions. *American Journal of Science* **258-A**, 80–97.

Kirkby, M. J. (1971) Hillslope process-response models based on the continuity equation. *Transactions of the Institute of British Geographers*, Special Publication No. 3, 15–30.

Kirkby, M. J. (1984) Modelling cliff development in South Wales: Savigear re-viewed. *Zeitschrift fur Geomorphologie* **28**(4), 405–426.

Kirkby, M. J., Bracken, L., Brandt, C. J. (2015) John Thornes and desertification research in Europe. In: Dykes, A. P., Mulligan, M., Wainwright, J. (eds.) *Monitoring and Modelling Geomorphological Environments*. John Wiley & Sons, Ltd, Chichester.

Phillips, J. D. (2006) Evolutionary geomorphology: thresholds and nonlinearity in landform response to environmental changes. *Hydrology and Earth System Science* **3**, 365–394.

Savigear, R. A. G. (1952) Some observations on slope development in South Wales. *Transactions of the Institute of British Geographers* **18**, 31–52.

Scheidegger, A. E. (1961) Mathematical models of slope development. *Bulletin of the Geological Society of America* **72**, 37–49.

Scheidegger, A. E. (1970) *Theoretical Geomorphology* (2nd edition). Allen & Unwin, London.

Schumm, S. A. (1977) *The Fluvial System*. John Wiley & Sons, Inc, New York.

Schumm S. A. (1979) Geomorphic thresholds: the concept and its application. *Transactions of the Institute of British Geographers* **NS4**(4), 485–515.

Schumm, S. A., Lichty, R. W. (1965) Time, space and causality in geomorphology. *American Journal of Science* **263**, 110–119.

Thomas, M. F. (2001) Landscape sensitivity in time and space – an introduction. *Catena* **42**, 83–98.

Thornes, J. B. (1978) The character and problems of theory in contemporary geomorphology. In: Embleton, C., Brunsden, D., Jones, D. K. C. (eds.) *Geomorphology: Present Problems and Future Prospects*. Oxford University Press, Oxford, pp. 14–24.

Thornes, J. B. (1982) Problems in the identification of stability and structure from temporal data series. In: Thorn, C. E. (ed.) *Space and Time in Geomorphology*. George Allen & Unwin, London, pp. 327–353.

Thornes, J. B. (1983) Evolutionary geomorphology. *Geography* **68**, 225–235.

Thornes, J. B. (1985) The ecology of erosion. *Geography* **70**, 222–235.

Thornes, J. B. (1990) Introduction. In: Thornes, J. B. (ed.) *Vegetation and Erosion: Processes and Environments*. John Wiley & Sons, Ltd, Chichester, pp. 1–3.

Thornes, J. B., Brunsden, D. (1977). *Geomorphology and Time*. Methuen, London.

CHAPTER 16

John Thornes and desertification research in Europe

Mike Kirkby[1], Louise J. Bracken[2] and Jane Brandt[3]

[1] School of Geography, University of Leeds, Woodhouse Lane, Leeds, UK
[2] Department of Geography, Durham University, Durham, UK
[3] Fondazione MEDES, Sicignano degli Alburni (SA), Italy

Introduction

In 1986, the European Commission embarked upon a new venture, the funding of collaborative research projects on themes of particular interest to Europe, with the aim of shared capacity building across the European scientific community. Inclusion of teams from the newly acceded member states of Spain, Portugal and Greece was encouraged. His long-term research interests and existing collaborations in Spain placed John Thornes in an ideal position to take advantage of the opportunities that this first framework programme (FP) offered. The 3-partner EPOCH Project was followed by Mediterranean Desertification and Land Use (MEDALUS) I, II and III, Concerted Action and DESERTLINKS, all of which John coordinated, as well as EFEDA I and II, MEDACTION and DESURVEY in which he was a partner or advisor. Undoubtedly, many have played their part in these and the other desertification-related projects. However, particularly in the first FPs, John's role was as significant as any in developing an understanding of the many aspects of desertification as it affects Mediterranean Europe and an interdisciplinary research community capable of tackling them.

This chapter does not attempt to review all the 20 or so European desertification research projects that have been completed to date. Instead, it looks at the evolution of the FPs and the key research issues and approaches that evolved through the successive projects that John Thornes was influential in planning and running.

What is desertification?

Although the intuitive view of desertification pictures the advance of sand dunes over previously fertile land and the burial and/or abandonment of settlements as the desert margin advances, the United Nations Convention to Combat Desertification (UNCCD) has defined desertification in a less dramatic way and one that may be important for many parts of southern Europe as well as for the Sahel. For the UNCCD, desertification means land degradation in arid, semi-arid and dry sub-humid areas resulting from various factors, including climatic variations and human activities. Dryland degradation, in turn, means the reduction or loss of biological or economic productivity and complexity for rainfed cropland, irrigated cropland or range, pasture, forest and woodlands. It is considered to result from land uses that have become unsustainable through the action

Monitoring and Modelling Dynamic Environments, First Edition. Edited by Alan P. Dykes, Mark Mulligan and John Wainwright.
© 2015 John Wiley & Sons, Ltd. Published 2015 by John Wiley & Sons, Ltd.

of one or a combination of processes, many of them arising from or aggravated by human activities and habitation patterns and potentially aggravated by climate change (UNCCD 2007). The three most important of these processes are soil erosion, loss of soil fertility and long-term loss of natural or desirable vegetation.

Soil erosion is the physical removal of the surface soil by wind, water or mass movements. In the most arid areas, wind mobilises material from the soil surface, usually redepositing coarser fractions locally and removing the fines (silt and clay) and organic matter that supports most of the nutrient in the soil and helps to store rainwater. The winnowed surface in turn grows only sparse vegetation, further increasing the intensity of wind shear at the surface. Where there is enough rain to generate significant runoff, erosion by water tends to become dominant, washing away fertile topsoil, so that less water is held in the soil to support vegetation growth. At progressively wetter sites, increasingly dense vegetation protects the surface from crusting that seals the surface and so decreases runoff. The conflict between increasing rainfall and increasing natural vegetation cover leads to a maximum of water erosion in semi-arid climates. However, cultivation generally exposes the soil at the beginning of the rainy season when crops must be planted, and water erosion can then be severe, irrespective of climate. Wetter sites can also be prone to mass movements where the parent material is rich in clays, and a few sites are also dominated by this type of erosion.

Loss of soil fertility can take a number of forms, and some of them are closely linked to soil erosion by water or wind that removes the organic matter and finer-grained topsoil. However, the most widespread form of desertification through loss of fertility is associated with salinisation of the soil. In a dry climate, much of the soil water evaporates near the soil surface, and any solutes in the water are deposited as a saline crust. Although the commonest constituent of these salts is calcium carbonate, which is relatively benign, the second most important components are sodium salts, which stunt the growth of most crop plants, damage the physical structure of the soil and are very hard to remove in severe cases. The conditions for salinisation are a dry climate and the availability of plentiful, poor-quality water, which combine to promote high levels of evaporation. Plentiful water in a dry climate is usually associated with shallow groundwater in areas of low relief and/or with irrigation.

Vegetation changes (Thornes 1990) can also degrade the potential land uses of an area, often due to the invasion of rangeland by unpalatable species, in many cases replacing grass with shrubs. Overgrazing exacerbates this process, giving a selective advantage to the unpalatable plants, but climate change and the introduction of non-endemic species may also play a part.

The history of semi-arid agriculture and civilisation, both in the Middle East and in Meso-America, has always been stalked by the threat of desertification, balancing sustainable production against water scarcity, erosion and salinisation. Today, combating desertification is generally one aspect of an integrated development approach that targets sustainable land use through a number of physical measures that aim to prevent or reduce land degradation with rehabilitation of partly degraded land and perhaps reclamation of more severely desertified areas. Desertification is generally defined in physical terms, but the root causes are intimately bound up with the availability of water and with socio-economic factors including the economic level and economic resilience of farmers, land tenure and societal values.

Desertification research in Europe

Within the European Community (EC), desertification has been a topic of active research since the 1980s. Table 16.1 summarises the evolution of the EC research agenda through successive FPs. The symposium held in Mytilene in 1984 (Fantechi and Margaris 1986) is often cited

Table 16.1 Desertification within the European research agenda.

Year	European research agendas	Projects	Themes
1984	Symposium on desertification, Mytilene (Fantechi, Margaris)		
1985			
1986			
1987	FP1: Impact of climate change/ variability on land resources	EV4C	
1988			
1989			
1990	FP2: Impact of climate and human action on the European environment		
1991		MEDALUS I	
1992			
1993	FP3: Causes and consequences of desertification in the Mediterranean area	EFEDA	
1994		MEDALUS II	
1995	UNCCD adopted, including northern Mediterranean Annex		
1996			
1997	FP4: Impact of desertification on natural resources and soil erosion in Europe		
1998		MEDALUS III	
1999			
2000	FP5: Scenarios and strategies for responding to land degradation and desertification. Tools to study and		
2001	understand changes in the environment and underpin EU policies (e.g. towards the UNCCD)	MEDACTION	
2002		DESERTLINKS	
2003			
2004	FP6: Detailed management strategies to monitor and combat desertification in the Mediterranean and worldwide		
2005		RECONDES	
2006		DESURVEY	
2007		LUCINDA	
2008	FP7: Tools and technologies to support sustainability and the UNCCD in Europe and worldwide	DESIRE	
2009			
2010			
2011		LEDDRA	
2012			

Themes (arrows, top to bottom): Desertification → sustainability; Biophysical processes → socio-economic processes; Top-down approaches → participatory processes as the birthplace of European desertification research as a specific topic in its own right. This workshop was sponsored by the EC and brought together scientists from across Europe and from a range of disciplines to promote research and establish an agenda. It was one of key activities responsible for the inclusion of desertification in the first and all subsequent research FPs.

Over more than 20 years, the emphasis of the FPs has changed significantly, both in its scientific stance and in a progressive shift from basic research to more applied objectives. In FP1 (1986), the EPOCH programme looked at the impact of climatic variability on land resources, desertification being considered a consequence of climate change. In FP2 (1989), a similar emphasis continued with the focus on climate impacts and climate-related hazards, together with the effects of climate change on the European environment. In FP3 (1991), the focus was on desertification in the Mediterranean area, assessing natural and human causes and mechanisms of desertification, its impact, history, causes and consequences. FP4 (1994) asked for a focus on the impact of climate changes and other environmental factors on natural land resources, in the context of desertification and soil erosion in Europe. At this time, FPs were being developed in parallel with the emergence of an international consensus on the global importance of desertification, and this led, in 1994, to adoption of the UNCCD, including a northern Mediterranean Annex (Annex 4) to reflect the importance of this issue in southern Europe. Consequently, FP5 (1998) included a requirement to make a European research contribution to the UNCCD. The focus shifted to scenarios and strategies for responding to the global issues of fighting land degradation and desertification. Researchers were asked to develop the scientific, technological and socio-economic methodologies and tools necessary to study and understand changes in the environment. Projects were expected to underpin EU policies relating to the UNCCD and the other environmental conventions.

In FP6 (2002), the focus was much more sharply defined, and there was an explicit requirement for projects to exploit their research results and to show the potential to deliver practical solutions. Combating desertification was mentioned specifically for the first time, together with management strategies and measures, development and demonstration of best practices, control measures, operational and analytical tools and techniques and harmonised data-information systems for the prevention, mitigation and control of land degradation. International cooperation with other relevant regions affected by desertification was also explicitly expected for the first time. The emphasis on development of practical tools and technologies was carried further in FP7 (2007) and focused on supporting sustainability with a less explicit concentration on desertification as a separate issue. There has also been an increased focus on international partnerships, so that research supports UNCCD activity both within Europe and worldwide.

Some broad trends can be discerned in the evolution of the FPs. Initially, climate change was clearly identified as the main driver of desertification processes, but the importance of considering human as well as climatic drivers was soon recognised. With the greater involvement of social scientists, there has also been an associated shift in emphasis from traditional top-down scientific approaches to a greater dependence on participatory methods, involving stakeholders at levels from the farmer to EU policymakers. In parallel, there has been a shift from treating desertification as a separable stand-alone issue to considering it as one aspect within the broader issue of sustainability.

A second important trend is in the study areas of interest. An initial focus on the pure science was quickly changed to a specific focus on the northern Mediterranean, helping to define the UNCCD Annex 4 countries and offering the potential for methodological support in the development of the National Action Plans. However, over time, this focus has broadened to encompass all areas impacted by the threat of desertification, asking researchers to engage with practical strategies for surveillance, mitigation and remediation worldwide.

As the FPs evolved towards improving the links between research and policy, their foci have also shifted to take account of research results and themes that emerged from the EU-funded projects themselves and other parallel work. At first, progress was hampered by a lack of already existing long-term data, in an environment that is defined by its irregular and high-magnitude natural events and commonly lacks the dense meteorological and hydrometric network of more humid areas. One important direction of initial work was therefore the establishment of common measurement methods and the establishment of measurement sites. It was also clear that results would eventually be needed at a regional scale if they were to be relevant to policy, so that although small measurement sites or plots were initially required for the basic scientific research, these sites were progressively incorporated into catchments and larger areal units so that results could be evaluated and applied at the scale of much larger administrative areas.

Two fundamental approaches to addressing regional scales have been developed, first through modelling and second through areal classification using indicators, together with some necessary crossover between these approaches. Modelling began by addressing the plot scale to engage with the underlying science, but it quickly became apparent that the data demands for directly extending this fine-scale approach to larger areas were prohibitive. Over time, therefore, models have tended to work at progressively coarser resolutions, allowing the use of new and pre-existing data sources but at the expense of fine detail and with some necessary simplification of the processes acting. Models have also increasingly been developed that begin to integrate biophysical and socio-economic processes and that acknowledge the importance of stakeholder participation in framing both the issues and the impacts, although there is still inevitable uncertainty about long-term trends. Models have also diverged into those that make specific forecasts about possible futures (or the range of possibilities) and those that look more generically at desertification syndromes.

Indicators have been derived from direct field observation or questionnaires, existing statistical and geographic (GIS) databases and remote sensing. In some cases they have been presented in the DPSIR framework (i.e. drivers, pressures, state, impact, response) and over time the number of proposed indicators has proliferated until almost 200 have been identified. It has become necessary to identify groups of indicators, looking at the extent to which they are related to one another as well as creating composite indicators, or indices, that give an overall view of the current state of desertification and/or its direction of change over time.

From these analyses of the state of current desertification and future trends, there has also been increasing attention on how best to combat desertification, through policy measures or through conservation measures to mitigate or reverse processes of desertification.

John Thornes' role and achievements in desertification research

Although the thread that is the contribution of any individual is always difficult to untangle from the string of concurrent and progressive advances in science, John Thornes has played a central role, perhaps the most central role, in the advance in understanding of desertification over the last 20 years. Although the term 'desertification' was perhaps first used by Aubréville (1949; Davis 2004), its emergence as a current theme can be traced most directly to the UN conference (UNSCD 1977) in response to recurrent drought in the Sahel and in Europe to the conference organised by Fantechi and Margaris (1986).

John was one of those attending the 1986 conference, and this led to a long series of EU-funded grants, principally but not only under the name of MEDALUS which ran in various phases from 1991 to 1999. His involvement always went well beyond simple participation in these FPs, which was still continuing at the time of his death. He provided a strong and clear leadership in coordinating the MEDALUS projects and, in so doing, created a Europe-wide research community with a sense of common purpose and formed a MEDALUS 'family' that is still influential in carrying forwards research into desertification. Although this is perhaps his most important and durable contribution overall, there are many more specific advances in which John's personal contribution and drive have played a crucial part, and of the issues

identified in Table 16.2, the direction of research was influenced by him in all but a very few cases:

• **Lack of field data – field manuals:** The first MEDALUS project established comparable field sites in seven locations throughout the northern Mediterranean, from the Alentejo in Portugal to Thessaloniki in Greece. A common monitoring programme provided both empirical and modelling activities with much needed data. A manual of measuring and monitoring techniques for specific use in semi-arid environments was written to support the field programme (Cammeraat 1998). Not all the study sites outlived the specific projects for which they were built. However, the Rambla Honda site in Almeria, Agri basin in Italy and the Greek Island of Lesvos all continued to be used for over 20 years in

Table 16.2 Some key issues in desertification research.

	EV4C	MEDALUS I	MEDALUS II	MEDALUS III	MEDACTION	DESERTLINKS	DESURVEY
Role of climate change	Analysis of change		Production of scenarios		Use of scenarios in regional modelling		
Modelling physical processes		2-dimensional and hillslope scale	Regional scale		Use of regional-scale output		Integrating with socio-economic processes
Use of target areas		Desertification Response Units – DRU	Target areas to integrate research	Environmentally sensitive areas – ESAs	Engagement of stakeholders in target areas		
Lack of data		Extensive field data collection, following MEDALUS I Field Manual				Mining of databases	
Integration of RS data		Analysis of land-use patterns, Analysis of RS time series					RS to monitor desertification
Integration with human dimensions		Dominated by the physical dimension			Progressive integration of social, economic and institutional dimensions		Integration within models
Engagement with stakeholders					Specific programmes for local and national engagement		
Dynamics of grazing patterns		Inclusion of grazing in models	Partnership with grazing experts				Ecological modelling for dynamics
Exploitation of results				Desertification Atlas		Application of indicators	Establishing surveillance systems

successive projects, enabling significant long-term data sets to be collected and a number of models to be developed and refined.

- **Use of target areas to integrate dimensions:** The small field sites established in MEDALUS I were soon expanded into larger target areas. This method of taking case study areas for an integrated investigation of desertification has had many benefits. In MEDALUS, an attempt was made to identify those areas in the landscape that would respond in a similar fashion to changing pressures. The areas were increased in subsequent projects to cover the whole municipality or river basin units in an attempt to integrate the social, economic and physical dimensions of desertification. This strategy was particularly successful in the method for determining environmentally sensitive areas (ESAs). By the time FP5 started, with its emphasis on interaction with stakeholders, the research teams were well placed to establish contact with extensive networks of local communities and to start to exploit the research results. Several of the sites originally established in the early 1990s are still active today, supported in various ways that were never envisaged in the original MEDALUS I project and providing a valuable springboard for subsequent research.

- **Modelling physical processes:** Starting with 2-dimensional and hillslope-scale models of physical processes, the modelling of desertification processes developed through increasing temporal and spatial scales until erosion and salinisation processes could be modelled at the European scale. This level of coverage and resolution has proved to be of particular use to those stakeholders working at national and international scales. Process models were initially developed at the finest hillslope scale, in parallel with the initial small field sites of MEDALUS I (Thornes and Brandt 1996). Although these were a useful test bed for developing concepts, it became apparent that their data requirements restricted their transferability and hence their practical utility.

In MEDALUS II, a catchment-scale model, MEDRUSH, was developed and applied within the larger test catchments, but even at this scale, the data requirements proved too onerous for wider application (Kirkby et al. 1997). Finally, in MEDALUS III, a coarser-scale model, finally operating on $1\,km^2$ grid cells, evolved through a parallel project into the PESERA model, and this has proved applicable to the whole of Europe and is still in widespread use (Kirkby et al. 2008). Other models used within MEDALUS included the SHETRAN hydrological and sediment transport model, which was successfully applied in MEDALUS I and II (Bathurst et al. 2003) and the conceptual Desertification Response Unit (DRU) model (Geeson et al. 2002).

- **Indicators and ESAs:** In MEDALUS II, assessment of catchment areas, many of over $1000\,km^2$, demanded the assessment of desertification risk using less intensive methods than the first small study sites. The most successful and widely applied outcome of this development was encapsulated in the concept of ESAs, which made use of a well-documented and widely used indicator system (DIS4ME 2004) to discriminate along the multidimensional scale between pristine and desertified environments, using a subset of indicators that were relevant within a particular environment and at a particular scale of interest (Kosmas et al. 2000; Salvati et al. 2009).

- **Role of climate change:** An analysis of climate change over the historical record was followed by the production of climate scenarios using general circulation models. While climate change was no longer specifically addressed by the projects after MEDALUS III, the scenarios and empirical data continued to be used in regional-scale and long-term modelling. However, it has been concluded that, over the timescale of 10–30 years, changes in land use due to economic pressures and migration are likely to have a much greater impact on desertification than expected climate changes, even though climate may have an increasing impact over longer time spans.

- **Engagement with local stakeholders:** Although the stated objective of all the projects since their start has been to produce results of use to decision-makers, it was not until FP5, in MEDACTION and DESERTLINKS, that external stakeholders were explicitly involved in the projects' work programme. This trend continued with the DESIRE project in FP6 in which stakeholders at all levels played a significant role in determining parts of the research programme and how they would like the results to be made available to them. It has gradually become clear that stakeholders operate at a very wide range of scales and styles. At the finest scale, stakeholders are directly involved in implementing remedial measures on the land, whereas at coarser scales stakeholders are providing advice or implementing and creating policies that may have local to supranational impacts. At the field scale, concerns are with physical feasibility and with the short- or long-term costs and benefits of proposed measures. At coarser scales, it is vital that policies, through subsidy, regulation or agro-environmental codes of practice, have sufficient flexibility to be applied over the full range of environments where they are relevant. Successful research projects provide tools for decision support, based, for example, on indicator systems or forecasting models, and these tools require training before adoption and application by technically qualified staff, whose role is generally to advise the politicians and senior administrators who ultimately create and modify policies. Technical advisers need to periodically re-brief their principals who, in many cases, retain a particular portfolio for only a few years. It follows that management tools, if they are to be used at all effectively, must be widely implemented through both a user community and within the scientific community, so that there is considerable value added where methodologies continue to be developed through a sequence of projects. Through MEDALUS and successor projects, this objective has at least partially been achieved for the indicator and ESA methodologies.

- **Integration of socio-economic and physical aspects of desertification:** Perhaps one of the biggest research challenges faced by the interdisciplinary teams in desertification research has been really effective integration of firstly different physical sciences and then the physical and social sciences. Until the end of MEDALUS III, the main efforts were in placed in integrating climate, hydrology, soil physics and chemistry and vegetation growth. One of the most influential outcomes of this process was the publication of the Desertification Atlas (Mairota et al. 1996) which synthesised the level of integration that has been achieved by the end of MEDALUS III. Thereafter, the research teams were expanded to include the economic and social sciences. Both MEDACTION and DESERTLINKS were effective in incorporating knowledge of land-use planning, policy and actor networks and decision-making in agricultural economics into an increasingly sophisticated understanding of a multifaceted problem. In the Desurvey (2011) project, models were constructed that integrated biophysical and socio-economic processes (e.g. van Delden et al. 2011). The LEDDRA (2014) project took this integrated approach even further with its goal of assessing how fit are the many types of responses (no action, prevention, mitigation, restoration/rehabilitation, adaptation; all within prevailing environmental, socio-economic and institutional conditions) that are made to land degradation and desertification in crop and grazing lands and in forest and shrublands.

- **Application of research results in mitigation and remediation strategies:** The exploitation of research results and their application to desertification management programmes has been included in the objectives throughout the series of projects. However, from FP5 onwards, the involvement of stakeholders in research projects meant that they were able to specify what they needed to know and in what form. In its Desertification Indicator System for Mediterranean Europe (DIS4ME 2004), DESERTLINKS made a significant effort to present its research results in a clear narrative that could be accessible and understood by non-specialist readers. This multilevel dissemination was further developed in DESIRE's Harmonised Information System (DESIRE 2013) and again in LEDDRA's Land and Ecosystem Degradation and Desertification Response Information System (LEDDRA 2014).

• **Grazing dynamics:** John Thornes had, for many years, been very strongly involved in researching the relationships between erosion, vegetation and ecology (e.g. Thornes 1990). In some of his last work, he had been developing a prototype model, GRAZER (Thornes 2005, 2007), to simulate grazing behaviour for rangeland, forest and marginal lands in Mediterranean and other semi-arid environments. It was developed within the DeSurvey (2011) project as a contribution to understanding the vulnerability of marginal lands with respect to desertification and provides one important conceptual underpinning to the development of the vulnerability model developed in DESURVEY. These land-use types are thought to be among the least resilient, lacking buffering capacity against land degradation due to centuries of misuse and abuse, through neglect and/or over-exploitation. GRAZER has been developed for the semi-extensive livestock production that is ubiquitous in Mediterranean rangeland systems, thus providing a fresh and relevant approach, in which the four main components interact dynamically: soil hydrology, plant growth, grazing and management.

The legacy

It is premature to judge the longer-term impact of John Thornes' work, but it is already clear that many of the concepts and approaches that

he pioneered in the MEDALUS projects, both personally and with the very many colleagues involved, are still the active currency of research in desertification. We can look at publications and citations (most notably Thornes and Brunsden (1977) and Brunsden and Thornes (1979)) and see that John's work has been widely and consistently cited, particularly since the beginning of the 21st century, as the impact of the various projects has been absorbed by the scientific community. However, the greatest, most lasting and pervasive legacy has probably been the community of researchers working on desertification that has been developed through the series of EU framework projects and through John's work at Bristol University and King's College, London. It is hard to make an exact count, but at a conservative estimate, at least 400 researchers have been involved over the last 20 years, including the 200 directly involved in MEDALUS and the many more through successor projects. The reach of this research is ultimately felt through its continuance and evolution through this community.

The three particular areas of MEDALUS-related desertification research that seem, at present, to have had the most long-lasting impact are the methodologies for defining ESAs, the potential to model the interactions between social and physical processes as one way of exploring the multiple impacts of policy changes and the fundamental importance of the links between degradation and ecology. All of these show the progress that has been made, away from simplistic analyses of degradation towards a fuller appreciation of the complex web of interactions that influence degradation, productivity and maintenance of sustainable livelihoods.

References

Aubreville, A. (1949) *Climats, Forêts et Désertification de l'Afrique Tropicale*. Société d'Editions Géographiques, Maritimes et Coloniales, Paris.
Bathurst, J. C., Sheffield, J., Leng, X., Quaranta, G. (2003) Decision support system for desertification

mitigation in the Agri Basin, southern Italy. *Physics and Chemistry of the Earth* **28**, 579–587.

Brunsden, D., Thornes, J. B. (1979) Landscape sensitivity and change. *Transactions pf the Institute of British Geographers NS* **4**(4), 463–484.

Cammeraat, L. H. (1998) *MEDALUS (Mediterranean Desertification and Land) Field Manual (Version 4)*. MEDALUS Project Office, Thatcham, 116 pp.

Davis, D. K. (2004) Desert 'wastes' of the Maghreb: Desertification narratives in French colonial environmental history of North Africa. *Cultural Geographies* **11**(4), 359–387.

van Delden, H., van Vliet, J., Rutledge, D. T., Kirkby, M. J. (2011) Comparison of scale and scaling issues in integrated land-use models for policy support. *Agriculture, Ecosystems and Environment* **142**, 18–28.

DESIRE (2013) DESIRE Project Harmonised Information System: providing local solutions to global sustainable land management problems. Available at: www.desire-his.eu (accessed 2 June 2015)

DeSurvey (2011) A Surveillance System for Assessing and Monitoring of Desertification. CSIC, Almería. Available at: www.noveltis.com/desurvey/ (accessed 30 May 2015).

DIS4ME (2004) Desertification Indicator System for Mediterranean Europe. DESERTLINKS Project. Available at: www.kcl.ac.uk/projects/desertlinks/indicator_system/introduction.htm (username: desertlinks, password: dis4me) (accessed 2 June 2015).

Fantechi, R., Margaris, N. S. (1986) Desertification in Europe. Proceedings of the Information Symposium in the EEC Programme on Climatology, Mytilene, Greece, 15–18 April 1984. Reidel, Dordrecht.

Geeson, N. A., Brandt, C. J., Thornes, J. B. (eds.) (2002) *Mediterranean Desertification: A Mosaic of Processes and Responses*. John Wiley & Sons, Ltd, Chichester.

Kirkby, M. J., Abrahart, R., McMahon, M. D., Shao, J., Thornes, J. B. (1997) MEDALUS soil erosion models for global change. *Geomorphology* **24**, 35–49.

Kirkby, M. J., Irvine, B. J., Jones, R. J. A., Govers, G., PESERA team (2008) The PESERA coarse scale erosion model for Europe: I – Model rationale and implementation. *European Journal of Soil Science* **59**, 1293–1306.

Kosmas, C., Danalatos, N. G., Gerontidis, S. (2000) The effect of land parameters on vegetation performance and degree of erosion under Mediterranean conditions. *Catena* **40**, 3–17.

LEDDRA (2014) Assessing the Fit of Responses to Land and Ecosystem Degradation and Desertification. Department of Geography, University of the Aegean, Lesvos. Available at: leddris.aegean.gr (accessed 2 June 2015).

Mairota, P., Thornes, J. B., Geeson, N. A. (1996) *Atlas of Mediterranean Environments in Europe: The Desertification Context*. John Wiley & Sons, Ltd, Chichester.

Salvati, L., Venezian Scarascia, M. E., Zitti, M., Ferrara, A., Urbano, V., Sciortino, M., Giupponi, C. (2009) Land degradation assessment in time and space: A review of bio-physical and socio-economic impacts in the Mediterranean basin. *Italian Journal of Agronomy* **4**, 77–90.

Thornes, J. B. (ed.) (1990) *Vegetation and Erosion*. John Wiley & Sons, Ltd, Chichester.

Thornes, J. B. (2005) Coupling erosion, vegetation and grazing. *Land Degradation and Development* **16**, 127–138.

Thornes, J. B. (2007) Modelling soil erosion by grazing: Recent developments and new approaches. *Geographical Research* **45**, 13–26.

Thornes, J. B., Brandt, C. J. (eds.) (1996) *Mediterranean Desertification and Land Use*. John Wiley & Sons, Ltd, Chichester, 303–354.

Thornes, J. B., Brunsden, D. (1977) *Geomorphology and Time*. Methuen, London.

UN Secretariat of the Conference on Desertification (1977) *Desertification: An Overview. In: Desertification: Its Causes and Consequences*. Pergamon Press, New York.

United Nations Convention to Combat Desertification (UNCCD) (2007) Climate Change and Desertification. UNCCD Thematic Fact Sheet No 1. UNCCD, Bonn, 2pp.

CHAPTER 17

John Thornes: An appreciation

Denys Brunsden

Vine Cottage, Dorset, UK

John Thornes was born on 27 December 1940, in Horbury, near Wakefield. He was the only child of Edwin Thornes, engineering foreman, and his wife, Gladys (née Carpenter). He attended a primary school in Horbury where he quickly showed academic promise and was transferred to Ossett Grammar School. There he gained good A-level grades. More importantly, because he was introduced to walking, scouting and school fieldwork, he acquired a lifetime commitment to the understanding of landscape and the challenge of interpreting 'great nature's open book'. This instilled in him a passion for the study of the form, processes, biotic activity and human interactions that shape the Earth's surface. He had a deep need to understand the influence of time and the changing environment on its evolution.

At Queen Mary College, University of London, John was awarded a first-class degree in Geography (BSc Special) with subsidiary Geology (1962). He was then awarded a Commonwealth Scholarship (1962–1964) to McGill University, Montreal. Under F.K. Hare, he studied advanced geomorphology, climatology and glaciology. Ken Hare strongly influenced his MSc thesis on the 'Late Glacial Stages in the Development of the Coaticook Valley, Southern Quebec'. In 1964, Hare took the Chair of Geography at King's College, London. John quickly followed to take up a DSIR Studentship. He completed his PhD in 1967 on 'Erosion and Sedimentation in the Alto Duero, Spain'.

To a geomorphologist, the academic environment of the 1960s was one of great excitement. The British Geomorphological Research Group (BGRG) had been established. In London, geomorphology was leaving behind the qualitative, Davisean, geological approach to landscape evolution established in King's College by S.W. Wooldridge, and the subject of the denudation chronology of the British landscape was being taken over by the multidisciplinary Quaternary Research Group. A new breed of young geomorphologists in the BGRG began to follow the ideas of the Columbia School of A.N. Strahler, Stanley Schumm, Mark Melton and Marie Morisawa. John was captivated by the work of Luna B. Leopold, M.G. 'Reds' Wolman, W.B. Langbein and R.E. Horton who were leading the way towards a new understanding of fluvial processes.

Monitoring and Modelling Dynamic Environments, First Edition. Edited by Alan P. Dykes, Mark Mulligan and John Wainwright.

All of the 'process geomorphologists' in the BGRG were deeply influenced by the 'general systems theory' paradigm being taught by R.J. (Dick) Chorley at Cambridge. We revelled in the field and lecture programmes where a brilliant group of young geomorphologists were seeking to establish a new process-based foundation for the subject. John debated vigorously with all the stars of that period, with perhaps M.J. (Mike) Kirkby at Bristol as his inspiration and Ron Cooke, Mike Thomas, Andrew Goudie, John Doornkamp, Ken Gregory, David Sugden, myself and many other members of the group as people he could not wait to greet and exchange ideas with at the inspirational meetings. He was also entranced by the exciting changes in human geography during the period 1960–1980, especially the ideas of Peter Haggett, Ron Johnston and David Harvey.

At the London School of Economics, John was mentored by Michael Wise and Emrys Jones, human geographers who he felt understood the importance of physical geography in the 'man and environment' equation. He was supported at the LSE by David Jones and Helen Scoging and greatly benefitted from the larger geomorphological staff at King's College who formed the other half of the Joint School of Geography. With their guidance, his career was meteoric. Appointed assistant lecturer at the London School of Economics (1966–1968), he became lecturer (1968–1978) and reader (1978–1981). He interspersed this with visiting scholarships at the Universities of Toronto and Heidelberg.

John became Professor and Head of Department (1983–1985), Dean of Science (1983–1984) and Deputy Principal (1984–1985) at Bedford College, London, followed by Professor and Head of Department (1985–1989) and Dean of Graduate Studies (1989–1992) at the University of Bristol and Distinguished Visiting Professor at the Catholic University of Leuven, Belgium (1988). Then in 1992, John returned to his *alma mater* at King's College as Professor and Head of Department, charged

with rebuilding a major research school after the depredations of university reconstruction and retrenchment that had taken place in London. He established the powerful team which still exists today and remained an active member of that team until his death in 2008.

Throughout his career, John maintained a very deep love for the Spanish landscape. His early work in the Alto Duero blossomed into a lifelong research interest into the origins, evolution and management of Spanish and other Mediterranean landscapes. One of my fondest experiences was to stand in a village cantina listening to John talking to the farmers and asking about erosion, flood control and terracing – always in the present tense and with a soberano and café cortado in hand! Although many British geomorphologists developed their own traditions of fieldwork in Spain, especially Adrian Harvey and Mike Kirkby, it is not an exaggeration to say that John began the fashion for Spanish undergraduate field trips taking advantage of the many cheap holiday tours then developing. Our first trip was in 1968 to Torremolinos and then Mojacar, Garrucha and Tabernas. John never missed a field trip for the next 40 years with the annual Joint School of Geography (KCL–LSE) fieldwork programmes and later with Bedford College and Bristol University students. Generations of students were inspired, research students flowered, and soon the Spanish university links began. Some of the studies were hugely original, such as the archaeology studies with Antonio Gilman and Stephen Wise at El Argar which demonstrated that many of the larger soil erosion features were due to catastrophic events that probably had only operated a few times since the Bronze Age!

John's wide ranging, almost furious reading and enthusiasm for new ideas and concepts led to other interesting themes. An early interest was palaeohydrology, carried out as an international commission with Leszek Starkel of the Polish Academy of Sciences, K.J. (Ken) Gregory and John Lewin. Generations of students experienced his enthusiasm for the auto-geometry of river channels, the channel changes with time, the processes of ephemeral streams and the influence of catastrophic floods. Almost all of the photographs of this period show him standing by or in streams usually with a survey pole or current meter in his hands.

Beginning with his membership of the spectacular Royal Geographical Society (RGS) Hovercraft Expedition to the Casiquiare and Orinoco Rivers, John also developed an interest in the rainforests of Amazonia from Manaus to the Xingu–Araguaia headwaters. With Stephen Nortcliff from King's College, he produced a series of innovative papers on the hydraulic geometry of the streams, seasonal variations in hydrology and cation exchange and the nutrient status of the soils, which greatly increased our understanding of forest ecosystems. This work flowered into an interest in the importance of vegetation and ecology to geomorphology, soil erosion and, later, desertification. He led the European Commission projects on Mediterranean Desertification and Land Use (MEDALUS) which in 1991–1999 helped to establish the modern methodology, models and data sets to evaluate and mitigate land degradation problems in the region. MEDALUS set a standard for modelling and monitoring that has seldom been bettered in geomorphological research. The programme was internationally acclaimed, and soon, academic colleagues from Spain and then many other European universities became involved in fruitful scientific exchanges and collaborations. Of course, this work developed out of his Spanish friendships. At first, this involved cooperation with F. Lopez Bermudez of the University of Murcia but spread to many other research schools using funding from the RGS, UK Research Councils and Europe. The attempt by Roberto Fantechi and the United Nations, in 1994, to develop a Convention to Combat Desertification coincided with the MEDALUS programme and became the formative research link between more than 40 European institutions and continues to influence research through the MODULUS, MEDACTION, DESERTLINKS and DESURVEY projects.

John was an innovative and strategic but pragmatic thinker who influenced research and teaching in all the departments, research groups, academic societies and international institutions with which he was associated. He travelled widely to disseminate this knowledge throughout the world. In addition to Europe, he worked in Arctic Canada, Iceland, Brazil, Argentina, Brunei, the Mojave Desert, Nepal, China, Mexico and South Africa. At Bristol and King's College, he developed research schools or themes which focused on clearly defined objectives and enabled successful research funding applications. He received nearly ECU 3,000,000 research funds over his career. Under John's leadership, these departments did not neglect teaching but placed it firmly within the chosen research interests. Students lived at the cutting edge of their subject. It was leadership by example with ideas, training and research questions as the driver. This, in turn, produced a stream of talented research students who in turn produced undergraduates, research projects and PhDs. Twenty-four PhD students directly benefitted.

Above all, John was a team player. He undertook a full role in student societies, departmental administration, examination boards and studentship grant committees. He served for the Natural Environment Research Council, the Academic Councils of Bedford and King's Colleges, editorial boards of several major journals, the European Society for Soil Conservation, the British Association and the Institute of British Geographers (IBG) of which he also became President. There, he encouraged the merger with the RGS which has proved to be so successful today. All the major British academic organisations in geography benefitted from his boundless enthusiasm and leadership. He was an assessor for Commonwealth universities and advisor to the Spanish Ministry of Education, the European Commission DGXII and the Hong Kong Universities Funding Council.

So the rewards came: the Patron's Medal of the RGS in 1996 and the Linton Award of the BGRG in 1998, the highest honour in the United Kingdom that his chosen subject could bestow. Then, in 2005, the Honorary Doctorate of the University of Murcia as Spain recognised his seminal contributions to semi-arid and Mediterranean research and his dedication to the establishment of Anglo-Spanish scientific links over four decades.

John was a prolific author of more than 130 academic papers and book chapters. His early work was summarised in his LSE monograph *Semi-arid Erosional Systems* (1977), which established a pattern that became a hallmark for his subsequent work. It drew together a clear exposition of an academic problem, relevant observation, description and measurement of the system components underpinned by innovative conceptual models. *Land Use and Prehistory in South-East Spain* (1985) and several classic papers with Antonio Gilman rewrote the way in which we thought about the influence of human activities on land use and soil erosion. Eleven major books and numerous research volumes followed.

John's conceptual interests were first expressed in *Geomorphology and Time* (1977) with Denys Brunsden, with whom he also wrote an important theoretical paper on *Landscape Sensitivity and Change* (1979). His teaching style is clearly shown in *Processes in Geomorphology* with Clifford Embleton, a compendium of the lectures given at KCL–LSE in the mid-1970s. His palaeohydrology work was summarised in *Palaeohydrology in Practice* (1987) and *Fluvial Processes in the Temperate Zone* (1991) with K.J. Gregory, John Lewin and Leszek Starkel of the Polish Academy of Sciences.

The seminal work on Mediterranean landscapes and his interests in geomorphology and ecology came together in *Vegetation and Erosion* (1990). Subsequently, *Mediterranean Desertification and Land Use* (1996) with Jane Brandt, *Atlas of Mediterranean Environments in Europe: The Desertification Context* (1997) with P. Mairota and N.A. Geeson and *Environmental Issues in the Mediterranean* (2002) with John Wainwright, together with numerous research monographs, drew together a remarkable lifetime's achievement.

Index

Monitoring and Modelling Dynamic Environments, First Edition. Edited by Alan P. Dykes, Mark Mulligan
and John Wainwright.
© 2015 John Wiley & Sons, Ltd. Published 2015 by John Wiley & Sons, Ltd.